Experiencing Famine in Fourteenth-Century Britain

ENVIRONMENTAL HISTORIES OF THE NORTH ATLANTIC WORLD

VOLUME 4

Experiencing Famine in Fourteenth-Century Britain

by

Philip Slavin

BREPOLS

British Library Cataloguing in Publication Data

A catalogue record for this book is available from the British Library.

D/2019/0095/184
ISBN: 978-2-503-54780-0
ISSN: 2565-8131
Printed in the EU on acid-free paper

In memoriam

John H. A. Munro (1938–2013)
and
John Langdon (1944–2016)

— mentors, colleagues and friends,
always *hungry* for knowledge and truth

Knock, knock, knock! Who's there, i' th' name of Beelzebub?
Here's a farmer that hanged himself on the expectation of plenty.

William Shakespeare, *Macbeth*: Act II, Scene 3

A dearth has never arisen [...] from any other cause but a real scarcity, occasioned sometimes perhaps, and in some particular places, by the waste of war

Adam Smith, *The Wealth of Nations*

Famines do not simply occur: they are organized by the grain trade.

Berthold Brecht

There is a woman in Somalia,
Scraping for pearls on the roadside,
There's a force stronger than nature
Keeps her will alive.
This is how she's dying:
She's dying to survive.

Sade (Helen Adu), 'Pearls' (1992)

Contents

List of Illustrations

Figures

Tables

List of Abbreviations

Archival

BAHPS	Birmingham Archives, Heritage and Photography Service (Birmingham)
BARS	Bedfordshire Archives and Records Service (Bedford)
BelCM	Belvoir Castle Muniments (Belvoir, Leicestershire)
BerCM	Berkeley Castle Muniments (Berkeley, Gloucestershire)
BL	British Library (London)
BoL	Bodleian Library (Oxford)
BRO	Berkshire Record Office (Reading)
CBS	Centre for Buckinghamshire Studies (Aylesbury)
CCA	Canterbury Cathedral Archives (Canterbury)
CCCA (O)	Corpus Christi College Archives, Oxford (Corpus Christi College, Oxford)
CoRO	Cornwall Record Office (Truro)
CrCA	Christ College Archives (Cambridge)
CUA	Cambridge University Archives (Cambridge)
CUL	Cambridge University Library (Cambridge)
DCA	Durham Cathedral Archives (Durham)
DALS (E)	Devon Archives and Local Studies, the Exeter Branch (Exeter)
DHC	Dorset History Centre (Dorchester)
EaCA	Eton College Archives (Eton)

ERO	Essex Record Office (Chelmsford)
ESRO	East Sussex Record Office (Falmer, Sussex)
ExCA	Exeter Cathedral Archives
GA	Gloucestershire Archives (Gloucester)
HALH	Haringey Archive and Local History (Haringey)
HantsALS	Hampshire Archives and Local Studies (Winchester)
HCA	Hereford Cathedral Archives (Hereford)
HereARC	Herefordshire Archives and Records Centre (Hereford)
HertALS	Hertfordshire Archives and Local Studies (Hertford)
HaHM	Hatfield House Muniments (Hatfield)
HoHM	Holkham Hall Muniments (Holkham, Norfolk)
HuntALS	Huntingdonshire Archives and Local Studies (Huntingdon)
KCA	King's College Archives (Cambridge)
KHLC	Kent History and Library Centre (Maidstone)
KUL	Keele University Library (Keele)
LA	Lincolnshire Archives (Lincoln)
LH	Longleat House (Warminster, Wiltshire)
LMA	London Metropolitan Archives (London)
LPL	Lambeth Palace Library (Lambeth)
LUL	Leeds University Library (Brotherton Library) (Leeds)
MagCA	Magdalen College Archives (Magdalen College, Oxford)
MCA	Merton College Archives (Merton College, Oxford)
NAI	National Archives of Ireland (Dublin)
NCA	New College Archives (New College, Oxford)
NLS	National Library of Scotland (Edinburgh)
NLW	National Library of Wales (Aberystwyth)
NorthantsRO	Northamptonshire Record Office (Northampton)
NRO	Norfolk Record Office (Norwich)

NYCRO	North Yorkshire County Record Office (Northallerton)
QCA	Queen's College Archives (Queen's College, Oxford)
RCBCIA	Representative Church Body of the Church of Ireland Archives (Dublin)
RHM	Raynham Hall Muniments (Raynham, Norfolk)
SAL	Society of Antiquaries of London Library (London)
SalopRO	Shropshire Record Office (Shrewsbury)
SARS	Somerset Archives and Record Service (Taunton)
SGCA	St George's Chapel Archives (Windsor)
SRO (BSE)	Suffolk Record Office, Bury St Edmunds Branch (Bury St Edmunds)
SRO (I)	Suffolk Record Office, Ipswich Branch (Ipswich)
SRO (L)	Suffolk Record Office, Lowestoft Branch (Lowestoft)
StRO (L)	Staffordshire Record Office, Lichfield Branch (Lichfield)
StRO (S)	Staffordshire Record Office, Stafford Branch (Stafford)
TCA (C)	Trinity College Archives, Cambridge (Cambridge)
TCL	Trinity College Library, Dublin (Dublin)
TNA	The National Archives (Kew, Surrey)
UCA	University College Archives (University College, Oxford)
UCL	University of Chicago Library (Chicago)
UNL	University of Nottingham Library (Nottingham)
WAAC	Worcestershire Archive and Archaeology Service (Worcester)
WCA	Worcester Cathedral Archives (Worcester)
WAM	Westminster Abbey Muniments (Westminster)
WCA	Winchester College Archives (Winchester)
WFM	Wisbech and Fenland Museum (Wisbech)
WSHC	Wiltshire and Swindon History Centre (Chippenham)
YML	York Minster Library (York)

Printed

CCR, 1313–18	*Calendar of Close Rolls Preserved in the Public Record Office: Edward II, AD 1313–1318* (London: Her Majesty's Stationery Office, 1893)
CCR, 1318–23	*Calendar of Close Rolls Preserved in the Public Record Office: Edward II, AD 1318–23* (London: Her Majesty's Stationery Office, 1895)
CDRS	*Calendar of Documents Relating to Scotland*, vol. III (*1307–57*), ed. by Joseph Bain (Edinburgh: General Register House, 1887)
CIM	*Calendar of Inquisitions Miscellaneous (Chancery) Preserved in the Public Record Office*, vol. II (London: His Majesty's Stationery Office, 1916)
CPR, 1313–17	*Calendar of Patent Rolls Preserved in the Public Record Office: Edward II, AD 1313–1317* (London: Her Majesty's Stationery Office, 1898)
CPR, 1317–21	*Calendar of Patent Rolls Preserved in the Public Record Office: Edward II, AD 1317–1321* (London: His Majesty's Stationery Office, 1903)
CPR, 1321–24	*Calendar of Patent Rolls Preserved in the Public Record Office: Edward II, AD 1321–1324* (London: His Majesty's Stationery Office, 1904)
RMEPS	*A Repertorium of Middle English Prose Sermons*, ed. by Veronica O'Mara and Suzanne Paul (Turnhout: Brepols, 2007)

Weights, Measures, Distance, Values, and Boundaries

Weight

1lb (pound *avoirdupois*) = 453.592 g

Volume (dry)

1 quarter = 8 bushels = 64 gallons = 2.819 hectolitres
1 Irish crannock = 1 quarter = 8 bushels = 64 gallons = 2.819 hectolitres
1 quarter of wheat, *c.* 1300 = 53 lb
1 quarter of maslin (wheat-rye mixture), *c.* 1300 = 52 lb
1 quarter of rye, *c.* 1300 = 51 lb
1 quarter of barley, *c.* 1300 = 46 lb
1 quarter of dredge, *c.* 1300 = 41 lb
1 quarter of oats, *c.* 1300 = 36 lb
1 quarter of beans and vetches, *c.* 1300 = 60 lb
1 quarter of peas, *c.* 1300 = 46 lb
1 tun of flour = about 6.67 quarters

Area

1 acre = 4 rods = 160 perches = 4046.86 square metres

Distance

1 mile = 1760 yards = 1.6093 kilometres
1 nautical mile = 1.15078 miles = 2025.37 yards = 1.852 kilometres

Value

1 pound sterling (£) = 20 shilling (s.) = 240 pence (d.)
1 mark = £0.66 = 13s 4d = 160d

Boundaries

All county boundaries are given as they existed before 1974.

Acknowledgements

My initial interest in the history of famine (which, subsequently, earned me the nickname 'Famine Phil', in some circles — which is, nevertheless, better than 'Dr Phil') goes back to 2008–2009, when, while doing my post-doctoral studies at Yale, I came across Amartya Sen's classic *Poverty and Famines* (1981). While reading and digesting this book, I was able to share my initial ideas with my colleague and friend Adam Franklin-Lyons, then a doctoral student at Yale, working on the 1374–1375 famine in Catalonia. Adam was quite instrumental in introducing me to the theory of famine and engaging me in countless conversations about famines, their causes, and consequences. Equally inspiring was the mentorship of Professor Timothy Guinnane, my post-doctoral supervisor (and an amazing scholar), who encouraged me to consider writing a monograph on the Great European Famine, even when I questioned the wisdom of the enterprise. It was not until I had landed a permanent position at the University of Kent in late 2013, however, that I could finally give the project proper attention, partially because of being preoccupied with my first monograph on Norwich Cathedral Priory and its food supply and consumption (2012).

The present study is based heavily on a large corpus of unpublished archival sources. In the course of research for this book, I have visited seventy-three archival repositories, both public and private, primarily in England, but also in Ireland, Wales, Scotland, and the US. I know I gave hundreds of archivists a hard time during my archival trips (sometimes commencing as early as 5AM, when catching the first train from Canterbury to Durham). All I can do is thank them for their remarkable professional commitment, helpfulness, and forbearance and apologize in advance for being a pest — realizing that more hard times are yet to come. One by one (and in an alphabetic order), I should offer equal thanks to the following archives: Bedfordshire Archives and Records Service (Bedford); Belvoir Castle Muniments (Belvoir, Leicestershire; my special thanks go to the Duke and Duchess of Rutland for their kind permission to consult and photograph their muniments); Berkshire Record Office (Reading); Birmingham Archives, Heritage, and Photography Service (Birmingham); the British Library (London); the Bodleian Library (Oxford); Cambridge University Archives; Cambridge University Library; Canterbury Cathedral Archives; Centre for Buckinghamshire Studies (Aylesbury); Christ College Archives, Cambridge; Cornwall Record Office (Truro); Corpus Christi College Archives, Oxford; Devonshire Record Office, the Exeter Branch; Dorset History Centre (Dorchester); Durham Cathedral Archives; Essex Record Office (Chelmsford); East Sussex Record Office (Falmer, Sussex); Eton College Archives; Exeter Cathedral Archives; Gloucestershire Archives (Gloucester); Hampshire Archives and Local Studies (Winchester); Haringey Archive and Local History (Haringey); Hatfield House (Hatfield; my special thanks go to the Marquess of Salisbury for his kind permission to consult and photograph his muniments); Herefordshire Archives and Records Centre (Hereford); Hereford Cathedral Archives; Hertfordshire Archives and Local Studies (Hertford); Holkham Hall (Holkham, Norfolk; my special thanks go to the Earl of Leicester for his kind permission to consult and photograph his muniments); Huntingdonshire Archives and Local Studies (Huntingdon); Huntington Library (Pasadena, California); Keele University Library (Keele); Kent History and Library Centre (Maidstone); King's College Archives, Cambridge; Lambeth Palace Library; Leeds University Library (Brotherton Library); Lincolnshire Archives (Lincoln); London Metropolitan Archives (London); Magdalen College Archives, Oxford; Merton College Archives, Oxford; the National Archives of Ireland (Dublin); the National Archives of Wales (Aberystwyth); the National Library of Scotland (Edinburgh); New College Archives, Oxford; Norfolk Record Office (Norwich); Northamptonshire Record Office (Northampton); North Yorkshire County Record Office (Northallerton); Queen's College Archives, Oxford; Raynham Hall (Raynham, Norfolk; my special thanks go to the Marquess Townshend of Raynham Hall for his kind permission to consult and photograph his muniments); Representative Church Body of the Church of Ireland Archives (Dublin); Shropshire Record Office (Shrewsbury); Society of Antiquaries of London Library; Somerset Archives and Record Service (Taunton); Staffordshire Record Office, Lichfield Branch; Staffordshire Record Office, Stafford Branch; St George's Chapel Archives (Windsor); Suffolk Record Office, Bury St Edmunds Branch; Suffolk Record Office, Ipswich Branch; Suffolk Record Office, Lowestoft Branch; Trinity College Archives, Cambridge; Trinity College Library, Dublin; the National Archives (Kew, Surrey); University of Chicago Library; University College Archives, Oxford; University of Nottingham Library; Westminster Abbey Muniments;

Winchester College Archives; Wisbech and Fenland Museum (Wisbech); Wiltshire and Swindon History Centre (Chippenham); Worcestershire Archive and Archaeology Service (Worcester); Worcester Cathedral Archives; and York Minster Library. In addition, I am grateful to Mr Charles Berkeley of Berkeley Castle and to the Marquess of Bath of Longleat House (Warminster, Wiltshire) for allowing me to consult their manorial muniments, in a microfilmed format.

The numerous visits to these archives and the digitization of a host of original documents would not have been possible without the generous support of the School of History, the University of Kent; the Mayers short-term Fellowship at the Huntington Library, Pasadena (California); and the Buckinghamshire Record Society. I am deeply grateful for their support.

In the course of research and writing, I was extremely lucky to have profited from the very helpful insights of my colleagues and friends. In particular, and again in an alphabetic order, I wish to thank Chris Briggs (University of Cambridge), James Davis (Queen's University of Belfast), Sharon DeWitte (University of South Carolina), Chris Dyer (University of Leicester), Maryanne Kowaleski (Fordham University), the late (and much missed) John Langdon (University of Alberta), Cynthia Neville (Dalhousie University), and Richard Smith (University of Cambridge) for taking the trouble of reading entire chapters or parts of chapters of this book and suggesting numerous comments and ways to improve it. I am also thankful to the following colleagues, who reflected on various aspects of the book, in person (over a pint or at a research seminar) or through email: Jelena Bekvalac, (the late and much missed) Richard Britnell, Bruce Campbell, Jordan Claridge, Samuel Cohn, David Dickson, Ben Dodds, John Hare, John Hatcher, Richard Hoffmann, William Chester Jordan, Amnon Linder, Francis Ludlow, Michael McCormick, Miriam Müller, Tim Newfield, Cormac Ó Gráda, Richard Oram, Kathlyn Pribyl, (the late and much missed) Alasdair Ross, Alexandra Sapoznik, Phillipp Schofield, Robert Swanson, Erik Thoen, Rob Wilson, and Michelle Ziegler. In particular, I am grateful to Chris Briggs, Ben Dodds, Phillipp Schofield, and Robert Swanson for being kind in sharing their own data. At the University of Kent, I would like to thank my colleague and friend Omar Nasim (now at the University of Regensburg) for engaging conversations over a coffee.

At Brepols, I am most grateful to Dr. Rosie Bonté, publishing manager, and to Dr. Lauren Davis, the copyeditor for the book, whose eagle-eyed reading of the manuscript spotted and slew a series of embarrassing typos and inconsistencies. Likewise, I am grateful to the two anonymous reviewers of the book, whose fruitful suggestions helped improve the quality of the book.

The book was written at various settings: my office on the Canterbury campus of the University of Kent (my then home institution, before my move to the University of Stirling in July 2019); the National Library of Israel (the Hebrew University of Jerusalem, my undergraduate *alma mater*); various airports and train stations; and on the train, in the air, and at hotels, conference venues, cafeterias, pubs, and libraries. But one particularly enjoyable location was Pasadena (California), where I was able to spend a month, thanks to the Mayers short-term Fellowship at the Huntington Library (April–May 2015). Whether at the library or walking down East Del Mar Blvd through the Caltech campus on my way to Trader Joe's, I felt absolutely happy in submerging into the inspiring, relaxing, and yet scholarly landscapes of Pasadena.

As ever, I am grateful to my dear friends, Mr Ilia Avroutine and Ms Maria Kozlov, for sheltering me on countless occasions in the course of my research trips, thus, making my life so much easier and more enjoyable (and cheaper).

I am thankful to Tanya, Yannai, and Maya, who was born half-way through the process of writing, and whose inimitably gorgeous and cute smile has always reminded me that there is more to life than just writing. As ever, I feel awfully guilty about being an absent-minded family member and feel a strong need to apologize for making all three suffer from the Great Famine, in the emotional sense. Last, but not least, I would like to thank Liz — who came into my life when this book was already in its late editorial phases — for her genuine love, care, and friendship, during the hard and perplexing times of change through which I have been going.

In the course of writing, I have lost two dear mentors, colleagues, and friends: Professor John H. Munro, my PhD supervisor at the University of Toronto (14 March 1938–23 December 2013) and Professor John Langdon of the University of Alberta (24 December 1944–31 December 2016), who happened to serve as an external examiner of my PhD dissertation (June 2008). Both Johns were amazing and most inspiring scholars, with uncompromisingly high scholarly and moral standards. But apart from that, they also became kind and close friends of mine — always ready to talk and support me during my professional and personal crises. It is to their memory — with much sadness and loss — that I dedicate this book. I can only hope that their mentorship of me was not in vain. This, however, I will leave to judge to the readers of this book.

PHILIP SLAVIN

On the train from London Waterloo to Exeter Central

7 May 2019/22 Iyar 5779

Prologue

Famines and their consequences are among the most pressing and complex economic and political issues that scholars, policy makers, and NGOs are struggling with today. While famine has largely disappeared from Europe and most of Asia (with the notorious exception of war-torn Yemen), it still prevails in parts of Africa. In particular, the harsh famine of 2011–2012 in Somalia claimed an unaccounted number of civilians. The current food crises in south-western Yemen, parts of Somalia, South Sudan, and north-eastern Nigeria is defined by FEWS NET (Famine Early Warning Systems Network) as 'emergency', while the central parts of South Sudan are recognized to be in the state of famine.[1] The concerns about famine conditions, which exist in some parts of the world and are imminent in others, were repeatedly expressed at the recent COP21 Climate Change conference in Paris (30 November–12 December 2015) and the COP22 Climate Change conference in Marrakesh (7–18 November 2016).

To understand why famines still happen today, despite the globalization of trade and the existence of various international NGOs dedicated to famine prevention and food security, it is essential to understand the nature of famines in a wider, historical perspective. Famines and food crises happen across space and time, in different political, socio-economic, and cultural contexts. Yet, despite these contextual differences, there are some overarching 'meta-structures' that are common to different famines, no matter the contexts in which they are set. The topic of historical famines has attracted much scholarly attention, and much work has been done by historians, development economists, political scientists, and environmental scientists alike. In studying various cases and aspects of historical famines, scholars raise what is perhaps the single most important

[1] *Famine Early Warning Systems Network*.

methodological and conceptual question: What creates famines? Much ink has been spilled on this question; yet, because of different approaches and theoretical frameworks, there is no scholarly consensus regarding the causes of famines. As we shall see in Chapter 1, one can distinguish between three grand schools of thought, each coming up with its own grand narrative or interpretation of famine causes. The 'environmentalist' school of thought sees famines as exogenous disasters triggered by the forces of Nature. The 'institutionalist' approach points its finger into human institutions and decisions as the main factors responsible for famines. Finally, the Malthusian approach focuses on the relationship between the size of a population and the available landed resources and contends that famine happens when the gap between overpopulation and the available landed resources becomes too wide.

The present study offers a humble contribution to the field of famine studies, by looking at what was arguably the single harshest subsistence crisis in the recorded history of northern Europe: the Great European Famine of the early fourteenth century. The crisis was devastating, in terms of its length, geographic span, proportions, and impact. Initially caused by the weather anomaly of 1314–1316 (torrential rain pouring virtually incessantly between the summer of 1314 and the summer of 1316), the crisis was omnipresent across a wide geographic span — from Scandinavian lands in the north to the Loire in the south, from Ireland in the west to Poland in the east. It has been estimated that the disaster claimed some 10–15 per cent of the north European population.[2] Indeed, as we shall see later, the demographic losses in England were *at least* of that magnitude and probably more in the area of 15–20 per cent. It would be wrong to assert that the Great European Famine has escaped the radar of scholars. In 1930, Henry Lucas provided the first synthetic overview of the disaster from a pan-European perspective, primarily on the basis of printed chronicles.[3] In 1959, Hans van Werveke published an article about the extent of the crisis in Flanders and neighbouring regions.[4] In 1973, Ian Kershaw studied the famine from the English perspective, drawing our attention to its long-lasting impact on the pre-Black Death economy and society. For Kershaw, the famine was just

[2] Poos, *A Rural Society after the Black Death*, pp. 106–07; Titow, 'Some Evidence of the Thirteenth Century Population Increase', p. 224; Razi, *Life, Marriage and Death in a Medieval Parish*, pp. 31 and 40.

[3] Lucas, 'The Great European Famine'.

[4] Van Werveke, 'La famine de l'an 1316'.

a part of the wider 'agrarian crisis' of 1315–1322.[5] In 1991, Meghnad Desai, an economist rather than historian, attempted to refute Kershaw's assessment of the crisis as a disaster with long-term consequences and suggested that the Great Famine was just a passing event caused by 'entitlement failures', which resulted from the adverse combination of harvest failures and livestock mortality.[6] Desai clearly attacked the Malthusian interpretation of late medieval economic history and attempted to introduce the concept of 'entitlement crisis/failure', as advanced by Amartya Kumar Sen in his classic *Poverty and Famines* (1981).[7] Unlike Desai, William Chester Jordan saw the Great Famine as a major catastrophe with far-reaching implications. In his award-winning monograph, *The Great Famine: Northern Europe in the Early Fourteenth Century* (1996), Jordan studied the catastrophe from the pan-European, rather than regional, perspective.[8] Jordan's monograph was followed by his 2010 piece, 'The Great Famine: 1315–1322 Revisited', where he, in effect, 'updated' some aspects of his 1996 monograph, on the basis of new primary and secondary sources that appeared in print between 1996 and 2010.[9] In 2004, Mike Davies and Jonathan Kissock showed that the famine had a strong impact on land market, as evidenced in the increase freehold land transactions in three counties in western England (Herefordshire, Shropshire, and Gloucestershire).[10] In 2013, Buchanan Sharp studied the impact of the famine on grain trade and royal attempts to mend the situation, which he characterized as 'royal paternalism'.[11] In the following year (2014), I published two articles on the topic, one dealing with the impact of market failure on the crisis, and the other one considering the influence of early fourteenth-century warfare on the catastrophe.[12] In 2015, Matthew Stevens published a local study of the famine in the marcher lordship of Dyffryn Clwyd in Denbighshire, north-east Wales, where he showed that

[5] Kershaw, 'The Great Famine and Agrarian Crisis in England'.

[6] Desai, 'The Agrarian Crisis in Medieval England'.

[7] Sen, *Poverty and Famines*.

[8] Jordan, *The Great Famine*.

[9] Jordan, 'The Great Famine: 1315–1322 Revisited'.

[10] Davies and Kissock, 'The Feet of Fines'.

[11] Sharp, 'Royal Paternalism and the Moral Economy'. Much of this study's ideas reappear in the second chapter of Sharp's most recent monograph, *Famine and Scarcity*, pp. 33–55.

[12] Slavin, 'Market Failure during the Great Famine', and 'Warfare and Ecological Destruction'.

the situation in Wales was similar to that in England.[13] Stevens's regional study of the famine was followed by the most recent study of the disaster, from an all-Welsh perspective, by Phillipp Schofield (2018).[14]

The present monograph is the first full-scale study of the early fourteenth-century famine in the British Isles, based almost entirely on the plethora of unpublished and largely unused archival sources. For this study, I have systematically consulted, digitized, and analysed a large number of surviving manorial records (demesne accounts, court rolls, and tithe receipts), sheriffs' accounts, port customs accounts, lay subsidy rolls, gaol delivery rolls, coroners' rolls, royal writs, feet of fines (registers of freehold land transactions), episcopal registers, ecclesiastical valuations, and household accounts of religious institutions from the famine period (in effect, for the entire decade of 1311–1321) (See Appendix 8). Taken together, this unpublished archival corpus amounts to close to 3500 documents. The abundance of the surviving material, brought together and analysed in the present study, makes it possible to reconstruct the causes, nature, manifestations, and impact of the famine in great detail. Although I did my best to divert from the Anglo-centric perspective that characterizes much scholarship of pre-Industrial economic history of the British Isles, the comparative thinness of Scottish and Welsh material, coupled with the disastrously poor survival of the Irish material (due to the infamous Four Courts Great Fire of 1922 that destroyed virtually all documents of the English colonial government), implies that England had to receive the disproportionally large share of discussion and analysis here. The remarkably abundant nature of English material is indeed unparalleled by any other European counterpart, and it would be fair to say that it is only England, where such a full-scale, detailed, and thorough study, based on meticulous analysis of thousands of surviving documents, is possible. Just as Ireland, Scotland, and Wales, so do other parts of northern Europe, too, suffer from a shortage of archival documents, and most of the information derives from textual rather than statistical sources. This has to do not only with the poor survival of archival materials, but also with relatively low levels of document production compared to England. In the early fourteenth-century, England could boast, arguably, the single most advanced level of bureaucracy in Europe, on both the state and local levels. In particular, the strong presence of the manorial system led to the issue of annual manorial accounts (reports recording financial and agricultural incomings

13 Stevens, 'The Great Famine in Dyffryn Clwyd, 1315–1322'.

14 Schofield, 'Wales and the Great Famine'.

and outgoings on a farm level) and court rolls (minutes of sessional manorial courts). This advantage places England in a unique position to act as a lab, in which one can isolate and analyse various aspects of the Great Famine and use these conclusions to shed light on the situation in other famine-struck regions of northern Europe, whose experiences are concealed because of the lack of surviving/available source material. It would not be an exaggeration to say that we can learn as much about the early fourteenth-century famine in England as we can about the mid-nineteenth-century famine in Ireland.

Chapter 1 sets the Great Famine within a wider environmental, demographic, and socio-economic context of the early fourteenth-century British Isles, and it places the disaster within a wider conceptual framework of famine theory. As it will argue, climatic shifts affected all parts of the British Isles. Yet, the intensity of the crisis varied across regions. While the famine ravaged England, the Anglo-Norman controlled Lordship of Ireland, and parts of Wales, it seems to have spared Scotland altogether. To understand this regional complexity, it is necessary to consider the different demographic and institutional aspects characterizing each region and to see to what extent these aspects transformed food shortage into a full-fledged famine. The chapter wraps these factors into conceptual models and theories of famine, to stress the advantages and shortcomings of each model, and concludes that one cannot explain the presence or absence of famine with just one theory or factor. Only by taking into account all factors together may one understand the incredibly complex nature of famine. This notion stands at the very heart of the present study, and it guides the discussion and analysis throughout the monograph.

Chapter 2 deals with the impact of the bad weather (the torrential rain of 1314–1316, in particular) on crop harvests and food production. There were three back-to-back harvest failures, with the respective *gross* crop yields (inclusive of tithes and seed) standing at about 25, 35, and 14 per cent below average, in 1315, 1316, and 1317, respectively. Most of the available data comes from manorial accounts covering the demesne sector. Thanks to the surviving tithe receipts for the famine years, we can also estimate that harvest shortfalls within the tenancy sectors were as bad as within the demesne. Although crop failure was an omnipresent reality, the degree of its harshness varied across regions. It appears that the northern counties suffered particularly badly because of the combination of the adverse weather and the environmental destruction caused by frequent Scottish raids. Although the downpour destroyed much of the grain and legume harvests, it also caused grass to grow in such abundance that its supply well exceeded the demand — to the loss of the local lords. Conversely, hay resources, instrumental in feeding livestock during the winter time, were catastrophically deficient because of the difficulties of mowing, drying, and preserving grass. The abnormal weather also depressed other, more minor food sectors, including orchard produce, honey, salt, and vintage. The omnipresent depression of food and fodder production had a pronounced impact on the health of both humans and domestic animals.

Chapter 3 shifts the focus from Nature into humans and human decision making and customs in exacerbating the crisis. Although *gross* yields were low, *net* crop yields (net of tithe and seed) were even lower, standing at 38, 50, and 26 per cent below average. This had to do with the fact that manorial officials (and possibly tenants, too) tended to invest excessively large proportions of harvested grain as seed corn, to match the pre-famine seeding levels in absolute terms. This came at the expense of the annual budget allocated to grain-allowance among local *famuli* (hired labourers), some of whom were laid off during the famine. The abysmally low crop yields reduced the ability of tenant holdings to supply their households. For instance, a typical holder of a ten-acre plot could extract only about 68 and 58 per cent of the required dairy calories in 1315–1316 and 1316–1317, respectively. Given that the vast majority of tenants held less than ten acres of land each, they had either to resort to local markets to procure the 'missing calories' or starve. The widespread poverty and high grain prices, as well as the malfunctioning of grain markets, implied that the latter was a more realistic option.

Chapter 4 looks at the behaviour of grain markets during the famine years. Overall, it is a story of market malfunctioning. First, the chapter establishes the seasonal movement of grain prices and shows that they went up uncontrollably fast and aggressively, not only compared to 'normal' years, but also to other historical famines. Moreover, they rose disproportionally high even in relation to the levels of the disastrous harvests of 1315 and 1316. One clear manifestation of the market malfunctioning was its segmentation, with the coefficient of variance (CoV) of prices rising across places. To make things even worse, there were signs of the rise of preferential trade, based on *particularized*, rather than *generalized*, trust, whereby some groups would enjoy more profitable trade terms than others. The situation was exacerbated even further by the presence of black markets and speculation and the breaching of trade regulations. A close analysis of court rolls reveals that the famine years saw a drastic rise in the number of fines paid by bakers for breaching the *Assize of Bread*, while the number of amercements imposed on local brewers declined (reflecting a relative diversion of calories from ale to bread). Although there can be little doubt that some bakers did bake contrary to the assize (out of necessity), the rise in baking amercements also implies a stricter control of bakers by local manorial authorities. Similarly, manorial authorities did their best to control local regraters of bread and ale. Unlike manorial authorities, the Crown did relatively little to fight the chaotic situation — apart from encouraging imports and prohibiting exports. Another blow to market functionality was the phenomenon of hoarding, practised by more substantial grain producers. Despite all these challenges, trade in foodstuffs was not fully paralyzed, thanks to limited activities of both foreign and domestic merchants.

The food insecurity was complicated further by transportation and storage issues, the subjects of Chapter 5. The abnormal weather of 1314–1316 reduced travel and communication conditions to a deplorable state and made food transportation a serious challenge. While inland transport costs may not to have changed, sea transportation costs went up because of a combination of storminess, piracy, and reduced shipping speeds. In some instances, considerable amounts of grain were lost to the sea and to pirates (usually alien citizens). River communication seems to have been affected, too, because of flooding and blockage. Storage costs likewise went up considerably during the famine years because of rising maintenance costs and increased spoilage rates, which were caused by the adverse weather conditions. Yet, a close analysis of storage costs indicates that grain hoarding was exceptionally profitable for those who could afford it.

Chapter 6 studies the impact of warfare and taxation on the famine. Within the 'war zone' of the British Isles (northern England, the Lordship of Ireland, south Wales, and south Scotland), war caused much environmental destruction and cut off local communities from their food resources. The long-term implications of the environmental destruction were harsh, while recovery was long. The crisis was aggravated by the royal purveyance policy, whereby local communities were commanded to sell their food to provision royal garrisons. In addition, some communities (and especially those situated in a proximity to frontier castles) suffered from food robberies and extortion by local corrupt officials. Both purveyance and food seizure were highly burdensome on those communities and individuals, falling victims to malfeasance by royal officials. Finally, to finance his wars with Scotland, Edward II taxed his English subjects both in 1315 and 1316. Although substantial sums were collected, the price that the taxed population had to pay was high, and it contributed to the intensification of the crisis even further.

Chapter 7 diverts its attention from the causes of the famine to its victims. Testing the Malthusian views, it estimates the relationship between the population levels and the available landed and food resources. On the eve of the famine, England was a bottom-heavy society, characterized not only by overpopulation but also by a large proportion of poor people holding small parcels of land (or no land at all). After tithe payment, seeding, spoilage, and processing, harvested grains yielded little, well below the calorific requirements of an average commoner to feed his household. This deficiency of grain calories prompted an unaccountable proportion of the population to resort to alternative 'famine foods', including, possibly (and in some extreme cases), human flesh. The consumption of repugnant comestibles and the disastrous malnourishment compromised the immune system of humans and made them susceptible to various famine-related diseases. As in other famines, there was a pronounced spike in mortality rates (as indicated in the number of inheritance dues) and a decline in fertility rates (as revealed by the number of marriage dues and fornication fines). There seems to be a degree of correlation between poverty and mortality rates. Mortality levels would reach their peaks in the later spring/early summer (during the 'hungry gap' period), in contrast with non-famine years, when the high mortality season was winter. The lowered fertility rates may be explained by a combination of economic insecurity, reduced libido, and poor health of potential grooms and brides. All this led to a decline in population; in some isolated instances, the crisis may have led to the depopulation of entire communities. Estimating the population losses is extremely tricky and difficult, but all the available evidence points out that the previous estimates of 10–15 per cent may be somewhat downplayed and that England may have lost about 15–20 per cent of her population.

Chapter 8 looks at the ways in which the famine victims coped with the crisis. The spike in crime rates (especially those related to food) indicates that many commoners were ready to take various chances to secure missing calories. If caught, however, felons would be delivered to local prisons and if found guilty, hanged. Although, in many instances, felons were acting on their own, in some cases a number of individuals would form gangs and act collectively. Sometimes these activities developed into food riots, with entire rural communities taking action. Engaging in theft and robberies was risky, and because of that some tenants chose to resort to other illicit, but quasi-decriminalized and quasi-acceptable activities. One such activity was gleaning, namely, the collection of grain leftovers from post-harvest fields. Another way to secure calories was to break the 'suit of mill', a manorial custom prescribing local tenants to grind their grain produce at lords' mills for a milling fine, called 'multure' (usually representing between 1/12 and 1/20 of the ground amounts). Despite the spike in crime rates and manorial fines, in the majority of cases, starving commoners resorted to peaceful and legal means of securing their missing calories. One strategy was to transfer lands (chiefly, through sales), in exchange for cash, to be converted into calories. Indeed, the volume of land transactions went up considerably during the famine. Another strategy was to rely on rural credit, extended both in cash and in kind. Despite the crisis and widespread dearth of food, credit was still being extended, especially during the first year of the famine, when the volume of credit transactions was unprecedentedly high. It declined during the second year and remained pronouncedly low in the aftermath of the famine. In the case of both land sales and credit transactions, the vast majority of transactions were done between non-familial parties. All the same, there were many instances when people surrendered their land or extended credit to help their relatives or close friends in need. Indeed, land transfers and credit transactions could be seen as both impersonal economic activities and acts of charity. The idea of charitable works was certainly followed, on a limited scale, by religious institutions (monastic houses and hospitals), which made some efforts to feed local paupers and infirm through organized food distribution. But these efforts were a drop in an ocean of need. Outside the church, charitable works were practised on an even more limited scale. Another coping strategy was to turn to existing liturgical practices, involving both clergy and lay people and consisting primarily of barefoot procession and celebration of votive masses against bad weather, famine, and mortality.

The epilogue surveys, perhaps in a somewhat tentative manner, the long-term implication of the Great Famine on the social and biological well-being of famine survivors in the pre-Black Death period. Although good harvests returned in 1318, the subsequent outbreak of bovine pestilence (1319–1321), the disastrous harvest of 1321, and the dry years in the late 1320s and the early 1330s meant that the food crisis was not over with the end of the famine. On the basis of palaeopathological evidence, the epilogue suggests that famine survivors and their progeny were frail individuals, with compromised immune systems and possibly with epigenetic alterations that later rendered them susceptible to the single most devastating pandemic in recorded history — the Black Death, whose first wave devastated England between 1348 and 1350 and claimed *at least* half of her population (and probably more than that).

Chapter 1

Framing the Great Famine: Concepts and Contexts

One vitally important question that has occupied both scholars and policy makers is: why do famines happen? Famines are incredibly complex and eclectic phenomena, and as such they lack a uniform cause or explanation — despite some claims otherwise. The conceptual framework of famine theory can be, roughly speaking, divided into three main schools of thought: 'environmentalist', 'demographic', and 'institutionalist'.[1] In order to appreciate these famine theories and their applicability to the Great Famine better, it is essential to examine each approach both theoretically and contextually, in conjunction with the historical reality of the early fourteenth-century British Isles.

Environmentalist Approach

The 'environmentalist' interpretation of famine is endorsed by both environmental historians and environmental scientists, and it focuses almost entirely on natural factors and correlates them with occurrences of famine. Natural disasters, short-term weather anomalies, and long-term climatic changes are the most commonly perceived triggers of famine. This view is ingrained into a much wider conceptual framework, which interprets the course of history (past and present) through the prism of environmental and climatic determinism. It places Nature at the very centre and considers it as the primary protagonist of

[1] I have dealt with the conceptual framework of famine theory elsewhere. See Slavin, 'Climate and Famines'.

history. Other factors, such as institutions, demography, and technology, tend to be secondary, as they depend on and are influenced by Nature.[2] This view has been put forth by a number of historians, including Mike Davis and Brian Fagan. In his controversially called monograph, *Late Victorian Holocausts: El Niño Famines and the Making of the New World*, Davis took a deterministic approach to the roots of European Imperialism and considers the El Niño events in the late nineteenth century as the triggers of famines across the Pacific, which aided Europeans to gain control of their Asian and African colonies.[3] In a similar vein, in his *The Little Ice Age: How Climate Made History, 1300–1850*, Fagan considered pre-Industrial famines and major subsistence crises in Europe to be caused by short-term climatic anomalies and long-term climatic changes, within the wider context of the Little Ice Age (LIA).[4] Most recently, Bruce Campbell, in his *Great Transition: Climate, Disease and Society in the Late-Medieval World*, offered a re-interpretation of Eurasian history (albeit concentrating heavily on England) from the twelfth to the late fifteenth century, through a bioenvironmental prism. In particular, the Great Famine, food shortages, and other economic crises of the late Middle Ages were influenced and shaped by a transition from the Medieval Warm Anomaly (MWA) to the LIA — a period marked by a high degree of climatic and biological instability.[5] In other words, for Davis, Fagan, and Campbell, food shortages and famines are exogenous crises. The exogenous explanation of famines is also backed by some environmental scientists, whose research establishes the correlation between climate change and food supply and forecasts the gradual decline in global food production levels over the course of the twenty-first century.[6] Likewise, the environmentalist interpretation of famines strongly prevails within public famine discourse. Thus, the 2011–2012 famine in Somalia is widely believed to have been caused by the 2011 East African drought, while the 2012 famine in West Africa is said to have been brought about by the 2012 Sahel drought.[7]

[2] The idea that Nature plays a dominant role as an historical trigger is expressed in Campbell, 'Nature as Historical Protagonist'.

[3] Davis, *Late Victorian Holocausts*.

[4] Fagan, *The Little Ice Age*.

[5] Campbell, *The Great Transition*.

[6] Norse, 'Multiple Threats to Regional Food Production'; Mendelsohn and Dinar, 'Climate, Water, and Agriculture'; Mendelsohn, Dinar, and Dalfelt, *Climate Change Impacts on African Agriculture*.

[7] NASA, 'Severe Drought Causes Famine in East Africa'; World Food Programme, 'Drought Affects Millions in the Sahel'.

To appreciate the environmentalist approach and its applicability to the Great Famine, it is necessary to establish a wider climatic context of the early fourteenth-century British Isles. The short-term weather anomaly of 1314–1316, which initiated the crisis, was itself part of a much wider context of climatic changes occurring from *c.* 1250/70 onwards. It is now generally accepted that by the second half of the thirteenth century (*c.* 1250–1270), some profound climatic changes were under way. After some two hundred years dominated by warm climate (the Medieval Climate Anomaly, or MCA), when average annual temperatures were similar to those of *c.* 2000 CE, the North Atlantic region entered a new climatic phase, transitioning from the MCA to the LIA. This transitional phase (*c.* 1250/70–1420), part of a much broader global climatic shift, manifested itself in a highly unstable climatic regime marked by high variability of year-to-year sea-surface and air temperatures and great variance in year-to-year precipitation levels. This was caused by a general weakening of the North Atlantic Oscillation (NAO), especially starting in the 1320s, to the point that by the 1430s it became strongly negative. This shift created stormy conditions and cold spells, which became predominant for the duration of the LIA.[8] These conditions stood in a sharp contrast to those of the MCA, dominated by a strong positive NAO, when strong winter Westerlies brought mildly wet and relatively warm weather to north-western Europe and arid conditions to the Mediterranean and North Africa.

In addition, there was a gradual reduction in solar irradiance, partially caused by the eruption of Samalas (Indonesia) in 1257, the single largest volcanic eruption in the last seven thousand years, which released between three hundred and six hundred megatons of sulfuric acid, and by five major volcanic eruptions in 1268, 1275, 1285/6, 1328/9, and 1342/3.[9] In particular, solar irradiance was depressed between *c.* 1280 and 1340, a period known as the Wolf Minimum. During this period, levels of solar irradiance were significantly lower than average for the period between the years 1000–1500.[10] Piecemeal weakening of the NAO, on the one hand, and reduced levels of sunshine, on the other, meant gradual cooling. Indeed, Greenland witnessed a period of severe cold spells, peaking in 1303, 1320, and 1353, while Iceland saw sea-ice formation along its northern coast in the 1310s and 1330s.[11] The period 1315–1318, overlapping

[8] Dawson and others, 'Greenland (GISP2) Ice'.

[9] Oppenheimer, *Eruptions that Shook the World*, pp. 263–67.

[10] Muscheler and others, 'Solar Activity during the Last 1000 yr'.

[11] Campbell, 'Panzootics, Pandemics and Climatic Anomalies', p. 186.

with the Great Famine years, coincided with unusually warm North Atlantic sea-surface temperatures, which created atmospheric conditions encouraging unusually wet, cold, and stormy weather all over northern Europe.[12]

What did these macroclimatic shifts mean for the environment and the organic economy of the British Isles? Obviously, they meant long-term and harsh implications for various sectors of the economy, particularly food production. This is reflected in the arable sector, dominating most of the English landscape (with the exception of south-west and north-west), south and south-east Scotland, manorialized marcher lordships in south and east Wales, and the Lordship of Ireland, overlapping, roughly speaking, with the eastern part of the island. As the chronology of English crop yields, meticulously reconstructed by Bruce Campbell, indicates, the year-to-year harvests were highly volatile, reflecting the wider climatic volatility. Apart from the disastrous harvests of 1315 and 1316 (and the mediocre one of 1317), analysed in Chapters 2 and 3, there were ten further mediocre-to-bad harvests between 1280 and 1351 (1283, 1293, 1304, 1321, 1328, 1339, 1346, 1349, 1350, and 1351).[13] The increasing storminess across the North Atlantic had some devastating consequences for coastal communities and their economies. Recent palaeoenvironmental investigations reveal several major sand-blowing and dune deflation events across coastal areas all over Britain, all the way from the Hebrides to south England, from the west coast of Ireland to east England. These events made the quality of soil in the affected areas less fertile and reduced their arable and pastoral production capacity, with some regions abandoning once-fertile lands altogether.[14]

The climatic shifts also had a pronounced impact on pastoral economy. Gradual cooling and the decreased solar irradiance have reduced the grazing season by approximately one month, with early May and late September being effectively eliminated.[15] The shortening of the grazing season meant that the annual biomass for grazing animals was now reduced by some 10–15 per cent. This, in turn, would have made livestock husbandry more challenging and costly than before, precisely in those areas where this type of agricultural enterprise formed the backbone of local economies. Although livestock husbandry played an important role in all parts of the British Isles, cattle- and sheep-based economies were the single most predominant type of economy in the upland

[12] Dawson and others, 'Greenland (GISP2) Ice', p. 431.

[13] Calculated from Campbell, *Three Centuries of English Crop Yields*.

[14] Oram, 'Worst Disaster'; Griffiths, 'Medieval Coastal Sand Inundation'.

[15] Oram and Adderley, 'Lordship and Environmental Change', p. 79.

communities of the Cambrian Mountains, covering much of Wales, the Gaelic parts of Scotland (Gàidhealtachd), and the Gaelic regions of western Ireland. Therefore, the reduction in the available vegetation biomass for grazing must have been particularly devastating in those upland Cymric and highland Gaelic areas practising almost exclusively pastoral regimes. It is equally essential to consider the influence of climate change on yet another sector of husbandry: fishing. Fishing played a particularly major role in coastal communities of Atlantic Ireland, Scottish settlements along the coastline of the North and Irish Seas, the Scottish Isles, the great lochs of the Highlands, as well as Welsh villages situated in proximity to the Irish Sea in the west. As some recent palaeoclimatological, archaeological, and palaeoichthyological evidence suggests, the post-1270 period was marked by increasing storminess in northern parts of Scotland. The cooling of sea-surface temperatures changed the migratory behaviour of herring and other deep-sea species, leading to a significant decline of commercial fisheries in Scotland.[16]

In other words, the climatic shifts had a devastating effect on the local environment, economies, and the availability of food resources, making local communities more vulnerable to food shortages and subsistence crises. This was especially true in 1315–1317, when the short-term weather anomaly and three back-to-back harvest failures created an acute food shortage and exposed local communities to an unprecedented degree of vulnerability. However, one should be careful and by no means consider the same short-term weather anomaly as a bringer of the famine. True, torrential rain did bring about three back-to-back failures, but it *did not* destroy the harvests entirely: the respective *gross* yields (inclusive of tithes and seed) stood, respectively, at 25, 35, and 14 per cent below average in 1315, 1316, and 1317, while the respective *net* yields (net of tithes and seed) fell by fell by 38, 50, and 26 per cent. Thus notwithstanding the harvest shortfalls, there still was some food.[17] In other words, Nature created food shortage, rather than famine. The same food shortage should be regarded as a tipping point, while famine is to be regarded as *one possible*, but *not inevitable* outcome of that tipping point. This is reflected in the geography of the early fourteenth-century famine: while famine ravaged, with various degrees of intensity, all over England, lowland (and possibly some upland parts of) Wales, and the Lordship of Ireland, there is no indication of famine in Scotland (pos-

[16] Oram, 'Worst Disaster', pp. 231–32; Serjeantson, *Farming and Fishing in the Outer Hebrides*, pp. 78–79.

[17] See below, Tables 2.3 and 3.2.

sibly with the exception of the English borderlands). In other words, to blame climate as the universal trigger of cataclysm means to reduce a historical argument to the oversimplified level of environmental determinism, which allows geographic and causational generalizations and does not account for other, no less decisive, factors. It is non-exogenous factors that would determine whether food shortage would become famine (as it did in England, parts of Wales, and the Lordship of Ireland) or whether food shortage would remain as such (as it did in Scotland and, most likely, western Ireland).

Demographic (Malthusian) Approach

An altogether different theory approaches famine from a demographic perspective. According to this view, any famine or food crisis is related to the interaction between population levels, poverty proportions, and the available food resources. The connection between population size, poverty, and vulnerability to famine has been already stated by Malthus in his classic 1798 *Essay on the Principle of Population*.[18] In a clearly oversimplifying manner, the main Malthusian argument can be summarized as follows. Economic growth is determined by the growth of two variables: human population and natural resources. Because of the passion between the sexes, humans reproduce at a fast pace and, as a result, human population grows geometrically, while natural resources grow arithmetically (Malthus used the analogy of the hare and the tortoise). It is inevitable, therefore, that human resources outstrip natural resources. This leads to overpopulation, demographic congestion, poverty, low living standards, socio-economic tensions, and — most importantly — vulnerability to shortages and famines. According to Malthusian logic, the more poor people there are, the more likely it is for a shortage to become an acute famine. Famines themselves, alongside wars and diseases, are regarded by Malthus as 'providential' or 'positive' checks that prevent populations from exceeding available landed resources, thus assuring that the problem of overpopulation and poverty is temporarily solved.

Malthusian views have influenced historians, sociologists, and economists alike. In particular, Malthusian theory was revived and expanded by Michael Postan in his work on late medieval English economy and society. According to Postan, incessant demographic growth forced more and more inhabitants to retreat to more 'marginal lands', marked by low fertility and depressed yields.

[18] Malthus, *An Essay on the Principle of Population*.

Conversely, high population levels mean that even fertile lands would become more marginalized: since grains represented the single most important dietary component of the human population, there would be a disproportional allocation of land-use to arable agriculture and the consequent over-exploitation of arable land at the expense of the shrinking livestock sector. Deficiency of pasturage and domestic animals means deficiency of manure, which is the single most important fertilizing agent.[19] For Postan, the Great Famine was a par excellence example of a positive check. In the same vein, Paul Ehrlich, in his 1968 *Population Bomb*, forewarned that an unchecked population will lead into environmental degradation, global poverty, food crises, and eventually mortality.[20] Although not adhering explicitly to the Malthusian theory, Amartya Kumar Sen, arguably the single most influential famine theorist of the twentieth century, has convincingly shown that poverty is a decisive factor, as far as famines are concerned: the bigger the proportion of poor people, the more vulnerable a society becomes to various socio-economic stresses.[21] One obvious consequence of harvest failures is a rise in food prices, while excessively high prices inevitably mean the decline of the collective purchasing power of commoners because of the lack of financial resources or any other means of exchange. Fewer and fewer people can now afford to buy food, while the gap between the commodity prices and the available funds to buy them becomes wider and wider. The degree of decline of the collective purchasing power is largely determined by the real wages and by the size of a deprived population, *in relation to available resources*.

Can Malthusian logic be applied to the Great Famine? Surely, if Nature can create a tipping point, demography can intensify the same tipping point, by adding more stress. This is especially true when thick population densities (high population levels in relation to the available land and food resources) and, consequently, large proportions of poor echelons within a population create the Malthusian scenario. One of the questions debated among economic historians is whether England was a Malthusian society in the pre-Black Death era. The answer depends on estimates of total population levels, proposed by different scholars. Up until now, no consensus has been reached and current estimates vary from 4.5–4.75 million proposed by Bruce Campbell and his colleagues to the higher figures (5.5–6 million), suggested by Richard Smith and other

[19] For instance, Postan and Titow, 'Heriots and Prices on Winchester Manors'.

[20] Ehrlich, *The Population Bomb*.

[21] Sen, *Poverty and Famines*.

scholars.[22] Engaging in this debate lies outside the scope of the present study, but it appears that the levels of the English population in the early fourteenth century may have been higher than those proposed by Campbell and his colleagues; as one recent study on the first outbreak of the Black Death has shown, population losses may have been higher than previously thought, and most certainly higher than 50 per cent.[23] Also, one should not forget the outbreak of the *pestilencia secunda* of 1361–1362, which may have claimed a further 10–15 per cent.[24] If there were about 2.5 million people by the time of the first poll tax of 1377, as estimated by Campbell and his colleagues, then there must have been at least five million people (and probably more) in the early fourteenth century. For the rest of the study, I shall stick to the figure of 5.25 million, acknowledging all caveats, drawbacks, and uncertainties of this estimate.

England's overpopulation is reflected in the relationship between her population size and the available land resources, measured here in population densities. With about 5.25 million people and about 12.72 million arable acres at their disposal, the average population density in England may have been in the area of just over 100 people per square mile (= 40 per square km).[25] The situation varied greatly across counties. Thus, it appears that Norfolk, the single most densely populated county, may have had approximately 263 inhabitants per square mile (= 102 per square km), followed by Huntingdonshire with some 200 inhabitants per square mile (= seventy-seven per square km). Middlesex had an even denser population of approximately 302 inhabitants per square mile (= 117 per square km), but this has to do with the presence of London, a home of some 60,000 inhabitants. On the other hand, Cornwall had only some 28 inhabitants per square mile (= 11 per square km), followed by Lancashire with about 36 people per square mile (= 14 per square km).[26] Taken

[22] Broadberry and others, *British Economic Growth,* pp. 20–27; Smith, 'Plagues and People'; for an even higher (but obviously unrealistic estimate) of 6.5–7 million people, see Hallam, 'Population Movements in England', pp. 536–37.

[23] Arthur, 'The Black Death and Mortality'.

[24] Mullan, *Mortality, Gender and the Plague of 1361–2.*

[25] It must be made clear that the estimate of 12.72 million arable acres was made by Broadberry and his colleagues on the assumption that there were 4.7 million people in England (Broadberry and others, *British Economic Growth*, p. 74). While suggesting that the total population of 5.25 with the total arable of 12.72 million acres may appear somewhat problematic, solving this issue lies outside the scope of the present study.

[26] Recalculated (pegged to the total of 5.25 million) from Broadberry and others, *British Economic Growth*, pp. 25–26 and 51–67.

as a whole, England definitely appears to have been heavily overpopulated by the standards of the pre-Industrial era. This is indicated not only by the sheer size of her population and population densities but also by her socio-economic composition and resource distribution.

A close analysis of social structure, income share, and land tenure reveals that around the year 1300 England was a fundamentally 'bottom-heavy' society, with the vast majority of the population comprised of poor commoners. According to one recent estimate, based on the total population estimate of 4.7 million, rustics constituted about 80–85 per cent of the total population, the vast majority of whom (about 75–80 per cent of all rustics, or about 64 per cent of the entire population) held disproportionately small holdings and captured only about 36 per cent of the country's income. Overall, about 69 per cent of the population held about 40 per cent of all income and some 33 per cent of all land.[27] The traditional Anglo-Saxon ideal of one household holding one virgate (that is, thirty acres of land) has long been gone. In fact, 60–65 per cent of England's population was living on less than a virgate (thirty acres or so), and perhaps as much as half of her population held only ten acres of land or less.[28] In more densely populated regions, such as Norfolk, a substantial proportion of population had to get by on less than five acres.[29] Because of the overpopulation and widespread poverty, the gap in inequality among tenants within the same communities was not that remarkable: as one recent estimate suggests, in *c.* 1290 the Gini coefficient stood in the area of 0.33–0.37.[30] According to Campbell and his colleagues' estimate, based on the estimate of the total population of 4.7 million, as much as about 41 per cent of England's population lived below the poverty line *c.* 1300, and it is possible that this proportion was higher still, if we assume that the total population was in the area of 5.25 million.[31]

In other words, mixed-farming regions of England (with the exception of the south-west, the Welsh Marches, and the northern counties) may well be a classic Malthusian society, prone to a considerable socio-economic stress and vulnerable to environmental and biological disasters. But the same cannot be

[27] Campbell, *The Great Transition*, p. 169.

[28] Broadberry and others, *British Economic Growth*, pp. 317–18; Dyer, *Standards of Living*, pp. 118–27.

[29] Campbell, *The Great Transition*, p. 185.

[30] Broadberry and others, *British Economic Growth*, pp. 308 and 328–29.

[31] Broadberry and others, *British Economic Growth*, pp. 317–18, and 321.

said of other parts of the British Isles; neither Scotland, nor Ireland or Wales can be classified as a Malthusian society. The population estimates for these lands is even less satisfactory than those for England, but the best guess would be that, taken as a whole, the respective populations of Ireland, Scotland, and Wales may have been in the area of 1.3, 0.75–0.8, and 0.3–0.33 million people, implying population densities of about forty-four, thirty-four, and forty people per one square mile, respectively.[32] In other words, there was between 2.4 and 3 times more land per capita available in Ireland, Scotland, and Wales than in England. True, a large proportion of land in Highland Scotland, the Welsh Uplands, and west Ireland, no good for arable cultivation, would be used for pastoral husbandry. Although our information regarding an average size of an Irish, Scottish, or Welsh husbandman's land is virtually non-existent, it was undoubtedly much larger than a land-plot of an average English tenant. Thus, Barrow has assumed that the average size of a Scottish husbandman's land-plot may have been around two oxgangs (twenty-six acres), but it appears as a rather generalized statement, based on scarce information from a limited geographic range.[33] In any event, the lower population densities in Gaelic-speaking lands may nevertheless imply larger land-plots than in England and, potentially, more generous per capita income. Indeed, Campbell has concluded that 'England by *c.* 1290 would have lost this per capita economic lead [over other parts of the British Isles] altogether'.[34] This is notwithstanding the relative economic autarky of those largely pastoral and almost exclusively rural societies. To put it differently, England was a backward economy (when compared to the Low Countries and Mediterranean Europe) and a Malthusian society, while her Gaelic and Cymric neighbours were backward economies, but non-Malthusian societies.

According to Malthusian logic, then, famine should have occurred in the mixed-farming regions of England, while other parts of the British Isles should have been either spared, or, at the very least, experienced some sort of food crisis, but not a full-fledged famine. Indeed, the same Malthusian logic may explain why food shortage became famine in densely populated areas of England and why Scotland seems to have escaped famine, but it fails to account for the presence of acute famine in thinly populated areas in northern England

[32] Campbell, 'Benchmarking'; Campbell and Barry, 'The Population Geography of Great Britain', pp. 59–60.

[33] Barrow, *Kingship and Unity*, p. 18.

[34] Campbell, 'Benchmarking', p. 41.

(especially in the north-west), the Lordship of Ireland, and parts of Wales. All this, in turn, indicates the limitations of the Malthusian interpretation of famines. Just as with Nature, pointing the finger at demography as a sole factor responsible for famines is conceptually dangerous because it pushes us into the domain of deterministic reasoning. But even taken together in tandem, Nature and demography provide only a partial appreciation of such a complex and eclectic phenomenon as famine. Both Nature and demography can be regarded as very slowly changing *longue-dureé* super-structures, serving as 'passive' contexts in which food shortages may or may not turn into famines. It is anthropogenic factors, related to humans, human decisions, and institutions, however, that play a much more active role in determining whether food crises transform into famines. It is only when human factors are considered together with the exogenous and demographic factors that we can appreciate why famines happen — or not.

Institutionalist Approach: Sen's Entitlements

Despite their theoretical and conceptual merits, the environmentalist and demographic approaches to famine are overshadowed by the influence of the third school of thought, the 'institutionalist'. This approach has been advanced by social and economic historians, development economists, and political scientists. One institutional aspect is the idea of 'food entitlements', first expressed in 1981 by the Nobel Prize–winning economist and philosopher Amartya Kumar Sen. In his classic *Poverty and Famines*, Sen blames humans and institutions for creating famines, by misallocating the scarce food resources and denying the poor masses of their basic entitlements to food. It is vitally important, according to Sen, to distinguish between the 'food availability decline' (FAD) and 'food entitlement decline' (FED). The former is the objective unavailability/non-existence of food supply, while the latter is an intentional displacement of food resources, whereby better-off elements mobilize their capital to appropriate the very scarce food resources, at the expense of poorer socio-economic strata, who, in turn, lose their entitlement to food resources. In other words, it is vital to distinguish between *real* and *artificial* deficiencies of food. *Real* deficiency (= FAD) is an objective shortage/unavailability/non-existence of the food supply, while *artificial* deficiency or non-existence (= FED) is brought about by intentional displacement of food resources from place A/group A to/by place B/group B.[35]

[35] Sen, *Poverty and Famines*.

Despite some criticisms, Sen's view of famine as a FED-induced, rather than a FAD-caused, phenomenon remains highly influential.[36]

Institutionalist Approach: Manorialism and Serfdom

Sen's views on the connection between poverty and famine were undoubtedly influenced by Karl Marx's theory of class relations. For both Marx and Sen, famine was a strictly class-discriminating phenomenon, implying the ability of better-off elements to mobilize their resources to deprive worse-off elements of their access to food. In his *Das Kapital*, Marx used the examples of the Rice Famine in India (1769–1770) and the Irish Potato Famine (1845–1852), to illustrate that famines are tools used by capitalists to get richer.[37] In the same vein, Sen contended that there is no such thing as a 'general famine'; elites never starve because they never lose their food entitlements.[38] Although neither the Great Famine, nor any other late medieval food crisis received their deserved attention by Marxist historians, the latter did discuss the impact of lord-tenant relationships on commoners' poverty in the context of the late medieval manorial system. The essence of the manorial system revolves around the seignorial land tenure of units called manors, comprised of lords' demesnes and peasant tenancies. These tenancies were being held in either freehold or in villeinage. Although both free and unfree tenants had to pay their rent in either labour or cash, there is no doubt that unfree peasants were at disadvantage. The main difference between free and unfree peasants was in their legal status. In addition to their inability to move freely, villeins were subjugated to various socio-economic and financial burdens, such as customary dues and fees. These included such dues as 'heriots' (land inheritance payments), 'merchets' (marriage licence payments), 'chevages' (a licence fee to reside outside of a lord's manor), and 'tallages' (extraordinary aid contributions). Manorialism was present, with various degrees of intensity, all over England. Outside England, it was found within the Lordship of Ireland, the marcher lordships in south and east Wales (especially in Monmouthshire and Denbighshire), and, to a smaller extent, along the east coast of Scotland. In the Welsh Uplands, western Ireland, and Highland Scotland, manorialism was largely unknown.

[36] For the criticism of Sen's 'entitlement theory', see, for instance De Waal, *Famine that Kills*, pp. 59 ff.; Howe and Devereux, 'Famine Intensity'; and Rubin, 'The Entitlement Approach'.

[37] Marx, *Capital*, I, Chs 25 (pp. 854–71) and 31 (p. 917).

[38] Sen, *Poverty and Famines*, pp. 39–44.

About half of English tenants were freeholders, while the other half was made up of unfree or servile tenants (interchangeably known as villeins). Obviously, the situation varied across counties. Thus, serfdom was very weak in south-east Sussex, most of Kent, east Essex, south-east Suffolk, and north-west Lincolnshire, as well as along the western line stretching from south Westmorland, through most of Lancashire, Cheshire, Shropshire, and west Staffordshire, to Herefordshire. Similarly, a fair proportion of free men was found in the east counties, all the way from north Lincolnshire, through most of Norfolk, Cambridgeshire, Huntingdonshire, Suffolk, east Essex, Hertfordshire, Bedfordshire, east Buckinghamshire, Middlesex, east Surrey, and east Sussex. Elsewhere in the country, within the 'central province' stretching diagonally all the way from the Scottish borders to Cornwall, the presence of unfree tenancy was more paramount.[39]

Although comparable statistical material is missing altogether for the Gaelic and Cymric regions, it seems that here the proportion of unfree tenants has been smaller than in England. This has to do with the fact that the manorial system in the Lordship of Ireland, south and east Wales, and east Scotland was less intense. Unlike their English counterparts, local lords, such as tenants-in-chief of the king of Scots, substantial Anglo-Norman lords in Ireland, and great marcher lords in south and east Wales, showed little interest in direct exploitation of their lands and, in many cases, leased them out to their tenants. The majority of Scottish tenants were free husbandmen, and only a minority of peasants was constituted by unfree Gaelic-speaking bondmen known as *neyfs*. A similar situation prevailed in the Lordship of Ireland, where two groups of unfree peasants, *betaghs* and cottars, seem to have constituted a minority of all tenants. Here, unfree peasants were almost invariably Irish, rather than Anglo-Normans. Although Scottish *neyfs* and Irish *betaghs* and cottars suffered from legal disadvantages compared to freemen, they were much better-off, both legally and socio-economically, than their English counterparts. Thus, labour services and customary payments of Irish *betaghs* appear to have been less burdensome than in England.[40] Likewise, in manorialized parts of Wales, serfdom was in decline during the high Middle Ages, and by *c.* 1300, serfs (*taeogion*),

[39] Campbell, 'The Agrarian Problem', p. 35.

[40] On Scottish *neyfs*, see Taylor, 'Homo Ligius and Unfreedom', pp. 103–08; Neville, *Land, Law and People*, pp. 66–68. On Irish *betaghs*, consult Smith, *Colonisation and Conquest*, pp. 78–79; Nicholls, *Gaelic and Gaelicised Ireland*, p. 70; and Down, 'Colonial Society and Economy'.

consisting of both Welsh and English tenants, constituted only a tiny minority of the population.[41]

The impact of the manorial system on English society and economy is a subject of much scholarly debate. For Marxist historians, such as Rodney Hilton, Robert Brenner, and Guy Bois, it was the manorial system that contributed to the widespread poverty and misery of peasants. Manorialism was both exploitative and inefficient in its nature. As such, it hindered the economic development of peasants, through a series of customary obligations and dues payable by tenants to their lords; these burdens can be seen as 'surplus extraction'. 'Surplus extraction' forced deprived peasants into smaller and marginalized holdings, which implied diminishing returns, reduced income, and lack of capital (and encouragement) to invest in land production and economy. This resulted in a 'productivity crisis', where yields were low, and many peasant families barely made ends meet. The dichotomy between the exploiting lords and exploited peasants (aka, the 'working class'), led, inevitably, to class contradiction and conflict.[42]

According to Marxian logic, one would expect to see a degree of correlation between the geography and intensity of manorialism and serfdom, on the one hand, and the geography of famine, on the other. But was that the case? As we shall see later in the book, various manifestation of crisis and famine (including low crop yields, high prices, hoarding, and speculation, as well as excess mortality and reduced nuptiality) can be detected in both 'free' Kent, Norfolk, Suffolk, Shropshire, and Cumberland, and in 'servile' Hampshire, Oxfordshire, and Warwickshire. Moreover, the presence of numerous and obvious signs of famine in south and east Wales (including the Dyffryn Clwyd Lordship in Denbighshire and a number of Monmouthshire manors) and the Lordship of Ireland, undermine this connection even further. Famine did not discriminate between free and unfree; it did, however, discriminate between the better-off and the worse-off. It should be borne in mind that serfs were by no means worse-off than free tenants. While there is no doubt that manorial customs and seignorial demands were in many instances burdensome, the impact of coercive seignorial power on poverty and backwardness seems to have been exaggerated by Marxist historiography. In many cases, serfs held larger plots of land than

[41] Davies, *Lordship and Society*, pp. 247, 383–87.

[42] Hilton, *Class Conflict*; Brenner, 'Agrarian Class Structure'; Aston and Philpin, eds, *The Brenner Debate*; Dyer, Coss, and Wickham, eds, *Rodney Hilton's Middle Ages*; Bois, *The Crisis of Feudalism*.

their free neighbours. Moreover, customary rents, paid by tenants, were fixed and amounted to only about one half of their real market value. Such arrangement was undoubtedly beneficial for that group of peasants and cumbersome for their lords. It is likely, as Campbell pointed out, that such scenarios encouraged tenants, both free and unfree, to be actively involved in the land market, which caused the land to be subdivided further and further (a process sometimes referred to as 'morcellation'), in order to meet the increasing demands of a growing population.[43] It was that process of subdivision that allowed the population to grow unchecked, and which, in turn, led to increasing impoverishment and declining living standards within English society. In other words, according to Campbell it was the limitations rather than the power of seignorial institutions, on the one hand, and the ability rather than the impotence of tenants, on the other, that contributed to widespread poverty in England. But what about other parts of the British Isles? Land morcellation, driven by demand for land, is a feature of arable, rather than pastoral, landscapes. As we have seen, Gaelic and Cymric lands were more pastoral than England, and this fact may explain why there is no evidence of aggressive land morcellation and, consequently, population pressure. Yet, famine is well documented in both the Lordship of Ireland and marcher lordships of Wales. This shows that the connection between overpopulation and the intensity of land market, on the one hand, and the intensity of starvation, on the other, is not to be taken for granted. People were starving in very densely populated Norfolk, where land subdivision was an omnipresent reality, as much as they were starving in thinly populated Cumberland, the Lordship or Ireland, and the Dyffryn Clwyd lordship, where the land market was very under-developed.

Institutionalist Approach: Grain Markets

Grain market performance and its impact on famines is yet another contested concept. In his monumental *Wealth of the Nations* (1776), Adam Smith distinguished between 'dearth' and 'famine' and contended that the former would not transform itself into the latter, as long as free markets and an efficient network of interregional and international grain trade function properly. Free trade, associated with 'good' (liberal) governments, saves the situation by allowing grain merchants to transfer crops from famine-free regions and countries to starving lands, thanks to light tariffs or duty-free trade. Empty

[43] Campbell, 'The Agrarian Problem'.

food shelves get replenished again, and grain trade carries on. The availability of foreign foodstuffs also forces grain hoarders to release their stored supplies into local markets because reasonable prices make grain storing unprofitable and uncompetitive.[44] Conversely, Amartya Sen and Robert Fogel held less optimistic views. According to Sen, the misallocation of resources and the collapse of food entitlement in the context of food stress are closely related to market malfunctioning. Here, Sen blamed crop producers and merchants for creating artificial panic by hoarding and driving up food prices, thus, creating famine.[45] In the same vein, Robert Fogel noted the negative effect of markets on food crisis. In particular, Fogel contended that harvest failures and food shortages become famines because of the inelasticity of demand, which cause food prices to go up well beyond their supposed level, to the point that worse-off elements cannot afford to buy their most basic commodities.[46]

The Smithian, Senian, and Fogelian models that focus on the impact of markets on famine relief and famine intensification have been applied and tested by a number of economic historians to historical famines. In particular, much work on the functioning of markets in the context of historical famines has been done by Cormac Ó Gráda, who found that markets could either function normally or malfunction altogether.[47] What about the Great Famine? In the context of the early fourteenth-century crisis, grain markets were closely related to the processes of population growth and land morcellation. The subdivision of land into smaller parcels meant that a certain proportion of smallholders were unable to feed their households with their own produce only and had to resort to local markets to take up the slack. Much work has been done on markets and commerce in England in the later Middle Ages and the overall consensus is that *c.* 1300 England, notwithstanding various barriers, managed to achieve remarkable levels of commercialization. Between the Norman Conquest of 1066 and *c.* 1300, there was a pronounced rise in the volume and velocity of money in circulation.[48] On the eve of the famine, the total amount of the currency in circulation in England was in the area of between £1.2 and

[44] Smith, *Inquiry*, Bk 1:1–3.

[45] Sen, *Poverty and Famines*, pp. 160–62.

[46] Fogel, *Escape from Hunger*, pp. 5–7.

[47] Most notably: Ó Gráda, 'Markets and Famines', and *Famine: A Short History*, pp. 129–58.

[48] Mayhew, 'Modelling Medieval Monetisation', pp. 55–77; Allen, 'Silver Production'; Latimer, 'Money and the English Economy'.

£1.5 million.[49] Population growth and a rise in the volume of money in circulation had a positive impact on the proliferation of trade hubs. On the eve of the Black Death (1348–1351) there were around 2100 settlements with 'formal' markets and fairs in England, in contrast with only around 500 such places in *c.* 1200.[50] In Norfolk, one of the most commercialized counties, there were 283 markets and fairs in some 175 vills (that is, some eight per 10,000 inhabitants and fourteen per ten square miles). But even in a more 'backward' Lancashire, there were eighty-two trade hubs (about fourteen per 10,000 inhabitants and five per ten square miles).[51]

The single most demanded product was crops (grains and legumes). The efficiency of grain markets is attested by a fair degree of market integration across different trade *loci* in England, as reflected in similar or nearly identical prices across regions (the 'Law of One Price') and by trans-regional market involvement, based on commercial collaboration and networking between different regions within the same county or region.[52] An active role in the grain trade was played by grain traders (corn-mongers), acting as middlemen between producers and consumers, both urban and rural.[53] Although much business was done in cash, credit transactions played an important role in different sectors of late medieval trade.[54] In some cases, lords were willing to sell their grain on credit to corn-mongers, while some grain traders were found to have served as creditors to both landlords and peasants.[55] Despite its relative efficiency, the English grain market was not without limitations. The most obvious flaw was the under-developed character of international trade. True, there were *some* imports of foreign grain by Hanseatic merchants, chiefly from the Baltic region, and by Italian businessmen from the Mediterranean.[56] The volume of grain trade, however, was pathetically meagre, and it is unlikely that the sheer volume of foreign grain imported in any given year *c.* 1300 exceeded 50,000

[49] These estimates derive from Allen, 'First Sterling Area', esp. pp. 86 and 91, on the assumption that England had about 82 per cent of all the currency within the Sterling Area (England, Wales, Ireland, Scotland, and the Isle of Man).

[50] Retrieved from Letters, *Gazetteer*.

[51] Slavin, *Bread and Ale*, pp. 2–4.

[52] Galloway, 'One Market or Many?'.

[53] Campbell and others, *A Medieval Capital*.

[54] Schofield and Mayhew, eds, *Credit and Debt*; Briggs, *Credit and Village Society*.

[55] Briggs, *Credit and Village Society*, pp. 37–39, 47–51, and 61–62.

[56] Hybel, 'The Grain Trade'.

quarters of grain, capable of feeding no more than 40,000 people.[57] As we shall see in Chapter 4, the figures were considerably lower during the famine years.

Outside England, however, the degree of monetization, commercialization, and marketization was much more limited. Here, officially chartered and recognized markets were limited to the Lordship of Ireland, more manorialized parts of south and east Wales, and south-east and east Scotland, but their density was not nearly as thick as in England. Judging by coin hoards from these regions, local economies did achieve a degree of monetization, but its extent was by far more restricted than in England.[58] In the Gaelic world of west Ireland and Highland Scotland, as well as the Welsh Uplands, local trade systems were based on barter rather than cash or credit. In the Gaelic and Cymric societies, cattle were regarded not only as signs of socio-economic prestige, but also as universal exchange currency.[59] This was not only because of the tribal, rather than centralized, socio-political system, but also because of the pastoral nature of local economies, where grain production and consumption played a much more limited role than in lowland anglicized regions.

According to the Smithian model, efficient grain markets and international trade would come to rescue during the crisis years and prevent shortage from transforming into famine. This, however, was certainly not the case in England, where grain markets were certainly sophisticated for their era. On the contrary, as we shall see in Chapter 4, there were widespread signs of market malfunctioning, which made access to grain stalls virtually impossible for the majority of commoners. At the same time, the very limited volume of imports of international grain certainly did not help the situation. Ironically, it was under-marketized parts of the British Isles, namely Gaelic Ireland, Highland Scotland, and the Welsh Uplands, that seem to have been avoided famine altogether. This may support the view that efficient markets may have negative effect on food crises, through transferring food from worse-off locations, communities, and individuals to those with sufficient purchasing power.[60] It is equally important to bear in mind that more optimized trade systems are made up of more com-

[57] This is based on the idyllic scenario that each year, each of eighteen trade ports would receive, on average, 2500–2800 quarters of grain. In reality, the figures seem to have been lower than that, as the surviving custom accounts reveal (see Chapter 4).

[58] Allen, 'First Sterling Area'.

[59] Britnell, *Britain and Ireland, 1050–1530*, p. 10; McCormick, 'Decline of the Cow'; Davies, *Wales in the Early Middle Ages*, p. 53.

[60] Drèze and Sen, *Hunger and Public Action*, pp. 89–90.

plex infrastructure mechanisms; once one part of these mechanisms is affected, other parts will be inevitably affected too, through some sort of chain reaction, causing the entire mechanism to malfunction. The effect of this collapse can be devastating and costly. Conversely, less efficient markets may not exacerbate food crises precisely because of their simple infrastructure.

Institutionalist Approach: Warfare

In his *Wealth of the Nations*, Smith also pointed out that warfare could be another mechanism reducing the available food resources, through plunder, crime, and import bans, and, thus, creating famine.[61] The impact of war on food crises is yet to receive its full-scale scholarly treatment, but in recent years there are more studies heading in this direction. In particular, scholars such as Richard Tucker, Lisa Brady, and myself pointed out the negative implications of warfare on environmental destruction and, consequently, food availability.[62] The role of war in intensifying food shortage (and thus, creating the FED conditions) is also a central theme in Sen's *Poverty and Famines*, where he places the Bengali famine of 1942–1943 in the context of the Pacific theatre of WWII. According to Sen, wars tend to misallocate the scarce food resources from communities to belligerents and conflict epicentres. This misallocation also has a negative effect on markets, as chaotic dearth leads to the decline in information value, price rise, and speculation.[63]

In the early fourteenth century, the British Isles were divided into the 'war zone' and 'war-free zone'. During the famine years, the 'war zone' encompassed England north of the Humber, the Lordship of Ireland, parts of Gaelic Ireland, and south Wales. In England, the northern counties were being ravaged by Scottish marauders, following the embarrassing English defeat at Bannockburn (23–24 June 1314). In Ireland, warfare involved Scottish and English armies, as well as Anglo-Norman forces and pro-Scottish Gaelic clans. South-east Wales was devastated by the revolt of Llywelyn Bren, a local anti-English leader, in early 1316. Within the 'war zone', there was much environmental destruction and the plundering of food resources, which undoubtedly intensified the crisis. Conversely, Scotland was spared of raids or invasions during the famine years,

[61] Smith, *Wealth of Nations*, Bk 4:5.

[62] Tucker and Russell, eds, *Natural Enemy, Natural Ally*; Brady, *War upon the Land*; Slavin, 'Warfare and Ecological Destruction'.

[63] Sen, *Poverty and Famines*, pp. 52–86.

and this fact undoubtedly helped this country to avoid famine. This, however, does not mean that Scotland did not experience some sort of food crisis: years of English invasion and the devastating campaign of King Robert I Bruce subjugating local chieftains in 1307–1309 resulted in environmental destruction with long-term and far-reaching consequences. Indeed, food crisis within the 'war zone' lasted beyond the famine years. Warfare was undoubtedly a decisive factor in aggravating the disaster within the 'war zone', but, just as with other factors, it cannot alone account for the famine.

Proposed Approach: 'Complexity Theory'

In other words, each famine model has its own advantages and limitations. While one variable *may* explain one *aspect* or one *locality* of famine, it certainly cannot provide a comprehensive and convincing explanation of famine as a whole. All it can provide is a partial answer. Thus, the environmentalist approach may explain *how famines are being initiated*, but not *why they happen*. The Malthusian approach highlights the connection between overpopulation and poverty, but it ignores the connection between exogenous shocks and the *availability* of per capita food supply in the context of food crises. The analysis of the impact of market performance on famine is equally important, but its sole focus on market forces does not allow us to consider the link between reduced harvests and empty stalls, on the one hand, and the relation between poverty and food trade disruption, on the other. Arguably, it is the exclusivist nature of these models that prevents a holistic and comprehensive approach to the study of famine and reduces any explanation to the oversimplified level of environmental/demographic or anthropogenic/institutional determinism.

Famines are highly complex and eclectic phenomena and as such, they cannot be caused or explained by just one uniform factor. Individually, environmentalist, Malthusian, or institutional explanations just cannot do it. To understand the high complexity of famines, a much more balanced and complex model is needed, which takes into account several factors and treats them as potentially equal and important causes of famine. Here, I propose to make use of the so-called complexity theory, pioneered and developed by economists and environmental and life scientists, but still largely under-appreciated by historians. Complexity theory stresses the inability to reduce any explanation to a single or few variables and the importance to treat each particular context as potentially unique. In the case of historical disasters, it is imperative to 'break down' the analysis into smaller and diversified variables, in accordance with economic and ecological complexity models, and it is only through internal variations within specific contexts rather than by singular macrodynamics that the causes, nature, and consequences of these disasters can be appreciated. Recently, this approach has been undertaken by Eoin Flaherty in his study of the geography of the Great Irish Famine of 1845–1852.[64] In the same vein, I propose to treat the Great Famine as a highly complex phenomenon, caused, created, and intensified by a combination of exogenous, endogenous, and demographic factors, acting together as a great three-headed hydra. To understand the nature and dynamics of each of the three 'meta-structures' (Nature, demography, and institutions), each 'meta-structure' is to be broken down into smaller variables contained within. For instance, Nature is not only 'Nature' in the abstract sense, but a series of complex variables, including weather, topography, elevation, and vegetation. 'Demography' is not just down to the absolute number of people, but to the relationship between population and landed resources and the proportion of poor elements within a given population. Institutions may be, in effect, anything related to various types of human political, socio-economic, technological, and cultural systems, as well as actions and decisions. Together, these variables make each 'meta-structure' work; while acting together, the three 'meta-structures' make famine.

The mechanism of famine may be likened to the mechanism of a clock, where a large number of small mechanical parts make the entire mechanism work. To understand how these variables operate within each 'meta-structure', it is essential to consider as many of these variables as possible. Environmental variables are to be considered first, since Nature *initiates* (but does not create) famine. How destructive was weather? How much did it destroy? How were different landscapes and types of land-use affected? Institutional and demographic variables are to be considered next. Did agricultural customs and strategies reduce the dwindled food supply further? Did trade and markets contribute to the disruption of the food supply? Did transportation challenges and storage faults decrease the existing amounts of food even further? What was the impact of armed conflicts (if there were any) on food supply and the ability of people living in a war zone to secure food? Finally, once the various natural and anthropogenic mechanisms of food depletion have been established, it is vital to consider the levels of population in relation to the available food supply and the purchasing power of various socio-economic strata to buy scarce food supplies (at excessively high prices). Once all these various mechanisms

[64] Flaherty, 'Assessing the Distribution of Social-Ecological Resilience and Risk'.

are considered, we may endeavour to answer the most pressing question: what creates famine? Such a complex approach stands in contrast with existing models of famine, which focuses on one meta-structure or one factor, providing insufficient or unsatisfactory explanations, and it shall guide us throughout the present study.

Chapter 2

The Malevolence of Nature: Floods and Harvest Failures

The Greatest Flooding Ever Remembered

The rain started in the late summer of 1314. As the biography of Edward II, *Vita Edwardi Secundi*, narrates, the rain was so heavy that the harvest of 1314 could hardly be gathered and stored at local barns.[1] Although heavy rain (or excessive drought) was nothing new to that generation, it was not before too long that people began to realize that the flooding of 1314 was in a different league, implying consequences of disastrous proportions. There is no doubt that the people who viewed it were impressed and terrified by the flooding spectacle. Some chronicles speak about the 'greatest flooding of rain'; some other merely remark on a 'great rainfall'. Some sources talk about *inundatio aquarum* ('inundation/flooding of water' in Latin), meaning specifically coastal/riverside inundation. For instance, a Durham chronicler tells of a great inundation of the River Wear, while a Thorney Abbey annalist narrates how flood water went over and covered the marshes of Cambridgeshire Fens.[2] It is essential to bear in mind that the chroniclers had a clear agenda to make a crisis appear particularly apocalyptic — as the divine retribution for human sins. Thus, they aimed to sound as striking and dramatic as possible. Also, they tended to use phrases and allusions to scripture or other writers to brag about their learning. Finally, they often were ill informed, basing their narrative on rumours and 'news' of events

[1] *Vita Edwardi Secundi*, ed. by Childs, pp. 110–11.

[2] *The Chronicle of Robert Greystanes*, ed. by Raine, pp. 96–97; *Thorney Annals*, ed. by Hart, p. 33.

of which they often had no personal knowledge. As such, chroniclers' descriptions should always be taken with a grain of salt.

A detailed chronology and seasonality of rain is difficult to reconstruct because of relative paucity and fragmented nature of references. Most information derives from manorial accounts, which recorded demesne crops and contained occasional references to weather. It should be emphasized that manorial accounts, too, could exaggerate the severity of the weather, here and there, to avoid the imposition of financial penalties on manorial officials by their lords. Equally, it should be borne in mind that manorial officials sometimes tended to manipulate the contents of their accounts, such as the figures for prices of grain sold, the amounts of grain harvested, or the wages paid in cash and kind. To avoid embezzlement, manorial lords hired auditors, who occasionally spotted the false figures.[3] While we have to accept the possibility of reeves reporting false figures, we can assume that this was done on a smaller scale during the short-term weather anomaly of 1314–1316, given the ubiquitous references to the rainy weather and abysmally low crop yields. With this in mind, we may attempt a tentative chronological reconstruction of weather conditions during the crisis. By harvest time, namely around August/September 1314, the rain was strong enough to disturb demesne harvesting. One manorial account from northern Norfolk indicates that demesne oats did not grow well because of the rain, but it is unclear if this can be taken as a hint that the rain started much earlier than the harvest of 1314, or if oat growth was hindered only closer to that harvest.[4] One Kentish account noted *inundatio aquarum* during harvest time in 1314.[5] As some accounts indicate, the harvest of 1314 was later (*serum*) than usual because of the rain.[6] Unfortunately, very few accounts record the length and chronology of demesne harvests, and those that do are confined to the Norfolk demesnes. In addition, there are a small number of accounts from other counties indicating the duration of demesne harvests, without specifying the dates. As far as we can tell from the surviving Norfolk accounts, the 1314 harvest began, on different demesnes, around mid-August and lasted until around mid-September.[7] Elsewhere, the demesne harvest started in early

[3] On the topic of the embezzlement by manorial officials, consult Carlin, 'Cheating the Boss'.

[4] NRO, DCN 60/18/19 (Hindolveston).

[5] CCA, DCc/BR/East Farleigh/10 (East Farleigh, Kent).

[6] Thus: NRO, DCN 60/18/19 (Hindolveston) and NRO, DCN 60/26/15 (Monks Granges), both in Norfolk.

[7] Thus: NRO, Hare 213x1/4277 (Wimbotsham); DCN 60/18/19 (Hindolveston); DCN 60/26/15 (Monks Granges); DCN 60/35/18(a) (Sedgeford); and DCN 60/35/17a (Taver-

September and continued until early October.[8] While the harvest indeed started considerably later than usual (in 'normal' years, the harvest would commence in very late July or early August), its average length was only marginally shorter than normal, by four days (Table 2.1).[9] With the exception of oats, the 1314 demesne harvest was bountiful, standing at some 22 per cent above average.[10] The manorial authorities undoubtedly appreciated the abundance of the harvest and did their best to collect it and shelter from it from the adverse weather conditions, as fast as possible. This put extreme pressure on local harvesting force, and in some instances, additional harvesters were hired to expedite the process. For instance, at East Meon (Hampshire), extra *famuli (hired labourers)* were employed to gather legumes and cart them away from fields because of the fear of rain (*pro dubio pluviae*).[11]

As some accounts from Somerset demesne of Glastonbury Abbey report, the heavy rain continued into the autumn, and it still poured during the time of wheat seeding (usually, early-to-mid-October).[12] The language of the accounts clearly reveals that the rain was of disastrous proportions. We are unaware whether the rain had temporarily stopped at some point later in the autumn or winter, or if it turned right into snow in the winter, as one Hertfordshire account suggests.[13] What it is clear is that during the winter time, demesne pastures rendered very little or nothing, to the point that animals could not graze.[14] Moreover, winter sowing of legumes was extremely challenging because fields

ham). The duration of Norfolk harvests and their correlation with annual season patterns has been meticulously analysed in Pribyl, 'Weather in Late Medieval Norfolk'. The famine years have been analysed on pages 23–24, 125–33, and 158–59.

[8] HantsALS, 11M59/B1/69, membr. 35v (East Meon Church, Hampshire).

[9] It should be noted that in normal years, an average harvest length is thirty-five days. The duration tends to be longer in years of abundance: see, Pribyl, 'Weather in Late Medieval Norfolk', pp. 155–201. The 1314 harvest lasted, on average, some twenty-eight to thirty days.

[10] Derives from Campbell, *Three Centuries of English Crop Yields*. The short duration of the harvest, notwithstanding its bountifulness, reflects the necessity to collect and store as soon as it was possible, on the account of the inclement weather.

[11] HantsALS, 11M59/B1/69, membr. 34r.

[12] Longleat House Muniments (henceforth, LH) 10666, membranes 9v, 33v (Pilton and Wrington); LH 10030 (Walton).

[13] Westminster Abbey Muniments (henceforth, WAM) 8802 (Kinsbourne, nowadays known as Harpendenbury). This has been studied in Stern, *A Hertfordshire Demesne*, pp. 90, 94, and 98.

[14] HantsALS, 11M59/B1/70, membr. 14r (Wargrave, Berkshire).

were covered with water, as some sources indicate.[15] It is clear that the rain fell with unprecedented intensity in May, around Pentecost (11 May 1315), pouring incessantly into the summertime and disturbing hay and harvest works.[16] One demesne account refers to the destruction of 'rowen grass' (that is, the second grass growing, usually between July and mid-Autumn) by the heavy rain.[17]

Table 2.1. Chronology and duration of East Anglian demesne harvests, 1314–1316.

Manor name	**County**	**Year**	**Start date**	**End date**	**Duration (days)**
Hindolveston	Norfolk	1314	25 August	25 September	31
Monks Granges	Norfolk	1314	11 August	8 September	28
Sedgeford	Norfolk	1314	15 August	21 September	37
Taverham	Norfolk	1314	10 August	8 September	28
Burnham Thorpe	Norfolk	1315	1 August	29 September	59
Scratby	Norfolk	1315	10 Augusts	14 September	35
Stanhoe	Norfolk	1315	1 August	?	?
Worstead	Norfolk	1316	28 July	?	?
Scratby	Norfolk	1316	1 August	29 August	29
Kempstone	Norfolk	1316	8 August	5 September	28
Hinderclay	Suffolk	1316	1 August	29 September	29
Newton	Cambridgeshire	1316	1 August	?	?

Source: Norfolk Record Office (henceforth, NRO), Hare 213x1/4277; DCN 60/18/19; DCN 60/26/15; DCN 60/35/18(a); DCN 60/35/17a; WAL 63/269x1; DCN 60/30/4; DCN 60/30/5; Wis Z/163x1; Raynham Hall Muniments (RHM), Raynham Box 21; and University of Chicago Library, Special Collections (henceforth, UCL), Bacon Roll 446.

According to some accounts, the rain was accompanied by severe storms.[18] On two East Anglian manors, a gale was so strong that it felled trees in the

[15] For instance, Bodleian Library (henceforth, BoL), DD Ch Ch Queens 251 (Werrore-Cosham (Hampshire) account for 1315–1316).

[16] The references are plentiful, but a few instances may suffice: The National Archives (Kew) (henceforth, TNA), SC 6/996/14, memr. 15ʳ and 7ʳ (Haughley and Thorndon, both in Suffolk); Northamptonshire Record Office (henceforth, NorthantsRO), FM 248 (Boroughbury, Northamptonshire); and TNA, SC 6/1011/4 (Byfleet, Surrey).

[17] TNA, SC 6/1202/7 (Llanblethian, Glamorganshire, Wales).

[18] LH, 10666, membranes 9ʳ and 39ʳ (Baltonsborough and Pilton, both in Somerset).

summer of 1314.[19] Several chronicles report thunderstorms. As one chronicle indicates — surely with some exaggeration — on 15 July 1315, Woburn Abbey, a Cistercian house in Bedfordshire, was destroyed by a thunderbolt.[20] Likewise, a number of manorial accounts speak about a 'great wind' (*magnus ventus*) and occasionally about a storm (*tempestas*) over the course of 1314 and 1315.[21] As one palaeoclimatic study on weather in late medieval Norfolk tentatively estimates, April–July temperatures in 1314 and 1315 were low, standing at, approximately, 11.9 and 11.8 degrees centigrade respectively — about one degree lower than the alleged average for 1250–1330.[22]

We can only imagine the gloomy and terrified faces of crop producers and harvesters during the harvest of 1315. The implications of the abysmally low yields of that harvest will be discussed at a greater length below. For now, it would suffice to mention that unlike the harvest of the previous year, which started later than usual, the harvest of 1315 commenced around 1 August, its normal time. The length of the harvest, however, varied from place to place, as the surviving accounts from Norfolk and other counties indicate (Table 2.1). Thus, at Scratby and Stanhoe, both in Norfolk, the harvest began in early August and lasted for five weeks.[23] At Burnham Thorpe, in the same county, on the other hand, the harvest is reported to have lasted from Lammas (1 August) to Michaelmas (29 September), making it fifty-nine days in total.[24] This information can be augmented by the undated duration of harvesters on other demesnes. Thus, at South and East Brent (Somerset), the harvest lasted for an overwhelming eight weeks.[25] Similarly, on some Hampshire demesnes of Winchester Bishopric, the harvest went on until late September and, in some cases, until early October.[26] Several accounts record 'a long harvest' (*autumpnus longus*).[27] The incessant rain caused fields to be inundated, which caused

[19] NRO, DCN 60/18/19 (Hindolveston, Norfolk) and NRO, DCN 60/7/6A (Denham, Suffolk).

[20] *Chronicon Abbatiae de Parco Ludae* ed. by Venables, p. 24; *Annales Paulini*, ed. by Stubbs, p. 278.

[21] For instance, WAM 25800 (a 1313–1314 account for Kelvedon, Essex) and WAM 82426 (a 1314–1315 account for Hardwick and Deerhurst, both in Gloucestershire).

[22] Pribyl, 'Weather in Late Medieval Norfolk', p. 296.

[23] NRO, DCN 60/30/5 (Scratby); RHM, Raynham MSS, Stanhoe accounts (Stanhoe).

[24] NRO, WAL 63, 269x1 (Burnham Thorpe).

[25] LH, 10666, membranes 43^r (Doulting) and 22^r (East and South Brent).

[26] HantsALS, 11M59/B1/70, membr. 21^v (East Meon Church) and 22^v (Hambledon Church).

[27] CCA, DCc/BR/Eastry/42 (Eastry, Kent).

the soil to become muddy and clogged, rendering challenging harvest conditions and making the enterprise long. That the harvest was so poor could also mean that there was no incentive to get the crops in fast, contrary to the previous year's abundant harvest, when every effort was made to shelter the crops from the heavy rain. Moreover, as we shall see in the next chapter, the economic constraints of the crisis prompted manorial authorities to dismiss some *famuli* because of the increased expenditure in other sectors. Again, this was in contrast with the previous year's harvest, when, as some accounts indicate, additional harvesters were hired, so as to collect the harvest as fast as possible. All these factors explain the long duration of the 1315 harvest.

The autumn of 1315 was excessively windy and rainy, and it is unclear when the rain stopped.[28] As several accounts indicate, both the winter and early spring were excessively rainy.[29] One account speaks about the 'incongruity of weather' during the Lent (28 February–11 April 1316).[30] As the Chronicle of St Mary Abbey of York states, the month of March was very rainy and windy.[31] The rainy weather continued into the summer, as several accounts of Winchester Bishopric report.[32] At Maids Moreton (Buckinghamshire), the rain was so strong that a part of its spring fields were submerged under water by harvest time.[33] However, it seems that during the harvest, the rain stopped, and the weather changed to dry. Thus, the account from Kinsbourne (Hertfordshire) reported a 'dry harvest'.[34]

The harvest itself started around its usual time, namely around 1 August, and lasted, on average, four weeks (See Table 2.1). Unlike the harvest of the

[28] WAM, 8803 (Kinsbourne, Hertfordshire). See also, Stern, *Hertfordshire Demesne*, p. 94.

[29] HantsALS, 11M59/B1/70, membr. 13v (West Wycombe, Buckinghamshire); WAM, 8803 (Kinsbourne).

[30] HantsALS, 11M59/B1/70, membr. 14v (Wargrave, Berkshire).

[31] *The Chronicle of St Mary's Abbey*, ed. by Craster and Thornton, pp. 68–69. There is a *lacuna* in the printed text ('Eodem anno, videlicet in Vigilia sanctarum Virginium...tempestas maxima ex occindentali ventus (*sic*) fuit' (In the same year, namely in the Vigil of the Holy Virgins... there was a most great storm [arising] from the western wind). The translation is mine). A close analysis of the manuscript (BoL, MS 39), however, reveals that the original text contained, in the *lacuna*, most likely, 'Felicitatis et Perpetuae' (that is, the vigil of the Virgins of Felicitas and Perpetua (6 March)).

[32] The examples can be supplied ad nauseam. Most of these references have been extracted and printed in Titow, 'Evidence of Weather', p. 385.

[33] BoL, Ch Ch DD 27.

[34] WAM, 8766.

previous year, there was not much variation in the duration of the 1316 harvest across the demesnes. The harvest of 1316 was the worst one recorded in late medieval England (and possibly the worst one in the last eight hundred years). Although the rain was undoubtedly a paramount factor, other aspects, such as the bad quality of seed sown in the late autumn of 1315 and spring 1316 and the increased growth of weeds, encouraged by the rain, should be taken into account, as we shall see later. It also appears that the April–July temperature in 1316 was considerably higher than in the two previous years, and it stood, according to one estimate, at approximately 13.1 degrees Celsius in East Anglia.[35] It is unclear when the rain stopped in 1316, but it clearly did sometime after the harvest, since there are only three references to rainy weather during the 1316–1317 account year.[36] At the same time, it appears that the winter of 1317 was cold, and there are several references to snow and frost.[37] Although the gross harvest of 1317 was still below average, it was the point when the short-term climatic anomaly that caused the harvest failures was over.

Although there is no doubt that occasional weather references in the chronicles and accounts provide invaluable and reliable information, it is instructive to juxtapose them against proxy data, in this case, deriving primarily from the dendrochronological record. As a preliminary remark, it should be noted that such exercise can be somewhat problematic because dendrochronology can be misleading in the English (and indeed, British) context, for three main reasons. First, there is regional climate variability in England (and other parts of the British Isles), because of which, in some regions, the trees tend to be less sensitive to precipitation and, thus, register weather as being drier than it really is — in contrast with the Continent. Second, one has to take into account the seasonality factor. The English and other British oaks would mostly register precipitation in spring and early summer (March to July), while during the late summer and early autumn, when harvesting was particularly vulnerable to wet weather, they generally 'switch off'. In other words, the trees would not register the wet weather in other seasons. Finally, trees are not particularly good at registering precipitation in most of the British Isles. This is largely because of the temperate climate of the islands; thus, the trees do not clearly respond to climatic stresses like trees in other parts of the world (for instance,

[35] Pribyl, 'Weather in Late Medieval Norfolk', p. 296.

[36] Titow, 'Evidence of Weather', p. 387. The references relate to the demesnes of Waltham St Lawrence, Billingbear (both in Berkshire), and Bishopstoke (Hampshire).

[37] WAM, 25423 (Birdbrook, Essex). See also, Stern, *Hertfordshire Demesne*, pp. 94 and 99.

trees in the arid areas of the Americas tend to respond to precipitation levels very markedly, whereas trees at high-latitudes of Scandinavia respond strongly to temperature).[38] Despite these limitations, however, it is still worthwhile looking at the dendrochronological record. For our purpose, the comparative data derive from fifteen sites in 'Western' counties (one in Shropshire, two in Warwickshire, seven in Oxfordshire, one in Gloucestershire, two in Wiltshire, one in Somerset, and one in Hampshire) and twenty-one sites in East Anglia (three in Cambridgeshire, six in Suffolk, and twelve in Norfolk). In all cases, the data are based on oak trees, both living and historical (Table 2.2).

Table 2.2. March-to-July (MAMJJ) precipitation levels in East Anglia and the 'Western counties', based on instrumental (1971–2000) and dendrochronological data (1271–1350).

	East Anglia		**Western Counties**	
Year	**precipitation (mm)**	**as % of 1271–1350**	**precipitation (mm)**	**as % of 1271–1350**
1971–2000	234.0	96%	249.2	100%
1271–1350	244.3	100%	249.0	100%
1314	270.2	111%	241.7	97%
1315	306.9	126%	291.9	117%
1316	305.6	125%	291.5	117%
1317	263.8	108%	276.3	111%

Source: Cooper and others, 'East Anglia 1,100 Year Spring–Summer Precipitation Reconstruction'.; Cooper and others, 'A Tree-Ring Reconstruction of East Anglian (UK)'; Wilson and others, 'Southern-Central England'; and Wilson and others, 'A Millennial Long March–July Precipitation Reconstruction'.

As Table 2.2 indicates, the 1271–1350 levels of March-to-July rainfall are based on the dendrochronological record and are calibrated with twentieth-century instrumental observations and, thus, are remarkably consistent with the 1971–2000 levels. As far as 1314 is concerned, the evidence is contradictory. On the one hand, East Anglian oaks register an 11 per cent increase in rainfall, while in the 'Western' counties the precipitation levels seem to have been on par with an average year. This might to do with the fact that East Anglian trees, growing in the driest part of the British Isles, might have reacted to rainfall and temperatures more pronouncedly than their western counterparts. Certainly,

[38] I wish to thank Kathleen Pribyl and Francis M. Ludlow for communicating and clarifying this important information.

the gaps between the normal and famine years were all the more marked in 1315 and 1316. Thus, the respective figures for those years stood at some 26 and 17 per cent higher than average in East Anglia and the western counties. It is perhaps surprising that the rise in tree-recorded precipitation levels was not that remarkable, as one would expect. Again, this might indicate the unsuitability of British oaks (especially those from the 'Western' counties) to register precipitation levels in an adequate fashion. All the same, there is no doubt that those figures still reflect the increased rainfall levels, indeed corroborating both the words of chroniclers and glosses of manorial officials. Equally surprising are the relatively high rainfall levels of 1317, despite the contrary evidence from the manorial accounts and chronicles. Again, all this points to the often-problematic nature of the dendrochronological record of English oaks.

What about other parts of the British Isles? Unlike English sources, Irish, Scottish, and Welsh documents (with the exception of occasional manorial accounts from the Welsh Marches) contain only short and often indirect references to bad weather. This is hardly surprising, given the laconic nature of Irish annalistic entries, on the one hand, and the preoccupation of local Irish, Scottish, and Welsh writers with contemporary wars, on the other. Nevertheless, several entries *sub anno* 1315 are worth mentioning. Thus, in Ireland, the *Annals of Loch Cé* speak about 'destructive bad weather'.[39] The *Annals of Clonmacnoise*, surviving only in an early seventeenth-century English translation, mention 'ugly and gross weather in the summer'.[40] The *Annals of Connaught* talk about 'intolerable storms'.[41] The same annals describe how the raiding parties of the Ó Conchubhair clan could not drive their loots through the boggy area of Coraidh-Chula-Cuirc, near Boyle (in Roscommon) because of the softness of the bog in the same year.[42] In addition, the *Annals of Ulster* note 'great snow', presumably in the winter.[43] The *Annals of Inisfallen* and the *Annals of the Four Masters*, both much later compilations, as well as a number of fragmentary Irish annals, do not mention bad weather at all. A 1315–1316

[39] *The Annals of Loch Cé*, ed. and trans. by Hennessy, p. 579 (an online edition is available courtesy of the *CELT* Project).

[40] *The Annals of Clonmacnoise*, ed. by Murphy, p. 275.

[41] *Annála Connacht*, ed. by Freeman, p. 241 (an online edition is available courtesy of the *CELT* Project).

[42] *Annála Connacht*, ed. by Freeman, p. 238.

[43] *Annala Uladh*, II, 434 (an online edition is available courtesy of the *CELT* Project).

royal memorandum roll for Ireland (destroyed in the Great Fire of 1922) notes 'the rainy season and the great flood of water', which occurred in that year.[44]

In Wales the situation is as elusive. The only two Welsh chronicles covering that period, the *Brut y Tywysogyon* (The Chronicle of the Princes) and the so-called *Epitome Historiae Britanniae*, mention neither rain nor famine at all for 1315–1316.[45] On the other hand, several manorial and administrative documents provide some glimpse into local weather conditions. The 1315–1316 account from Usk (Monmouthshire) reported inundation of the River Usk, ruining both arable fields and pasturage.[46] Similarly, a coroner's roll from Dyffryn Ardudwy (Merioneth) recorded five people drowning, presumably because of the coastal inundation, in the same year.[47] Moreover, *c.* 1320, Einion Sais, Bishop of Bangor (1309–1328) petitioned Edward II to relieve him of his dues, in part because he had been 'impoverished by bad years', presumably referring to the famine, as well as cattle murrain.[48] The bishopric of Bangor held lands primarily in Caernarvonshire and Anglesey.[49] Anglesey, a low-lying island off the northern coast of Wales and Dyffryn Clwyd, a mixture of lowland and upland settlements, relied on both arable and pastoral types of farming. Most importantly, there is an exceptional series of court rolls from Dyffryn Clwyd, a marcher lordship in Denbighshire (in north-east Wales). During the famine years, one witnesses a spike in food-related crime, as well as exceptionally high grain prices, all pointing to the dire circumstances of that region.[50] Conversely, the upland belt of the Cambrian Mountains, covering much of Wales and practising almost entirely cattle- and sheep-based husbandry, have higher levels of annual precipitation than the lowland parts of Wales. It is possible that the excessive rainfall here would hardly be perceived by those observing it as an anomaly.

[44] Only excerpts of the memorandum rolls survive, in a transcribed version: National Archives of Ireland (NAI), RC 8/10, pp. 416–22. Printed in *Historic and Municipal Documents of Ireland*, ed. by Gilbert, pp. 366–71. The English translation can be accessed on the *CIRCLE* website.

[45] *Brut y Tywysogyon*, ed. and trans. by Jones, p. 123 and Luft, 'The NLW Peniarth 32 Latin Chronicle'. The only thing mentioned in both sources under 1315–1316 is the revolt of Llywelyn Bren.

[46] TNA, SC 6/927/18.

[47] TNA, SC 6/1170/10.

[48] TNA, SC 8/331/15659.

[49] As it derived from TNA, SC 6/1143/25.

[50] Stevens, 'The Great Famine in Dyffryn Clwyd'.

Surprisingly, no Scottish source mentions either flooding or failed harvests and famine. It does by no means imply, however, that Scotland was miraculously spared by the torrential rain. First, both Northumbrian and Cumbrian sources mention the flood, and their proximity to the Scottish borders implies that the situation was undoubtedly similar, at least in southern Scotland. Second, the references of the Irish annalist to bad weather may be used as a proxy to the Gaelic parts (the Gàidhealtachd) of Scotland. Finally, it should be borne in mind that Scotland (and especially the Gàidhealtachd region), just as the Welsh Uplands, has higher levels of precipitation than England, and it is possible that just as in the Cambrian Mountains, the heavy rain in the Scottish Highlands or Islands would not have been regarded as an anomaly. Unfortunately, this assumption could not be checked against dendrochronological record: unlike in England, Scottish trees, because of altitude, are sensitive to summer temperature rather than precipitation, and, thus, reconstructing precipitation levels using Scottish tree-rings would be extremely challenging, if not impossible. There is, however, a long-term precipitation data based on speleothem evidence from Uamh an Tartair cave (in Sutherland), which does not indicate any anomaly in precipitation index between 1300 and 1330, apart from 1321, when precipitation levels stood at 13 per cent below average.[51] One problem with a speleothem record is that it tends to contain a mixture of temperature and precipitation signals, rendering 'noisy' results, based on unclear relationship between the two types of signals. Thus, the quality of this data can be questioned. There is, however, much progress in our knowledge about historical temperatures from Scotland, thanks to the meticulous work of Miloš Rydval and Rob Wilson, who have, most recently, reconstructed Scottish summer temperature chronology, based on Scots pine and going back to 1200. The annual figures are indexed on the average summer temperatures for 1961–1990 (12.57 degrees Celsius). According to their reconstruction, the only summers standing out as cold in the period 1300–1325 were 1317 (–2.56), 1319 (–2.09), 1322 (–2.16), and 1323 (–2.74), but not 1314 (–0.29), 1315 (–0.42), nor 1316 (–0.24).[52]

[51] Proctor and others, 'A Thousand Year Speleothem Proxy Record'; Proctor, Baker, and Barnes, 'A Three Thousand Year Record'.

[52] I am grateful to both authors for being so kind and sharing their pre-publication data on Scottish temperature, 1200–2000. These are revised data of Rydval and others, 'Reconstructing 800 Years of Summer Temperatures'.

Harvest Failures: Gross Demesne Yields

The most obvious reflection of the disastrous weather was epic harvest failures. As we have seen above, manorial officials tended, at times, to report inaccurate and even false harvest figures to justify poor estate management. The calculations above are based on the assumption that these false figures were reported only on a small scale and that the weather anomaly indeed depressed the yields in an unprecedented manner. For the purpose of the present discussion, crop yields shall be measured per seed rather than per acre. Here, it is important to distinguish between three methods of expression. Most manorial accounts record yields net of tithe and gross of seed, without accounting for one-tenth collected as tithes by local parish priests or vicars. Moreover, while the accounts do not factor in the tithed part of the harvests, they do record the seed corn portion of the harvest, to be invested in seeding. For technical purposes, I shall refer to these yields recorded in manorial accounts as 'recorded yields'. These 'recorded yields' are a faulty indicator of arable productivity, since they account neither for the total harvest (but rather for about 90 per cent of it, with tithes deducted), nor for the true availability of arable produce, since they do not deduct the relative share of each crop used as seed corn after harvest. To appreciate the crop yields better, and especially in the context of the famine years, it is necessary to convert the 'recorded yields' into 'gross yields' and 'net yields'. Gross yields are the true ratio between the present year's harvest *inclusive of tithes* (that is, dividing the total harvest by the factor of 0.9) and the amounts of crop sown in the previous year. In other words, seeding 100 quarters of wheat in 1313 and harvesting 400 quarters of the same grain, inclusive of tithes, in 1314 gives the fourfold yield, expressed as 4:1. The gross yields inclusive of tithes provide the best 'environmental' indicator of crop growth and weather conditions. Net yields are the ratio between the present year's harvest *gross of tithe and net of the harvest share invested in seeding* and the amounts of crop sown in the previous year. In other words, net yields account only for the disposable share of the harvest, which could be consumed, sold, or stored. As such, it provides the best indicator of food availability.[53] However, since this chapter is dealing with the natural forces standing behind the crisis, our discussion shall be limited to the gross yields only. The net yields will be discussed in the next chapter.

[53] Wrigley, 'Some Reflections on Corn Yields'.

To calculate the gross yields, the following formula was used:

$$Gy^i \text{ (Gross Yield of individual crop)} = Ry^i \text{ (Recorded Yield of individual crop)} \div 0.9$$

A systematic collection of all the surviving manorial accounts for the years 1314–1318 rendered it possible to calculate crop yields for 111 demesnes in 1315, 84 demesnes in 1316, and 105 demesnes in 1317. These figures are lower compared to all the surviving accounts for the famine years (so far, I have been able to identify and analyse 140, 165, and 134 accounts for 1314–1315, 1315–1316, and 1316–1317, respectively). This discrepancy derives from two reasons. First, far not every account specifies the ratio between this year's harvest to the previous year's seed, and in many instances, it is necessary to have two consecutive accounts to calculate the yields internally. Second, manorial accounts record the harvest of the previous, rather than the present, year. Thus, the account of 1315–1316, running between the two Michaelmases, would record the harvest of 1315 rather than that of 1316.

The geographic distribution of manorial accounts, for both famine and non-famine years, tends to be scattered unevenly towards the southern and eastern counties, while the western and northern counties, where manorialism and manorial institutions were comparatively weak, have produced comparatively small number of records. Thus, about 31 per cent of all the surviving crop yields for 1314–1317 come from a single county of Hampshire, despite the fact that that county covered about 3.6 per cent of the total arable land in England in the early fourteenth century.[54] On the other hand, there is only one surviving crop yield from the county of Dorset, whose share in the total arable land was about 2.2 per cent around that period.[55] To make things even worse, there are no surviving accounts for several counties. That disproportional over-representation, on the one hand, and significant lacunae, on the other, can potentially create a misleading picture when calculating the average of all observations for each crop. One has to bear in mind regional variations in agricultural patterns and structures, as well as environmental and topographic conditions. For instance, south-western, western, and north-western counties tended to be more pastoral than the eastern parts of the country. Furthermore, relatively backward agricultural methods in some upland areas in the north, coupled with the increased precipitation levels, implied lower yields, compared

[54] That is about 460,000 acres out of the total 12.72 m acres. See Broadberry and others, *British Economic Growth*, pp. 66–67.

[55] Broadberry and others, *British Economic Growth*, pp. 66–67.

to other parts of the country. Conversely, because of comparatively progressive arable techniques and efficient seignorial arrangements, crop yields tended to stand out in East Anglia, Kent, and Somerset.[56] Averaging all the available observations does not take into account these differences and renders a somewhat distorted picture when talking about the 'national' average. To eliminate those potential noises, it is essential to convert the 'manorial average' into the 'national average', taking into account each of those structural differences and each county's relative share in the total national arable acreage. The method is closely based on that devised by Campbell and his colleagues and is outlined in a detailed manner in Appendix 1.

Likewise, it is important to distinguish between 'individual' and 'composite' yields. Individual yields relate to yields of each individual crop, whereas composite yields are, in effect, the weighted average of all crop yields, weighted according to each crop's share in land acreage in each county. Around 1315, the respective share of crops in the total arable acreage was 33 per cent for wheat, 7 per cent for rye and maslin (wheat-rye mixture), 2 per cent for winter barley, 6 per cent for legumes, 13 per cent for barley, 3 per cent for dredge (barley-oat mixture), and 36 per cent for oats. Winter oats, accounting for less than 0.2 per cent of all arable land, were accounted for together with spring oats, while oat-legume mixtures, representing a tiny fraction of 1 per cent, were combined with legumes.

With those adjustments in mind, we can now turn to quantifying the approximate extent of crop failures during the famine years (Table 2.3). In the harvest of 1315, the 'national' composite gross crop yields stood at about 25 per cent below average (individually, the below-average figures stood at about 31 per cent for wheat, 28 per cent for rye, 40 per cent for winter barley, 48 per cent for legumes, 11 per cent for barley, 8 per cent for dredge, and 21 per cent for oats, respectively). In other words, it was winter grains (wheat, rye, and winter barley) and legumes that failed particularly miserably. On some demesnes the situation was truly alarming. Thus, at Ingthorpe, How, Riddings (all situated on the infertile soils of the Yorkshire Dales), Kings Worthy (Hampshire), and Walworth (Surrey), wheat yields stood below, at, or just above the ratio of 1:1.[57]

[56] On Norfolk and Kent agriculture in the later Middle Ages, consult Campbell, 'Arable Productivity'; Campbell, 'Agricultural Progress'; Campbell and Overton, 'A New Perspective'; Campbell, 'Agriculture in Kent'.

[57] The Yorkshire demesne yields are calculated from the Bolton Priory account book: *The Bolton Priory Compotus*, ed. by Kershaw and Smith, pp. 391–96 (crop seeding in 1314) and pp. 411–16 (crop harvesting in 1315). The King's Worthy calculations derive from HantsALS, W/HI/2-3 (1314–1315 and 1315–1316 account rolls for the St John's Hospital, Winchester). The Walworth accounts are CCA, DCc/BR/Walworth/6–8.

All in all, on 18 per cent of all the demesnes, wheat yields stood at more than 50 per cent below average, while on 41 per cent of the demesnes, wheat yields stood at between 25 and 50 per cent below average. Only on some 9 per cent of the demesnes were wheat yields higher than average (Table 2.4). Thus, yields of almost 5:1 are reported at Littleton (Hampshire); at Westwood (Wiltshire), the yields of 5.75:1 were achieved, while Alton Priors (Wiltshire) saw the overwhelming yield of 6.6:1.[58] These, however, were clear anomalies, and it is unclear what accounted for that: out of these four manors, only Westwood is situated on exceptionally fertile soil, while the fertility of the other demesnes is mediocre.[59] Also, while Kent may experience lower levels of precipitation compared to other parts of England, both Hampshire and Wiltshire tend to have 'average' rainfall levels.[60]

It was legumes (beans, peas, and vetches), however, that failed most epically. Bean- and pea yields stood at more than 50 per cent below average on over half of the demesnes, while three quarters of the demesnes reported yields of more than 25 per cent below average. The figures for vetches are even gloomier. The average vetch yields were 0.77:1:00 (namely, harvests fell well short of seed), with almost one third of vetch-growing manors collecting no harvest at all, and with not even one demesne achieving the average 'non-famine' yield. The failure of leguminous crops undoubtedly had a negative impact not only on humans, but also on the nutrition and health of domestic animals. I will be returning to the issue of animal health in the end of this chapter.

Unlike the winter grains and legumes, spring grains (that is, barley, dredge, and oats), were faring remarkably well. As far as barley is concerned, only 7 per cent of all the demesnes experienced yields more than 50 per cent below average, while on almost half of the demesnes, the yields varied between 75 and 100 per cent of the normal level. Higher-than-average yields were produced on as many as 24 per cent of the demesnes. In the case of oats, 19 per cent of all the demesnes saw yields of more than 50 per cent below average; 38 per cent of the manors fell into the range of 51–75 per cent of the normal level, while 21 per cent of the demesnes achieved higher-than-average yields. As Table 2.4 indicates, the percentile distribution of the 1315 yields was quite different from that in normal years (with 1308 used as an example).

[58] HantsALS, DC/J1/14, membr. 13^{v} (Littleton); HantsALS, DC/J1/14, membr. 15^{v} (Westwood); and HantsALS, DC/J1/14, membr. 20^{v} (Alton Priors).

[59] The soil quality data derive from the *Soilscapes* database, courtesy of the Cranfield Soil and Agrifood Institute.

[60] MET Office, *UK Climate — Historic Station Data*.

Table 2.3. National, aggregated gross crop yields in England and south-east Wales (Monmouthshire), 1315 (111 demesnes), 1316 (84 demesnes), and 1317 (105 demesnes), indexed on 1270–1429 (1270–1429 = 1.00).

Harvest	Each crop's share in total acreage	Gross yields, 1315	Gross yields, 1316	Gross yields, 1317	Coefficient of variation, 1315	Coefficient of variation, 1316	Coefficient of variation, 1317
Wheat	33%	0.69	0.59	0.96	0.374	0.444	0.336
Rye	7%	0.72	0.48	0.95	0.649	0.487	0.493
Winter barley	2%	0.60	0.51	0.59	0.446	0.526	0.437
Legumes	6%	0.52	0.52	1.18	0.772	0.825	0.525
Barley	13%	0.89	0.78	0.89	0.380	0.446	0.320
Dredge	3%	0.92	0.75	0.88	0.289	0.328	0.423
Oats	36%	0.79	0.70	0.71	0.477	0.412	0.359
Composite		0.75	0.65	0.86			

Source: Slavin, Manorial accounts database (see Appendix 8); Campbell, *Three Centuries of English Crop Yields*.

Note: The figures are indexed on average gross yields (inclusive of tithes and seed) for the period 1270–1429. These are per-seed yields, calculated as the ratio between a previous year's seed and the current year's harvest. Average gross yields in normal years are 4.18:1 for wheat, 4.28:1 for rye, 4.06:1 for winter barley, 3.16:1 for legumes, 3.89:1 for barley, 3.5:1 for dredge, and 3.12:1 for oats (the first figure is the harvest and the second figure is the seed). The indexed figures in the table are expressed in relation to the average yields for the period 1270–1429. For instance, the indexed gross wheat yield in 1315 is 0.69, which is meant to be 2.88:1 (that is, the harvest was 2.88 times its seed). The coefficient of variation is a degree of deviation from the average yield achieved in a given year. The higher the coefficient figure, the higher the variety of yields attained by different demesnes, in relation to their average yield.

The harvest of 1316 was even worse, with 'national' composite gross yields standing at about 35 per cent below average (Table 2.3). The below-average figures were about 41 per cent for wheat, 52 per cent for rye, 49 per cent for winter barley, 47 per cent for legumes, 22 per cent for spring barley, 25 per cent for dredge, and 30 per cent for oats, respectively. Again, the degree of failure varied from place to place. For instance, at Morton (Buckinghamshire), Halton (in the uplands of the Yorkshire Dales), and Wymondham (Norfolk), wheat yields stood below the ratio of 1:1, while other crops either failed as badly or stood slightly higher than that.[61] Unlike the harvest of 1315, the proportion of

[61] The Halton calculations derive from *The Bolton Priory Compotus*, ed. by Kershaw and Smith, pp. 411–16 (the 1315 seeding) and pp. 430–34 (the 1316 harvest); the figures for Morton and Wymondham derive from Campbell, *Three Centuries of English Crop Yields*.

Table 2.4. Disaggregated indexed crop yields divided into percentile groups during the famine years and a 'normal' (non-famine) year (1308).

	<0.33	0.33–0.50	0.51–0.75	0.76–1.00	1.01–1.25	>1.25	No accounts
Wheat, 1315	9%	9%	41%	31%	7%	2%	108
Wheat, 1316	14%	31%	26%	21%	6%	1%	77
Wheat, 1317	1%	2%	23%	36%	16%	21%	99
Wheat, 1308	2%	3%	27%	27%	21%	20%	66
Peas, 1315	28%	25%	13%	13%	11%	10%	61
Peas, 1316	34%	18%	24%	10%	10%	4%	50
Peas, 1317	5%	2%	17%	17%	15%	45%	66
Peas, 1308	1%	2%	22%	37%	21%	18%	57
Barley, 1315	1%	6%	22%	47%	12%	12%	95
Barley, 1316	6%	16%	33%	20%	17%	8%	64
Barley, 1317	0%	5%	31%	36%	13%	14%	83
Barley, 1308	0%	2%	16%	22%	39%	22%	51
Oats, 1315	3%	16%	38%	23%	13%	8%	102
Oats, 1316	6%	22%	31%	31%	5%	5%	81
Oats, 1317	4%	14%	49%	22%	7%	5%	101
Oats, 1308	0%	2%	19%	43%	21%	16%	63

Source: Slavin, Manorial accounts database (see Appendix 8); Campbell, *Three Centuries of English Crop Yields*.

Note: There are 6 percentile groups (<0.33, 0.33–0.50, 0.51–0.75, 0.76–1.00, 1.01–1.25, and >1.25); each percentile group signifies the indexed yields in relation to the average yields for the period 1270–1429 (1270–1429 = 1.00) (see Table 2.3). Each row establishes a relative proportion of demesnes whose indexed yields fall into each percentile group. For instance, 'Wheat 1315' row indicates that during the harvest of 1315, 9 per cent of all demesnes achieved less than 0.33 the average wheat yields (namely, their wheat yields stood at more than 67 per cent below average). A further 9 per cent of all demesnes achieved between 0.33 and 0.5 the average wheat yields (namely, their wheat yields stood anywhere between 50 and 67 per cent below average); 41 per cent of all demesnes achieved between 0.51 and 0.75 the average wheat yields (namely, their wheat yields stood anywhere between 25 and 49 per cent below average), etc.

demesnes whose yields stood well below average was quite remarkable (Table 2.4). In the case of wheat, on 45 per cent of the demesnes the yields were more than 50 per cent below average, and almost three quarters of all the demesnes did not achieve 75 per cent of the normal level. Only 7 per cent of the demesnes saw slightly higher-than-average yields, but these were, just as in the case of the 1315 harvest, clear exceptions. For instance, at Malham (Yorkshire), Eastry (Kent), Fareham (Hampshire), and Brandiston (Norfolk), wheat yields ranged between 4.5:1 and 5:1, while at Copton (Kent), they stood slightly over 5.6:1.[62] Again, it is unclear what explains such high yields; none of those manors could boast exceptionally fertile soil or ecological advantages in the context of the torrential rain. Barley and oats fared much better. In the case of former, only on some 13 per cent of the demesnes were the yields lower than half their normal level; almost half of the demesnes saw yields between 50 and 100 per cent of their normal level, while 38 per cent of the demesnes managed to attain higher-than-average yields. As far as oats are concerned, crop yields were more than 50 per cent below average on 28 per cent of the demesnes. The majority of the yields (62 per cent) stood between 50 and 100 per cent their normal level, and a further 10 per cent were slightly above average. Just as in the case of the 1315 harvest, the percentile distribution of the 1316 yields was quite different from that in normal years (Table 2.4). Within the legume sector, the yields were as abysmally low as they were in 1315. Again, as in 1315, vetches were faring particularly bad, yielding, on average, 1.2:1. that is about 62 per cent below normal level. Out of a total of thirty-four observations, four manors harvested nothing at all, twenty-three did not achieve the yield of 1:1, and only on two demesnes did the yields stand above average.[63] Although legumes occupied a meagre share of the total arable acreage (about 6 per cent, with vetches occupying no more than 2 per cent), their failed harvests undoubtedly had some far-reaching implications. After all, apart from being a vital fertilizing agent and a source of human food, legumes were the single most important component of swine fodder, widely used for their fattening.

The gross composite yields of the 1317 harvest were 14 per cent below average, a situation far from being ideal. Although wheat and rye yields were just slightly below average, and legume yields were 19 per cent above average, other crops did not perform that strikingly. The figures for barley and dredge stood, respectively, at 11 and 12 per cent below average; oat yields were 29 per cent below average, while winter barley achieved only 59 per cent of the normal levels. Given that winter barley occupied only some 2 per cent of the national

[62] The data for Malham derive from *The Bolton Priory Compotus*, ed. by Kershaw and Smith, pp. 411–16 (the 1315 seeding) and pp. 430–34 (the 1316 harvest); the figures for other demesnes derive from Campbell, *Three Centuries of English Crop Yields*.

[63] The figures for Fareham (Hampshire) and Bishopstone (Wiltshire) were, respectively, 4.47:1 and 5:85:1. Calculated from HantsALS, 11M59/B1/71, membr. 6[v] and 33[v] and HantsALS, 11M59/B1/72, membr. 5[v] and 24[v].

acreage, it was the failure of oat harvests that had the most pronounced short-term consequences on agrarian economy. Just as legumes, oats were an important component of both equine fodder and the human diet and, as such, were a highly versatile and demanded crop, whose dearth could have a strong impact on the feeding patterns of both humans and domestic animals. One could, perhaps, add the fact that oats can increase serotonin levels, vital for human psychological well-being.

As we have seen, winter grains and legumes performed much worse than spring grains in 1315 and 1316. What accounts for that? On the one hand, it might do with the fact that wheat, the single most capricious grain, is highly vulnerable to inclement weather. On the other hand, rye, by far more tolerant grain, capable of withstanding various climatic challenges, failed as much as wheat did.[64] Moreover, while winter barley failed epically, spring barley fared reasonably well. Despite the distinction made in the accounts between 'winter barley' (*ordeum hiemale*) and 'Lenten barley' (*ordeum quadrigesimale*), the two were, in fact, the same type of barley (*Hordeum vulgare L.*), sharing exactly the same botanic traits. Both types of barley functioned as a drink grain in normal years, but because of the widespread failure and scarcity of wheat and rye, both winter and spring barley were diverted into baking during the famine years to ensure improved calorific efficiency, as we shall see in Chapter 4. Does it, therefore, have to do something with the seasonal patterns of growth rather than with the botanic peculiarities of each crop? It is likely that the weather conditions were particularly harsh in the winter time, when wheat, rye, and winter barley germinate and legumes are being planted (usually in late January/early February). As we have seen, there are references to snow, implying negative temperatures in the winters of 1315 and 1316. Negative temperatures can seriously impede the germination and growth of crops and result in harvest failures. In some cases, one frosty night can ruin an entire harvest.[65] At the same time, there is no evidence that the winters of 1315 and 1316 were *especially freezing*; there is no reference to frozen rivers or excessive frost. Instead, there are plenty of references to incessant rain and cold (and occasionally snowy) winters.

[64] Comet, 'Technology and Agricultural Expansion', p. 17.

[65] Myllyntaus, 'Summer Frost', pp. 77–78.

Regional and Local Trends?

Can we talk about geographic trends in crop failures in 1315–1317? Overall, it appears that there was a good degree of consistency, with the highest composite yields achieved in Somerset and Wiltshire, followed by East Anglia in all three harvests. Thus, on Somerset-Wiltshire demesnes, the wheat and rye yields for 1315 stood at only 12 per cent below average, while the figures for oat and barley were above average. In the following year, the wheat yields dropped to 41 per cent below average, and the oats figures fell to 17 per cent below average, while the barley ones stood slightly above average. The 1317 yields, with the exception of rye, were about or higher than average. In East Anglian demesnes, the composite crop yields for 1315 and 1316 were 14 and 31 below average, respectively, while the 1317 ones were about average. The lowest yields, in all three cases, are reported on Yorkshire demesnes. Here, the composite yields were about 50 per cent below average in 1315 and 1316 and some 25 per cent below average in 1317. This dichotomy between the relatively high demesne yields of Somerset-Wiltshire and East Anglia, on the one hand, and the abysmally low yields in Yorkshire (and, most likely, other northern counties), on the other, is hardly surprising, given the environmental, technological, and economic peculiarities of those regions. It would not be an exaggeration to say that the south and East Anglia were the two most agriculturally advanced regions, achieving higher-than-average yields in non-famine years. Thus, parts of the Somerset Levels and Moors were characterized by exceptionally fertile soils and, usually, a lower precipitation index.[66] Conversely, the East Anglian counties of Norfolk, Suffolk, and Cambridgeshire were marked by comparatively dry climate and advanced techniques of intensive arable husbandry.[67] Both regions achieved yields higher than national yields in non-famine years, as it is evident from local manorial accounts. Parts of Yorkshire and other northern counties, on the other hand, suffered from a series of environmental and technological disadvantages. In many instances, local soils tended to be less fertile than those in other regions. This is especially true of high-altitude settlements, whose local topographic conditions favour pastoral husbandry, rather than arable farming. Moreover, high altitude usually implies a higher precipitation index and colder temperatures.[68] To make thing even worse, however, from 1314 onwards, that

[66] Natural England, 'Sheet 142: Somerset Levels and Moors'.

[67] Campbell, 'Agricultural Progress'.

[68] On various agricultural challenges and adjustments of the northern agriculture, see Waites, *Monasteries and Landscape*, pp. 101–57; Winchester, *Harvest of the Hills*, pp. 52–102.

Table 2.5. Grain yields of 1315 (111 demesnes), 1316 (84 demesnes), and 1317 (105 demesnes), correlated against explanatory variables.

Yields	Wheat/barley/oat specialization	Legume specialization	Acreage size	Horse density	Ox density	Cattle density	Sheep density	Swine density	Total livestock density	Elevation	Soil fertility	Lay/royal demesne	Religious demesne
Wheat, 1315	0.089	-0.026	0.199	0.036	-0.036	-0.020	-0.009	0.082	-0.025	-0.154	0.098	0.070	-0.070
Barley, 1315	0.062	-0.190	0.088	-0.110	-0.052	-0.043	0.008	-0.008	-0.058	0.082	0.141	-0.121	0.121
Oat, 1315	-0.152	0.187	0.174	0.265	-0.035	0.034	-0.072	0.264	-0.003	-0.278	0.050	0.096	-0.096
Wheat, 1316	-0.112	0.241	-0.068	0.072	-0.084	-0.087	-0.096	0.207	-0.121	-0.227	-0.072	0.045	-0.057
Barley, 1316	0.103	0.054	0.116	0.054	-0.122	-0.204	-0.142	0.071	-0.237	-0.114	0.125	0.128	-0.128
Oat, 1316	-0.078	0.415	0.059	0.070	-0.175	0.011	-0.158	0.286	-0.147	-0.382	0.097	0.067	-0.072
Wheat, 1317	0.029	0.088	0.228	0.156	-0.050	-0.150	0.246	0.192	0.009	-0.144	-0.010	-0.135	0.135
Barley, 1317	-0.016	-0.190	0.060	0.036	0.034	-0.079	0.080	0.003	0.007	-0.024	0.208	-0.090	0.090
Oat, 1317	-0.065	0.131	-0.041	0.124	-0.034	0.027	0.095	0.085	0.041	-0.262	0.255	-0.143	0.143

Source: Slavin, Manorial accounts database (see Appendix 8).

region became an unfortunate victim of Scottish raiders plundering the countryside in the course of the First Scottish War of Independence (1296–1328).[69] This fact had some harsh regional implications, as we shall see in Chapter 6.

[69] McNamee, *Wars of the Bruces*, pp. 72–122; Slavin, 'Warfare and Ecological Destruction'.

In order to appreciate why some demesnes performed better or worse than others, it is essential to correlate individual crop yields from a substantial number of demesnes (111 demesnes in 1315, 84 in 1316 and 105 in 1317) against a series of exogenous and endogenous variables (Table 2.5). In essence, the table establishes the impact of different factors (explanatory variables) on the degree of the success or failure of crop yields on different demesnes (dependent variable). The impact is expressed in terms of 'correlation of coefficient', that is, the measure of a degree of the strength and direction of the linear relationship between two variables. As the table indicates, in most cases, the correlation coefficient between two variables was weak and, seemingly, coincidental. Thus, the degree of specialization of each crop (measured in the share of each crop's acreage within the total arable acreage) does not seem to have had any impact on the yields. The only exception here seems to have been wheat and oat yields in 1316 that were higher on those demesnes that allocated larger shares of their arable acreage to legumes (possibly thanks to their nitrogen restoring properties). But then again, the same correlation between legume specialization and grain yields is seen neither in 1315, nor in 1317. Also, with the possible exception of the 1317 wheat yields, the size of demesne did not play any role here. In other words, grain yields could either fail miserably or get away with lighter failures on both great and small demesnes. Likewise, the type of lordship had no impact on the yields. Perhaps surprisingly, the correlation between the quality and fertility of soil, on the one hand, and the yields, on the other, was very low, with the possible exception of barley and, especially, oat yields of 1317. It should be kept in mind that in that year, with the exception of winter barley, oats failed more than any other crop, and it was in that context that soil fertility might have impacted its yields. In most instances, livestock density had either limited or no impact on the yields. One exception was horse-specializing demesnes, which saw higher oat yields in 1315. Also, there seems to have been a positive correlation between sheep density and wheat yields in 1317, possibly thanks to high densities of sheep, which functioned as 'fertilizing machines'.[70] It is elevation that seems to have most paramount impact on crop yields, and the inverse correlation between the two variables appears to be especially visible in the case of oats. In other words, upland settlements tended to yield lower yields, and vice-versa. This undoubtedly reflects the positive correlation between the altitude of settlements and the precipitation index: as a rule of thumb, more rainfall is expected on high-altitude terrain.

[70] Newman, 'Medieval Sheep-Corn Farming'; Campbell, *Seigniorial Agriculture*, p. 182.

Harvest Failures within the Tenancy Sector: Evidence from Tithe Receipts

The crop yields extracted from the manorial accounts refer to the demesne sector, whose contribution to the agrarian economy *c.* 1315 may have been in the area of 20 per cent.[71] What about the peasant sector, accounting for some 80 per cent of the agrarian economy? The biggest challenge in studying the peasant sector is a frustrating lack of sources similar to the plentiful corpus of demesne accounts. Still, we are extremely lucky to have a small number of tithe accounts, recording approximately one-tenth of the agricultural produce, usually crops, paid by local communities to priests or vicars of parish churches.[72] For the famine years, I was able to identify thirteen parishes whose tithe accounts have come down to us. The parishes in question are Skipton and Kildwick (both situated in the Yorkshire Dales and encompassing nine and eight hamlets, respectively), Billingham (County Durham), Birchington and Monkton (both in Kent), Feering (in Essex), Scratby and Sedgeford (in Norfolk), the Hampshire parishes of East Meon, Hambledon, and Wootton, and the Wiltshire parishes of Overton and Enford.[73] In order to appreciate the situation at these parishes, it is necessary to juxtapose their tithe receipts against the demesne accounts pertaining to the same manor. Unfortunately, unlike the demesne accounts, the majority of the tithe receipts survive with considerable gaps. Thus, there are only five manors for which both the demesne accounts and tithe receipts survive, with few gaps for the period *c.* 1300–1325 (Skipton and Kildwick, both juxtaposed against the neighbouring Bolton, and Hambledon, East Meon Church and Feering, juxtaposed against their respective demesnes).

[71] Campbell, *Seigniorial Agriculture*, p. 59.

[72] The single most comprehensive study of peasant agriculture, based on the tithe material is Dodds, *Peasants and Production*.

[73] The references of the relevant tithe accounts, utilized in the present study, covering *c.* 1300–1320, are as follows: *The Bolton Priory Compotus*, ed. by Kershaw and Smith (Skipton and Kildwick); Durham Cathedral Archives (henceforth, DCA), Billingham Accounts, 1300–1320; CCA, DCc/BR/Monkton/22, 30, 33, 37–38, 40, 47, 50 and 54 (Monkton and Birchington); WAM 25602, 25604, 25607, 25609, 25611, 25613, 25615, 25617, 225619, 25621, 25623, 25625, 25627, 25629, 25631; TNA, SC 6/841/5; WAM 25633, 25635 and 25637 (Feering); NRO, DCN 60/30/3–7 (Scratby): NRO, LEST/IC 2, LEST/IB/79–80 (Sedgeford); HantsALS, 11M59/B1/57–76 (East Meon and Hambledon; those figures have also been extracted and tabulated, in a written format, by Jan Z. Titow and later deposited at the Hampshire Archives and Local Studies, see, HantsALS, 97M97); and HantsALS, DC/J1/6–16 (Wootton, Overton and Enford). I am deeply thankful to Ben Dodds for his kindness in sharing his tithe data for the Durham Cathedral Priory manors, the tabulated version of Hambledon tithes, and the images of Feering tithe receipts.

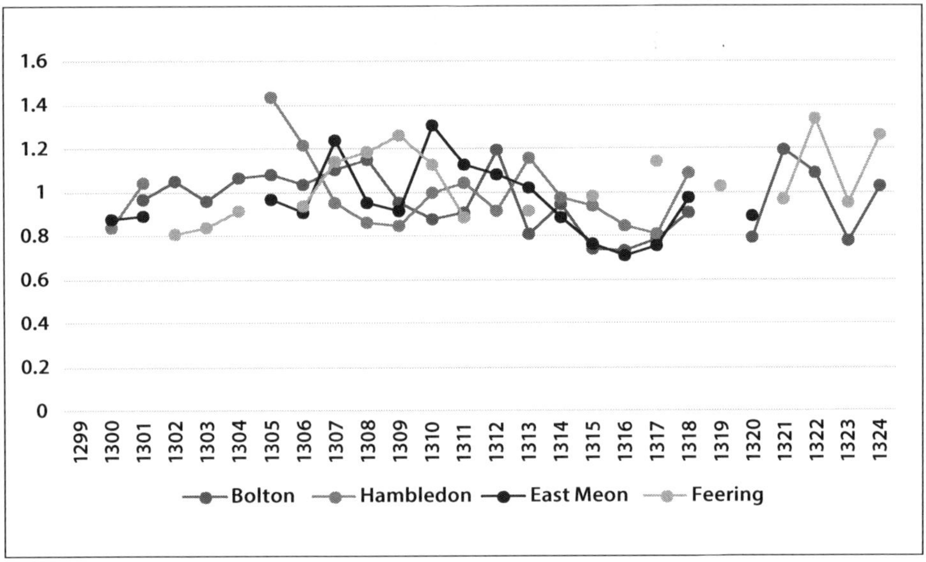

Figure 2.1. Grain production levels on five manors, expressed as the indexed ratio between tenancy and demesne production levels, 1299–1324. *Source*: Slavin, Tithe receipts database (see Appendix 8).

Note: When the production level = 1.0 (indexed on 1299–1314), it means that the demesne and parish were equal; an indexed production level of 1.2 means that levels of production were 20 per cent higher in parish than the demesne

Unlike the demesne accounts, the tithe receipts do not report yields per se. Instead, they record the quantities of crops paid by local parishioners, representing about one-tenth of their sheer annual produce. In other words, tithe receipts represent only an indirect indication of arable production rates, which, nevertheless, are telling. To get a better sense of comparison between demesne and tenancy, I have indexed the sheer annual quantities of crops harvested on both the demesnes and the tenancies, based on their average for 1299–1314. The indexed figures for tenancy production levels were divided by those of the demesne within the same manor, resulting in a ratio between the two sectors. This ratio is perhaps the most revealing indicator of similarity and difference between the two sectors.

As Figure 2.1 indicates, the peasant sector seems to have been even more depressed than the demesne one during the crisis years. Almost invariably, the relative production levels were lower within tenancy. With the exception of Feering, the composite crop production (that is, the weighted average of all crop yields, weighted according to each crop's share in land acreage on each parish) levels at parishes were 15–30 per cent below those on the demesne. At Kildwick and Skipton, the figures remained depressed into the 1320s, reflecting depopulation and land abandonment, related to poverty and the constant threat of Scottish marauders who pillaged local countryside.[74] Wheat yields were particularly low: at Kildwick and Skipton, the tenant production levels were 24, 54, and 50 per cent below the demesne ones in 1315, 1316, and 1317, respectively. At Hambledon, the tenancy production levels of wheat stood at 33, 47, and 9 per cent below the demesne ones, respectively. At Feering, the situation was a bit better: in 1315, the composite crop production levels within the tenancy sector were only 4 per cent below the demesne ones, while in 1317 the tenancy figures stood at 12 per cent above the demesne ones. Unfortunately, the 1316 demesne account for Feering does not survive, so no comparison between the two sectors is possible for that year for that particular manor.

Although the tithe accounts from those five parishes, alongside with the demesne accounts from the same manors, provide a unique glimpse into a virtually uninterrupted comparative chronology, additional tithe receipts related to all twelve parishes, despite their chronological paucity, still represent an invaluable source for studying the harvest failures within the peasant sector. As Table 2.6 demonstrates, there is unambiguous evidence that the peasant sector of agriculture was at least as depressed as the seignorial one — and apparently, even more so. On average, the tithe receipts of 1315, 1316, and 1317 stood at about 36, 52, and 42 per cent below average (with the respective figures for wheat, legumes, spring barley, and oats standing at 49, 55, 14, and 40 per cent below average in 1315; 62, 74, 47, and 47 per cent below average in 1316; and 24, 32, 26, and 49 per cent below average in 1317).[75] True, the sheer levels of tithed crops do not provide a direct measure of their yields. Most importantly, because of the lack of information on the amount of land under cultivation on a parish by parish level (with the possible exception of Feering), it is impossible to calculate or estimate peasant yields per acre. Also, it is unclear if the lack of any reference to tithed rye in the Monkton and Skipton accounts, during some famine years, indicates the total destruction of this crop by the disastrous weather or a possible evasion from paying rye tithes (or possibly, clerical negligence). In any event, this does not present any methodological issue, given the

[74] On the Scottish raids on Bolton, see Kershaw, *Bolton Priory*, pp. 14–16, 173; Kershaw, 'Note on the Scots'.

[75] Because of the difference in size of each manor, and hence its production capacity, the average figures were calculated as weighted averages, taking into account the size of each crop's receipts on each parish, in normal years.

Table 2.6. Annual levels of tenant arable production on twelve parishes, as indicated by tithe receipts, 1315–1317 (indexed on 1299–1314).

Parish	Wheat	Rye/ maslin	Winter barley	Leg umes	Barley	Dredge	Oats	Total
Monkton, 1316	0.34	0.00	0.00	0.00	0.92		0.00	0.59
Monkton, 1317	0.91	0.13	0.38	0.80	0.81		0.73	0.78
Birchington, 1316	0.57	0.32	0.75	0.71	0.26		1.95	0.54
Birchington, 1317	0.77	0.41	0.65	1.35	0.89		1.00	0.94
Billingham, 1315	0.41			0.82	0.66		0.61	0.53
Skipton, 1315	0.40	0.59		0.14	0.44		0.54	0.52
Skipton, 1316	0.11	0.00		0.25	0.72		0.55	0.50
Skipton, 1317	0.50	0.00		0.66	1.06		0.43	0.44
Kildwick, 1315	0.39	0.55		0.34	0.56		0.56	0.55
Kildwick, 1316	0.03	0.03		0.16	0.47		0.44	0.42
Kildwick, 1317	0.38	0.23		0.57	0.70		0.52	0.52
Hambledon, 1315	0.41		0.85	0.27	0.85		1.16	0.79
Hambledon, 1316	0.51		0.85	0.16	0.60		0.91	0.67
Hambledon, 1317	0.79		1.04	0.62	0.63		0.61	0.68
E. Meon Church, 1315	0.47		0.33	0.44	0.66		0.58	0.54
E. Meon Church, 1316	0.24		0.15	0.11	0.29		0.55	0.34
E. Meon Church, 1317	0.68		0.43	0.21	0.44		0.70	0.60
Wootton, 1315	0.52		0.72	0.00	1.40		0.59	0.67
Alton Priors, 1315	0.80	0.35		0.31	0.83	0.18	0.57	0.72
Enford, 1315	0.75		0.35	0.08	0.97		0.95	0.82
Scratby, 1315	0.61				0.98			0.84
Feering, 1315	0.71	0.15		0.62	0.70		0.51	0.62
Feering, 1316	0.76	0.49		0.41	1.08		0.74	0.72
Feering, 1317	1.08	0.37		0.32	0.70		0.84	0.92
Sedgeford, 1315	0.30	1.13	0.00	0.72	0.89		1.03	0.89
Weighted avg., 1315	0.51			0.45	0.86		0.60	0.64
Weighted avg., 1316	0.38			0.26	0.53		0.53	0.49
Weighted avg., 1317	0.76			0.68	0.74		0.51	0.58

Source: Campbell, Bruce M. S., *Three Centuries of English Crop Yields, 1211–1491* Slavin, Tithe receipts database (see Appendix 8).

Notes: All figures are relative to the average quantities of tithes paid by tenants between 1299 and 1314. Because of the paucity of evidence and, presumably the occasional omission, rye, winter barley, legumes, and dredge were not included in the average columns.

fact that the scale of rye production on both demesnes was rather limited, ranging between 1 and 4 per cent of all the tithed crops. There is no doubt, however, that the trends of tithe receipts directly reflect similar trends in productivity levels, however inaccurately. In other words, a 40 per cent decline in tithe receipts *does not mean* exactly a 40 per cent decline in yields, but it certainly means that yields drastically declined, possibly close to that figure.

Fodder Produce Depression

So far, we have focused on the arable sector. Arable crops, however, were not the only type of vegetation resources that were devastated by the adverse weather conditions. To appreciate the impact of the torrential rain on land produce better, it is essential to consider the situation within the grassland sector. Just as arable land, grassland, too, was a crucially important type of land-use, in terms of its contribution to agrarian economy. Grassland fodder, deriving from both pasture grass and meadow-mown hay, was an essential cornerstone of the late medieval economy, which depended on healthy animals to ensure food supply for humans. Grass fodder was the single most important energy generator within the working animal sector, consisting of draught horses and oxen. Roughly estimating, pasture grass and hay contributed, collectively, about half or slightly more to an average daily calorific intake of farm horses, and it appears that the figures were considerably higher for bovids and sheep, perhaps around 75 per cent.[76] It should be noted that the fodder crisis caused by the torrential rain of 1314–1316, was, in effect, the peak within a much longer climatic and environmental changes, related to biomass production. As we have seen in Chapter 1, the transitional period from the Medieval Climatic Anomaly and the Little Ice Age and chiefly the Wolf Minimum of *c.* 1280–1340, affected vegetation growth a great deal, by shortening the growing and, consequently, the grazing season by approximately one month, mainly in upland regions of the British Isles. The reduced growing season could have produced less biomass for animal grazing than it would have during the Medieval Climatic Anomaly.[77]

While the torrential rain of 1314–1316 destroyed much of the arable crop land, it produced grass in such abundance that the supply well exceeded demand, and the value of pasturage declined a great deal. There are numerous

[76] Estimated from various pre-Black Death account rolls, on the basis of grain and legume allocation to draught animals.

[77] Oram and Adderley, 'Lordship and Environmental Change', p. 79.

references to an over-abundance of grassland and the difficulties of making any profit out of it. Thus, at Byfleet (Surrey), and Eye and Thorndon (both in Suffolk), the local manorial accounts state that very little pasture was sold because of the abundance of grass on common pastures and heaths. To make things even worse, however, between 1313 and 1317 there was an outbreak of sheep murrain in England and Wales (and possibly other parts of the British Isles), caused seemingly by a combination of liver fluke and scab, and it targeted primarily young animals. At the same time, there was an excess of bovine mortality in 1315–1316 in England, which, however, was nothing compared to the devastating outbreak of bovine pestilence that decimated European stocks between *c.* 1315 and 1321 (ravaging the British Isles in 1319–1321).[78] As we shall see later, the corrupt nature of mown forage undoubtedly contributed to livestock mortality. The decline in livestock numbers decreased the demand for permanent pasture. Some accounts, indeed, bemoan the lack of demand for grassland on the account of animal shortage.[79]

Contrary to permanent pasture, available in abundance, hay produce, deriving from mowable grassland, was severely depressed. That domesticates were deprived of hay supplies during the famine years is implied ubiquitously in both literary and statistical sources. According to the sources, both narrative and statistical, the rainfall was so destructive that very little or no meadow could be mown and converted into hay. But even if demesne workers managed to mow meadows, they did not necessarily succeed in drying the hay. The torrential rain of 1315 and 1316 prevented meadows from drying and, as a result, doomed them to rot and get infected with parasitic fungi, bacteria, worms, and gastropods, posing a dangerous threat to the health of domesticated animals. Indeed, several accounts refer to the great shortage of forage, and in some cases, hay remained unmown and thus, destined to rot.[80] Some accounts describe the bad condition of unmown grass.[81] These references are supported by literary evidence. As John de Trokelowe (d. after 1330) narrates in a gloomy fashion, animals suffered from the putrefaction of herbage as much as humans suffered from the ruining of food.[82]

[78] Newfield, 'A Cattle Panzootic'; Slavin, 'The Great Bovine Pestilence'.

[79] Thus, LH 10688 (Longbridge Deverill, Wiltshire) and TNA, SC 6/996/14 (Eye Castle and Thorndon, both in Suffolk).

[80] Thus, Maids Moreton (Buckinghamshire), in 1316 (BoL, Ch Ch DD27) and Longbridge Deverill (Wiltshire), in the same year (LH, 9764).

[81] Thus, Castlemorton and Longdon (Worcestershire): TNA, SC 6/1109/5.

[82] *Johannis de Trokelowe Annales*, ed. *by* Riley, p. 94; see also, *Thomae Walsingham Historia Anglicana*, ed. by Riley, p. 147.

According to the anonymous composer of *Vita Edwardi Secundi*, grass lay so deep under water that it could not be mown.[83] The nearly contemporary poem *The Simonie, or Poem on the Evil Times of Edward II* (composed *c.* 1320–1325) bemoaned the starvation of beasts.[84] A similar situation prevailed in the Lordship of Ireland. According to the extent of lands of the Archbishopric of Dublin, enrolled into the now-lost 1315–1316 royal memorandum roll for Ireland, no hay could be mown, and no turf could be dug from turbaries of several archiepiscopal manors in County Dublin because of the rainy season and the great flood of water which occurred in that year.[85]

Numerous glosses inscribed into manorial accounts reveal some insightful details about the length of mowing and hay making. Some accounts justify the high expenses related to meadow mowing because of the difficulty to do so and the unusually long duration of the process. Thus, at Raynham (Norfolk), mowing and haying lasted as long as nine weeks in 1316, between 27 May and 1 August.[86] This was a truly remarkable duration, given the fact that in many instances, mowing and haying on the demesne would not last more than two weeks, while common field hay could be gathered and stacked in just one day in June or July.[87] Two explanations can be offered here. First, the soggy conditions of the grassland implied it would take a longer time to mow meadows, make hay, and bring it in — just as in the case of arable harvest. Indeed, several accounts refer to the destruction of pasture and hay resources by flooding.[88] At Wallingford Castle (Berkshire), pasturage sales in 1316 commanded unusually low revenue because seven acres of pasture were lost to the Thames inundation.[89] Getting hay dry would certainly have added more time than usual because of the lack of solar irradiance. Second, we have to account for the state of enervation among manorial labourers around the time of the 1316 harvest. A year of

[83] *Vita Edwardi Secundi*, ed. by Childs, pp. 110–11.

[84] *The Simonie*, in *Medieval English Political Writings*, ed. by Dean, lines 43–46. A fully searchable online version is available courtesy of the University of Rochester-based *TEAMS Middle English Text Series*.

[85] National Archives of Ireland (NAI), RC 8/10, pp. 416–22. Printed in *Historic and Municipal Documents of Ireland*, ed. by Gilbert, pp. 366–71. The English translation can be accessed, in English translation, on the *CIRCLE* website.

[86] RHM, Great Raynham Account, 1315–1316.

[87] Ault, *Open-Field Farming*, pp. 26–27.

[88] For instance, Tickhill Castle (Yorkshire), in 1316 (TNA, SC 6/1088/1) and Pyrford (Surrey), Castlemorton, and Longdon (both in Worcestershire), in 1315 (TNA, SC 6/1109/5).

[89] TNA, SC 6/1096/3.

starvation, which was particularly severe around the harvest time, undoubtedly left a negative mark on workers, whose physical exhaustion must have created a severe energy deficit.[90] One marginal gloss is particularly illuminating. At Wroughton (Wiltshire), in 1316, an additional 2s 8d had to be spent on carting hay because of tenants' weakness (*propter impotenciam customariorum*).[91] The inability to gather harvests because of malnourishment and the physical weakness of tenants is documented in other historical famines. During the Great Irish Famine of 1845–1852, labour gangs, consisting primarily of starving men, were shuffling along from one government-subsidized work project to the next. This created a severe energy deficit that would have hurt the efficiency of any task the same Irish labourers were asked to do.[92] The physical debilitation of mid-nineteenth-century Russian serfs, unable to collect harvest, is echoed in Nikolai Nekrasov's poem 'Unharvested Parcel' ('*Neszhataya Polosa*') (1855).[93]

Although the accounts do not provide, in the vast majority of cases, direct data on grass yields (contrary to crop, wool, and, in some cases, dairy yields), there is a number of indirect indications about that.[94] First and foremost, the harshness of the crisis is revealed in the rates and costs of meadow mowing. It should be noted that in most cases, hay mowing was one of customary dues owed by manorial tenants, and because of this fact, the vast majority of the accounts did not record meadow mowing costs.[95] At the same time, however, there were a number of demesnes which either employed additional labour force, or relied entirely on hired mowers, to shear their meadows and make haystacks. Since different manors were differently structured in terms of their land-use and size of fields, sheer sums expended on mowing and haying will tell little, if anything. To appreciate the implications of the crisis of pasture produce on grazing animals, it is instructive to convert the sheer figures into mowing costs per one animal unit. This can be achieved by dividing the total sums spent on mowing by the total number of grazing and hay-consuming livestock units.[96]

[90] Thus, the 1314–1315 account for Werrore-Cosham (Hampshire) talks about 'bad herbs' (*male herbe*) (Oxford, Queen's College Archives (henceforth, QCA), Roll 246).

[91] HantsALS, DC/J1/14, fol. 16r.

[92] Ó Gráda and Mokyr, 'What do People Die of during Famines', p. 341.

[93] Nekrasov, *Polnoe Sobranie Sochineniy v 15 Tomakh*, I, 135–6.

[94] Although there are indeed some rare and scattered data on grass yields, I could not find a single reference from the famine years.

[95] On the customary hay services, consult Jones, 'Harvest Customs'.

[96] Livestock units are calculated, following the methodology proposed by Campbell, on

Once we have adjusted the figures to give mowing costs per animal unit supported on the demesne, we may now index these figures on non-famine years (in this case, 1301–1314). This renders the relative costs of hay production per animal, during both famine and non-famine years. As Figure 2.2 indicates, the relative hay-producing costs per grazing animal rose during the famine years to unprecedentedly high levels, being about twice as high as in the normal years. At Werrore and Cosham (two Hampshire demesnes, belonging to God's House Hospital, Southampton and managed collectively), the hay-producing cost per animal stood at 0.8d and 0.9d in 1315 and 1316 (compared with the average 0.5d per animal in normal years). At Bolton, the figures rose from 2.1d to 3.9d and 5d per animal in 1315 and 1316. In 1317, the costs were still higher than usual, standing at 2.8d per animal, but clearly, the figures were in decline. The gap in relative expenditures on hay per animal between Werrore-Cosham and Bolton can be partially explained by the fact that Bolton had much higher stocking densities of animals (that is, livestock units per crop acres), which required local authorities to seek hay and pasturage outside the home farm, as it is indeed implied from various meadow-related entries into its account book.[97] In any event, it is apparent that the increased costs of hay production may have been caused by two factors. First, the abundance of grass and the abnormally wet conditions implied a longer duration of mowing and hay harvesting. As we have seen, in some cases, mowing could take as long as nine weeks. Second, a prolonged exposure of hay to rain and humidity made it infested with parasitic fungi, bacteria, worms, and gastropods. Consequently, good-quality hay became deficient, and this undoubtedly commanded higher prices of that crop. It should be noted that the increased sums spent on mowing by no means reflect the possibility that the wages of mowers rose during the famine years. In fact, the *nominal wages* of agricultural workers remained unchanged during the famine year, while their *real wages* reached their single lowest point ever in 1316, with the possible exception of 1597 (See Chapter 4).[98]

Unlike hay, whose prices went up considerably, the value of unmowed (permanent) pasture declined because of the abundance of grass, on the one hand,

the basis on weighting each animal group according to its relative financial value (horses = 1.0; oxen and mature cattle = 1.2; immature cattle = 0.8; and sheep and swine = 0.1). See, Campbell, *Seigniorial Agriculture*, pp. 104–06. Poultry units are calculated following my methodology, outlined in Slavin, 'Late-Medieval Goose Farming', p. 6.

[97] *The Bolton Priory Compotus*, ed. by Kershaw and Smith.

[98] Campbell, 'Nature as Historical Protagonist', p. 285.

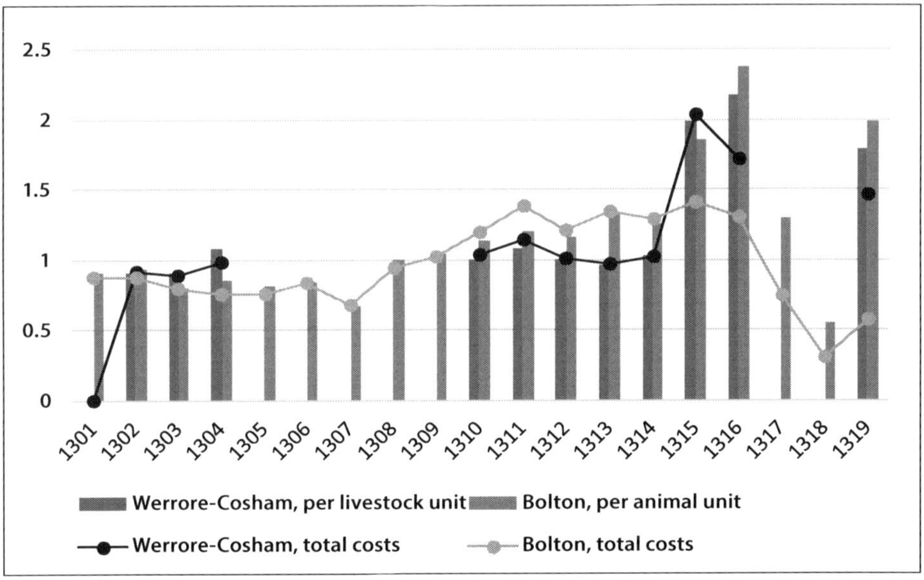

Figure 2.2. Total and relative expenditure on hay production at Werrore-Cosham (Hampshire) and Bolton (Yorkshire), 1301–1319. *Source*: *The Bolton Priory Compotus*, ed. by Kershaw and Smith; QCA, Rolls 241–50; BoL, DD Ch Ch Queens 251–55.

Note: The respective figures indexed on 1301–1319 (1301–1319 = 1.0), appearing on the vertical axis, are as follows: (1) Werrore-Cosham, total mowing costs, 1.0 = 2s 5d; (2) Bolton, total mowing costs, 1.0 = £6 10s; (3) Werrore-Cosham, relative mowing costs per livestock unit, 1.0 = 0.49d; and (4) Bolton, relative mowing costs per animal unit, 1.0 = 2.11d.

and animal murrains, on the other. This is reflected in drastically decreased revenue from pasture sales. As Figure 2.3 shows, the levels of pasturage reached their lowest point during the famine years. On the estates of the Winchester Bishopric, consisting of over fifty demesnes, pasturage sales stood at 16, 45, and 16 per cent below average in 1315, 1316, and 1317, respectively, while the 1315 and 1317 figures for Hinderclay, a Suffolk demesne of Bury St Edmunds Abbey, stood at 34 and 28 per cent below average, respectively. The figures for 1318 and 1319 were still low, perhaps reflecting the decreased demand for pasture, resulting from a reduced sheep population, hit by outbreaks of disease, during the famine years. Similar contraction in pasturage sales revenue is found elsewhere. At Mitchinhampton (Gloucestershire), local manorial officials received 12s 4d in 1316, compared with 29s 7d and 25s 3d in 1306 and 1311.[99]

[99] TNA, SC 6/856/15–17.

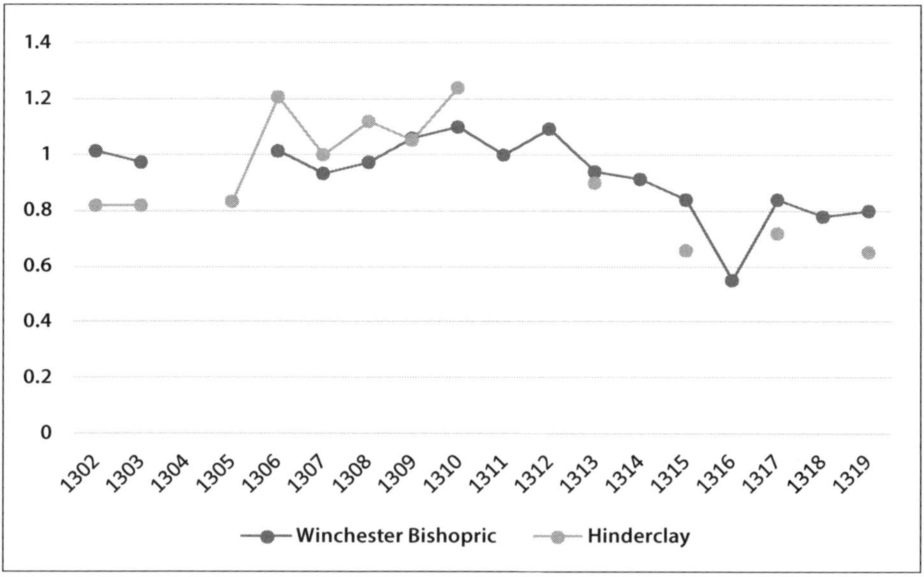

Figure 2.3. Levels of pasture sales, 1302–1319 (indexed on 1302–1314). *Source*: HantsALS, 11M59/B1/57–76 and University of Chicago Library, Special Collections (henceforth, UCL), Bacon Rolls 434–50

At Wheatley (Nottinghamshire), the total revenue from pasturage and hay sales amounted to only £2 10s and £1 6s in 1316 and 1317, compared with the average of £12 for 1313–1315.[100] Similarly, at Byfleet (Surrey), annual sales of pasturage and hay contributed, on average, about £8 between 1310 and 1313, while the respective figures for 1315, 1316, and 1317 declined to £3 12s, 6s 2d, and £4 12s.[101] Conversely, the income from pasture sales at Longbridge Deverill and Monkton Deverill, two Wiltshire demesnes belonging to Glastonbury Abbey, remained, more or less, steady between 1309 and 1319, averaging at 25s 7d in the former and 22s 10d in the latter.[102] This may possibly be because both demesnes got away relatively lightly with sheep mortality during the famine years. But these were clear exceptions, and in the vast majority of cases, little grass could be sold because of the decline in the demand for this commodity.

[100] TNA, SC 6/954/19–23.

[101] TNA, SC 6/1011/1–7.

[102] Longleat House 9641–45, 10688, and 9646–47 (Longbridge Deverill) and Longleat House 9758–67 (Monkton Deverill).

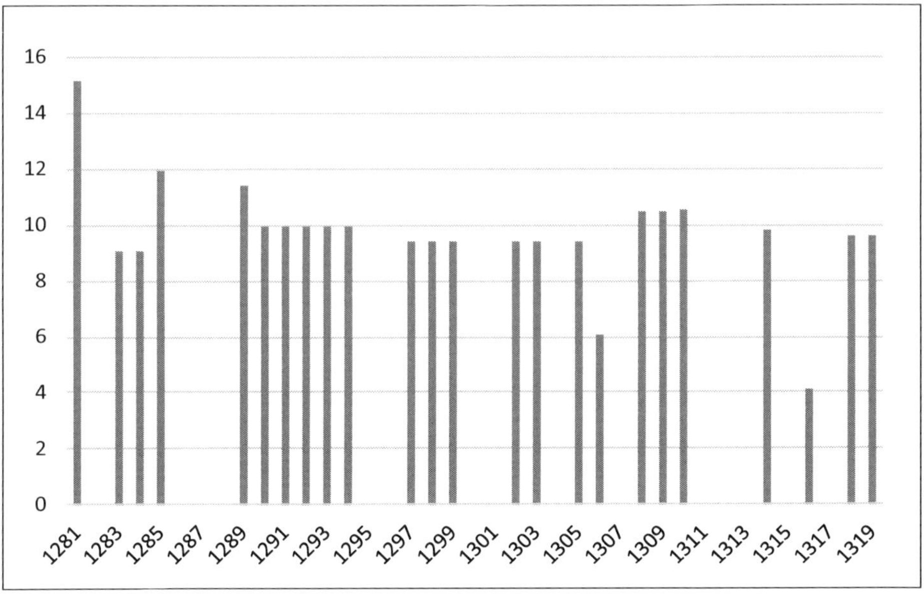

Figure 2.4. Approximate parish hay-produce levels (in £ sterling) at Oldington (Worcestershire), 1281–1319. *Source*: TNA, SC 6/1070/5–24

The glimpse into the tenancy sector is trickier, given the extremely scarce nature of evidence. In the vast majority of cases, tithes were paid in crops or young animals rather than hay. To make things even worse, most of the few surviving hay tithe receipts pertain to the post-Black Death era, in conjunction with the increasing tendency to sale tithes.[103] The only exceptional run of hay tithe sale receipts, with reasonably full chronology and partially covering the famine years, survives (to my best knowledge) for the manor of Oldington in Worcestershire (near Kidderminster) (Figure 2.4). As its 1315–1316 account indicates, in that year the total value of sold hay tithes was 8s 2.5d, compared with the annual average of 20s for 1281–1314. In other words, it appears that in 1315, the total value of hay produced by Oldington tenants may have been about £4 2s, which is about 79 per cent below average, compared to the period 1281–1314. This data has to be approached with caution. In many instances,

[103] For instance, Lytham St Anne (Lancashire) hay tithe sales accounts cover 1381–1426, while those of Murton (Dalton-le-Dale) (Durham) extend from 1350 to 1456. See, Durham Cathedral Archives, Lytham tithe accounts, and Dalton tithe accounts. Again, I am grateful to Ben Dodds for making this data available to me.

the annual hay tithe value was static for several years, standing at 19s 10d in *c.* 1290–1295 and 18s 10d in *c.* 1296–1305. This reflects a commonplace managerial practice to 'standardize' an approximate tithe value in cash, rather than calculate the exact levels of produce.[104] Still, the abysmally low figures for 1315–1316 undoubtedly reflect the depressed levels of hay production, but it appears that hay production levels at Oldington returned to normal in 1317–1318.

Minor Food Sectors

Although arable and pasturage were the two most important sectors of organic production, one has to account for the impact of the crisis on more minor food produce sectors. One such sector is fruit horticulture. Fruit consumption contributed only a fractional share to the total calorific intake. Also, fresh fruit and vegetables were traded on a small scale and informal matter.[105] Still, fruit, chiefly apples, pears, and cherries, did supplement a few calories, in both hard and liquid (pressed) form, and because of that fact, such an exercise is worth a try. If finding direct evidence, textual or statistical, for grassland production is difficult, doing so for fruit produce is nearly impossible. To make things even worse, those few accounts containing information on orchard produce tend to be rather general, and in many instances they seemingly under-record the real produce yields. This was especially true when fruits were consumed domestically during the picking season rather than collected for pressing or selling, owing to their perishable nature. All those constraints make our task to estimate the impact of the inclement weather on orchard produce a very difficult, if not an impossible, one.

Year-to-year per-tree yields can vary a great deal, depending on local weather conditions, age, size, cultivation density, pollination levels, and physical well-being of each tree. I was unable to find a single direct reference to fruit yields per-tree in late medieval England (or anywhere else in Europe), but some fragmentary insights from later pre-Industrial periods can be instructive. For instance, in his 1750 *Country Housewife's Family Companion*, a Hertfordshire-based farmer, William Ellis (*c.* 1700–1758), claimed to have harvested 20

[104] A similar practice existed also for garden tithes. See Dyer, 'Gardens and Garden Produce', pp. 32–33.

[105] The evidence on gardening and fruit and vegetable consumption in late medieval England is masterfully surveyed in Dyer, 'Gardens and Garden Produce', pp. 27–40.

Table 2.7. Reported apple production levels (in quarters) on three south English demesnes, 1303–1334

Year	Shapwick (Somerset)	Westonzoyland (Somerset)	East Meon Church (Hampshire)
1303	5.0	*11.9*	
1305	8.5	32.0	
1311			*10.6*
1312	9.8	*21.2*	*26.5*
1313	16.5	*26.5*	5.5
1314	16.0	*26.5*	4.0
1315	2.5	*10.6*	1.0
1316			0.0
1317			*26.5*
1334	32.9	37.0	

Source: HantsALS, 11M59/B1/59–72; LH, 11246, 11271, 11215, 11216, 10655, 10656, 10766, 10761, 10632, and 10633.

Notes: 1 quarter = 8 bushels = 336 lb of apples. The italicized figures have been inferred by converting the tuns of pressed cider into their hard equivalent (using an approximate ratio of 10.6 quarters = 1 tun (*c.* 240 gallons)).

bushels (that is, some 840 lb) of apples and pears per tree.[106] These, however, are exceptionally high, if not fanciful figures, undoubtedly related to mature and carefully managed trees. All the same, it appears that 10 to 15 bushels per tree was an expected figure in the early nineteenth century, and it is unlikely that the figures would have been much lower in the late medieval period.[107] Two main conditions are essential for good fruit tree yields: sufficient solar irradiance and sufficient pollination levels carried out by pollinating insects, primarily bees and butterflies. Cold and rainy springs hinder fruit tree blossom and prevent insect pollination. Also, bees and butterflies (or any other insects, with the exception of mosquitoes) cannot fly in the rain. Furthermore, cold and rainy winters increase mortality rates of insects because of their inability to secure food. Finally, the torrential rain of 1314–1316 undoubtedly encouraged

[106] Ellis, *The Country Housewife's Family Companion*, pp. 39–41; this reference is discussed in Muldrew, *Food, Energy and the Creation of Industriousness*, pp. 111–12.

[107] Jackson (*The Arithmetical Catechism*, p. 31) reckoned that each tree produces 15 bushels of apples.

the activity of gastropods — that is, slugs and snails — both notorious destroyers of tree leaves. Given all these factors, it is hardly surprising that those few manorial accounts recording orchard produce reported abysmally low figures.

As Table 2.7 hints, the torrential rain of 1314–1316 undoubtedly had a pronounced impact on apple produce levels. Although the reported figures may be far from accurate (because of a possibility that a part of the annual harvest may have been consumed domestically and remained unaccounted for), they, nevertheless, provide a good idea of general trends. Indeed, the figures for the famine years were remarkably low, as local demesne managers managed to pick very few apples compared to non-famine years. Moreover, the accounts also reaffirm the impression that produce levels varied a great deal in non-famine years. Thus, at East Meon Church, produce level trends seem to have been already in decline by 1313. As far as other fruits are concerned, I was able to find only one reference to pears during the famine years: according to the 1315–1316 account from Harston (Cambridgeshire), only 1 bushel of pears had been harvested.[108] Naturally, this figure might have been downplayed, but it may still reflect the fact that other fruit trees suffered in a similar fashion.

The low fruit yields were related, first and foremost, to the bad physical state of pollination agents. Although comparatively few accounts include a portion on bee-keeping, it is still possible to have a general idea about the impact of the weather crisis on this hitherto neglected sector of late medieval agrarian economy. Beehives were usually kept in gardens, to produce honey and wax, both highly commercial products.[109] Liquid honey could be used either for making mead (an alcoholic drink made by fermenting honey with water) or as a sweetening ingredient for wine, a beverage reserved primarily for better-off echelons. In addition, some quantities of honey must have been eaten either spread on bread or as an ingredient in dishes. According to one late medieval estimate, 1 gallon of honey could produce 5 gallons of mead.[110] It appears that mead was falling out of fashion by the time of the famine, and there is no evidence about large-scale production and consumption of this beverage for that period.

Those few accounts that record bee-keeping provide some clear indication about the physical state of beehives and their production depression during the crisis years. As the accounts indicate, the rainy and freezing winters of the famine years increased the mortality rates of swarms a great deal. Thus, at

[108] British Library (henceforth, BL), Add. Roll 18522.

[109] Dyer, 'Gardens and Garden Produce', p. 29.

[110] *Walter of Henley*, ed. by Oschinsky, p. 474.

Werrore and Cosham (Hampshire), five out of eight swarms died during the winter of 1315–1316 'because of excessively rainy time'.[111] Similarly, at Pilton (Somerset), thirteen out of seventeen swarms died over the course of the winter of 1314–1315.[112] Those manors that saw excessive bee mortality seem to have suffered from depressed yields of honey and wax production. At Werrore-Cosham, a total of eleven swarms produced only 1 gallon of honey and about 4.3 lb of wax in 1314–1315 (that is, approximately 0.09 gallons [= 1.07 lb] of honey and 0.39 lb of wax per swarm), while in the following year, a total of eight swarms produced no honey or wax at all.[113] The 1314–1315 figures were approximately 27 per cent below the average for period 1302–1314. Unfortunately, I was unable to find any comparable series of accounts recording honey and wax production for 1315–1317.

Salt-making was yet another food industry that was severely depressed by the inclement weather. In the late medieval period, salt-making depended much on the natural evaporation of brine (created by sand washing during the high tide period in the late spring).[114] Clearly, the flooding of 1314–1136 would either hinder or make the evaporation of brines impossible. Stoking fires around the salt pan to quicken evaporation would have been extremely wasteful and inefficient.[115] Although I was unable to trace a single manorial account containing a section on salt production during the famine years, there are a number of direct references to the crisis in the accounts. Thus, in 1315–1316 at Great Chart (Kent), 38s were expended on 18 bushels of salt because of the 'great dearth and deficit of salt'.[116] Similar comments on the catastrophic deficiency of salt are found in the 1315–1316 accounts from Stockton (Wiltshire) and Werrore and Cosham (Hampshire).[117] Just as with grains, salt prices trebled and quadrupled, and this fact made fish and meat processing and preservation, as well as

[111] BoL, DD Ch Ch Queens 251.

[112] LH 10766, membr. 40^{v}.

[113] QCA, Roll 246 and BoL, DD Ch Ch Queens 251. It is clear that the Werrore-Cosham accounts should be approached with caution, as they seem to drastically under-record the total rates of honey production. According to late-thirteenth century treatise *Husbandry*, each swarm ought to yield 1 gallon of honey — about ten times more than reported in the Werrore-Cosham accounts (*Walter of Henley*, ed. by Oschinsky, pp. 432–33).

[114] Hurst, 'Fuel Supply'.

[115] Jordan, *The Great Famine,* p. 52.

[116] CCA, DCc/BR/Great Chart/29.

[117] BoL, Ch Ch Queens 251.

making dairy products, increasingly difficult. This fact had some far-reaching implications on food security, as we shall see in Chapter 4.

The torrential rain had an equally devastating effect on vineyards. In England, viticulture was practised on a very limited scale, confined primarily to the southern and western counties.[118] There are no accounts for vintage yields, and we can only speculate that local vineyards were damaged at least as badly as on the Continent. The destruction of the wine harvest is reported in Normandy, the Paris region, and other parts of northern France, with vintage yields declining by 50 per cent in 1315 and 1316 and by some 80 per cent in 1317. Similarly, the bad weather ruined vineyards of the Rhineland. In those instances, when grapes were harvested and processed, the quality of wine was described as deplorably sour. Also, making wine was impossible in parts of northern Italy because of the rainy weather in the summer of 1315.[119]

Wider Implications

There is no question that the torrential rain of 1314–1316 created a catastrophic situation by depressing various sectors of food production. Most importantly, the arable sector, for which unusually detailed, reliable, and robust statistics exist, could not provide the starving population with the required calories. As we have seen, the composite gross crop yields are reported to have stood at 25, 35, and 14 per cent below average in 1315, 1316, and 1317, with the first two years being particularly bad for winter crops and legumes, and with oats and winter barley failing during the third harvest failure. On the eve of the famine, an English commoner derived about 70 per cent of his/her calorific intake from crop-based products (both hard and liquid), and there can be no doubt that the impact of the grain failures on food supplies were catastrophic.

The torrential rain, moreover, did not only deprive humans of their most basic foodstuff. Other types of food were destroyed too. Take fruit, for instance. Never a paramount ingredient of the human diet, fruits (primarily, apples and pears) were, nevertheless, consumed in modest quantities by all social strata. Although not as omnipresent as ale, apple (or pear) cider was still produced and consumed in non-negligible quantities, especially in the southern and western counties.[120]

[118] Campbell and Bartley, *England on the Eve of the Black Death*, pp. 105–06; Dyer, *Standards of Living*, p. 62.

[119] Lucas, 'The Great European Famine', p. 373.

[120] McLean, *Medieval English Gardens*, pp. 24, 25, 27; Dyer, 'Gardens and Garden Produce'.

Across the world, which at that time lacked any bacteriological knowledge any bacteriological knowledge, the consumption of fermented beverages was life-saving. With the epic harvest failures, the availability of drinking grains shrank a great deal, and cider became the most obvious substitute for ale. The stormy weather, accompanied by the torrential rain and possibly ice storms, destroyed much of the apple tree produce and deprived humans of their access to cider. With the decreased availability of both ale and cider, the only drinkable alternative left was fresh water. This would have had the most devastating impact on human health: flooding and high winds associated with torrential rain tend to pollute water a great deal, by discharging waste and mud into rivers and other reservoirs of water. Although drinking fresh water was always dangerous, it has never been as hazardous as it was during the Great Famine years.

Finally, the catastrophic dearth of salt implied that preserving and curing fish and meat became a serious challenge. The fact that fresh meat had to be consumed shortly after slaughter undoubtedly caused much inconvenience for peasants, who were used to storing salted meat over extensive periods of time.[121] The same can be said about dairying. In effect, dairying can be regarded as milk preservation, with salt being an essential ingredient in making cheese and butter. Just as with meat, the salt deficit may have prompted many commoners to switch from processed dairy products to raw milk, which had to be consumed swiftly due to its highly perishable nature. All these dietary complications had profound implications on the food security of the populace of England and other parts of the British Isles.

Humans were not the only victims of the disastrous weather and the ensuing food crisis. The scarcity of good-quality hay and the infestation of hay with parasitic fungi, bacteria, worms, and gastropods posed a dangerous threat to the health of domesticated animals. As we have seen, contemporary and nearly contemporary chroniclers reported the consumption of corrupt hay by the domesticates. Also, cold and wet weather encourages the mating activity of scab-mites and, consequently, the spread of sheep scab.[122] Furthermore, domestic animals also suffered from wet weather and cold temperatures, which likewise had a negative effect on their immune system and health. Exposed to colder temperatures and deprived of the most basic kinds of fodder, animals had to waste more energy to maintain body heat. This, in turn, was likely to have lowered resist-

[121] Scully, *The Art of Cookery*, pp. 54–55; Dyer, *Standards of Living*, p. 160.

[122] Slavin, 'Landscapes of Panzootics'.

ance to pathogens within a short period of time.[123] The outbreaks of sheep and bovine murrain, as indicated in both chronicles and manorial accounts, were on a modest scale. Thus, in 1314–1315, 1315–1316, and 1316–1317, the annual losses of sheep through murrain stood at 10 per cent for each year. Similarly, in the course of 1315–1316, England and Monmouthshire lost about 9 per cent of their bovids, which was nothing compared to the massive losses during the 1319–1320 outbreak, when the sheer losses stood at about 62 per cent.

Paradoxically, however, it was the survivors, rather than the victims, that were affected by the long-term impact of the 1314–1316 weather crisis. This is particularly true in the case of bovids. The cold weather and malnourishment, especially in young cattle, may have had harsh consequences for their health and physical growth, chiefly the development of muscles. Consequently, bullocks were more likely to have grown into sterile bulls and weak oxen, while heifers were doomed to become infertile and prone to abortion. It should be borne in mind that the period of growth in cattle and sheep is significantly shorter than in humans, and this fact undoubtedly had some far-reaching implications for the well-being of stocks. Normally, it takes about eighteen months for calves to develop into sexually mature animals. Therefore, a six-month deprivation and malnutrition in a calf, bullock, or heifer, related to the cold weather and unavailability of good-quality fodder, is undoubtedly much worse than a year-long deprivation in a sub-adult human. A close analysis of manorial accounts for the period of 1310–1319 indicates that about 87 per cent of the English and Welsh bovine stocks that succumbed to the disastrous pestilence of 1319–1320 (most likely caused by rinderpest) were survivors of the 1314–1316 crisis. Almost half of the stocks consisted of animals that either were maturing, or were *in utero*, during the earlier crisis years. The immune systems of calves were probably compromised, and it is hardly surprising that a large proportion of mature bovids (especially cows and bulls) died in such large numbers during the pestilence. The bovine pestilence, in turn, had a long-lasting impact on human health and vulnerability. As I shall argue in the Epilogue, the biological connection between the three disasters of the early fourteenth century — the Great Famine, the Great Cattle Pestilence, and the Black Death — should by no means be overlooked.

[123] Newfield, 'A Great Destruction of Cattle', p. 64.

Chapter 3

Declining Food Security: Harvest Production and Management

Famine as an Anthropogenic Phenomenon

There is no doubt that the weather anomaly of 1314–1316 had a pronounced impact on various food sectors — arable produce, first and foremost. The torrential rain, strong winds, and cold temperatures brought about a very rare scenario of three back-to-back harvest failures in 1315, 1316, and 1317. Yet, even such a remarkable weather anomaly could not destroy the annual harvests of 1315, 1316, and 1317 in their entirety. As we have seen, all it did was reduce the respective *gross* crop yields by approximately 25, 35, and 14 per cent, respectively, thus creating a food shortage. Yet, fair proportions of food were still available for human consumption. With some stress and malnourishment, these harvests were, at least in theory, still sufficient to supply the vast majority of the English population with food and avoid widespread starvation and, consequently, a mortality crisis. Yet, the same food shortage did become famine; and not just any famine, but arguably one of the worst (and perhaps the worst one) in recorded European history. It was not Nature, however, that saw the transformation of shortage into famine: it was humans, human factors, institutions, and decisions that did so.

Because of their dependence on food availability and supply, humans inevitably react to food shortages. In theory, human intervention, through a system of political, legal, socio-economic, and cultural means and values, can either mitigate a food crisis and thus prevent famine or, on the contrary, aggravate a food crisis and thus create famine. Students of famine, both historians and social scientists, appreciate the impact of anthropogenic factors on subsistence crises. As we have seen, this is especially true since the publication of Amartya Sen's monumental *Poverty and Famines* in 1981. For Sen, famine is a man-made phenomenon. Using the example of the Bengali famine of 1942–1943, Sen contended that in many cases famines occurred not because of a lack of food resources, but because of the decline in 'entitlements' to (depleted) food resources. In other words, what creates famine is not an actual 'food availability decline' (FAD), but the lack of the 'exchange entitlements' to ensure steady access to food, or 'food entitlement decline' (FED). According to Sen, famines take place when lower social strata lose their entitlement to food, while the better-off, at the expense of the rest, increase their own supply of food. The misallocation of food resources is especially true in the context of the free market, whose behaviour in dear years tends to cater for people of means, while poorer elements do not have the purchasing power to secure their dietary needs. In other words, the link between poverty and famines is apparent.[1]

Notwithstanding some criticisms and contrary views, Sen's view of famine as a FED-induced rather than FAD-caused phenomenon is supported by countless examples of historical famines, past and present.[2] Perhaps the most obvious historical example is the Irish Potato Famine, or *An Gorta Mór* in Irish Gaelic (1845–1852). It is true that the catastrophe started as a climatic and biological crisis, but it was the disastrous policies of the British Government and rent-seeking tendencies of local lords that caused Ireland to lose some 10–15 per cent of her population to hunger and hunger-related diseases and a further 10 per cent to emigration.[3] Similar scenarios can still be seen today in some African countries. Thus, the 2005–2006 famine of the African Horn and the 2011–2012 famine in East Africa may have started with drought, but they were aggravated by tribal military conflicts, failed state governance, human rights abuses, and the international humanitarian aid failure. Taken together, those anthropogenic factors prevented local populations from accessing what was naturally theirs, thus, creating a FED scenario, par excellence. A similar

[1] Sen, *Poverty and Famines*.

[2] De Waal, *Famine Crimes*; De Waal, *Famine that Kills*; Howe and Devereux, 'Famine Intensity and Magnitude Scales'; Rubin, 'The Entitlement Approach', and Rubin, 'A Collapse of Entitlements'.

[3] Ó Gráda, *The Great Irish Famine*.

Table 3.1. Shares of annual harvests used as seed corn in 'normal' and famine years, expressed as a percentage of total annual harvests, 1270–1349.

Harvest	**1270–1349 (excluding the famine years)**	**1315**	**1316**	**1317**
Wheat	20%	34%	38%	25%
Rye	22%	33%	43%	26%
Winter barely	23%	35%	46%	38%
Legumes	27%	54%	47%	37%
Barley	23%	29%	30%	26%
Dredge	20%	40%	40%	32%
Oats	29%	43%	39%	39%
Composite	24%	38%	38%	31%

Source: Slavin, Manorial accounts database (see Appendix 8).

situation can be seen in Haiti. It is true that much agricultural produce was destroyed by a mighty 7.0 magnitude earthquake on 12 January 2010, but it was a series of anthropogenic factors, including the inefficient humanitarian response, the inability of the Haitian government to address the most basic concerns of public health, sanitation, and housing, and the persistence of a black market, as well as remarkable levels of political corruption and high rates of violence in various earthquake-stricken regions, that were responsible for the inability to contain the outbreak of cholera and famine, both lasting at least until 2013.

Was the Great European Famine yet another classic example of a FED crisis? As we have seen, the three back-to-back harvest failures had already created FAD conditions, but not FED conditions, because non-negligible quantities of food were still available for consumption. If three back-to-back harvest failures were not bad enough, however, various socio-economic, political, and demographic factors contributed to the misallocation of already very scarce food resources and turned the FAD situation into a FED crisis. Several aspects were of particular importance: harvest management, market performance, transportation and storage of grain, an ongoing military conflict, and social poverty. A separate chapter will be devoted to each of these five factors.

Table 3.2. National, aggregated net crop yields in England and south-eastern Wales (Monmouthshire), 1315, 1316, and 1317, indexed on 1270–1429 (1270–1429 = 1.00).[4]

Crop	Each crop's share in total acreage	Net yields, 1315	Net yields, 1316	Net yields, 1317	Coefficient of variation, 1315	Coefficient of variation, 1316	Coefficient of variation, 1317
Wheat	33%	0.57	0.43	0.87	0.498	0.604	0.435
Rye	7%	0.62	0.32	0.86	0.809	0.630	0.449
Winter barley	2%	0.51	0.32	0.45	0.494	0.712	0.389
Legumes	6%	0.34	0.37	0.95	1.248	1.232	0.714
Barley	13%	0.81	0.66	0.81	0.523	0.512	0.407
Dredge	3%	0.69	0.52	0.71	0.599	0.438	0.487
Oats	36%	0.63	0.56	0.56	0.789	0.576	0.527
Composite, indexed on 1270–1429		0.62	0.50	0.74			
Composite, harvest-to-seed ratio (gross of tithe, net of seed)		1.68	1.32	2.08			

Source: Slavin, Manorial accounts database (see Appendix 8).

Note: The figures are indexed on average net yields (inclusive of tithes, but exclusive of seed) for the period 1270–1429. These are per-seed yields, calculated as the ratio between a previous year's seed and the current year's harvest *minus* seed corn. Average net yields in normal years are 3.26:1 for wheat, 3.25:1 for rye, 3.04:1 for winter barley, 2.21:1 for legumes, 2.92:1 for barley, 2.73:1 for dredge, and 2.12:1 for oats. Average net yields for 1270–1429 have been estimated by converting the gross yields (deriving from Campbell's database of crop yields, as in Table 2.2) into their net equivalent, using the formula $Ny^i = Gy^i - (Gy^i \times S^i)$ (see previous page and Appendix 1). The indexed figures in the table are expressed in relation to the average net yields for the period of 1270–1429. For instance, the indexed net wheat yield in 1315 is 0.57, which translates to 2.06:1 (that is, the net harvest was 2.06 times its previous year's seed). The coefficient of variation is a degree of deviation from the average yield achieved in a given year. The higher the coefficient figure, the higher was the variety of yields attained by different demesnes, in relation to their average net yield.

[4] My results differ from those in the database by Campbell, *Three Centuries of English Crop Yields*. This discrepancy in calculations is due to four reasons. First, my calculations are based on a somewhat bigger sample of demesne accounts. Second, in converting 'raw' yields into 'net' yields, Campbell utilized the following general formula: [(Yield per Seed ÷ 0.9) – 1]. Third, since Campbell's database covers a much longer period (1211–1492), the regression method had to be applied in order to splice many individual and discontinuous manorial yield series into a single master yield series. Fourth, Campbell's calculations of net yields on a regional (as opposed to manorial) basis accounted only for wheat, barley, and oats, while legumes and minor crops were left out. According to Campbell's database, the composite yields of those three grains for 1315, 1316, and 1317 stood at 39, 63, and 10 per cent below average, respectively.

Net Demesne Crop Yields

As we have seen in the previous chapter, the composite gross yields of 1315, 1316, and 1317 stood at approximately 25, 35, and 14 per cent below average. But did it mean that the available crop resources for human and animal consumption were cut by approximately 25, 35, and 14 per cent? It is essential to bear in mind that, as Tony Wrigley has shown, the relative supply of food and market prices were determined *not* by *gross yields* (that is, how much was reaped in relationship to a previous year's seed), but rather by *net yields* (yields gross of tithes and net of seed, namely, how much was left for consumption, *after the deduction* of a certain proportion of harvest, which was used as seed corn for next year's harvest).[5] In other words, food supply was determined not by how much harvested, but how much was left after post-harvest seeding. Conversion of gross yields into their net equivalents is given, in a detailed manner, in Appendix 2. Overall, some 38 per cent of harvest was allocated to seeding in both 1315 and 1316, while in 1317, the figures declined to 31 per cent. These were remarkably high figures, compared with those from non-famine years from the pre-Black Death period, when about 24 per cent of 'composite harvest' was invested in seeding. The figures rose, especially within the winter grain and legume sectors — namely those crops that failed miserably in 1315 and 1316. At the same time, however, the relative proportions of harvest used as seed rose within the spring grain sector too. Although figures varied across manors, there appears to be a clear negative correlation (varying between -0.40 and -0.60) between gross yields and the proportion of harvest used as seed. In other words, those demesnes that achieved miserably low yields tended to allocate disproportionally high shares of harvest as seed, and the other way around.

The investment of a disproportionally large share of harvest as seed corn had a negative impact on net crop yields and, hence, on the theoretical crop supply. As Table 3.2 indicates, net crop yields (that is, gross of tithe and net of seed) were abysmally low, much lower than gross yields. The respective figures for 1315, 1316, and 1317 were approximately 38, 50, and 26 per cent below average. Just as with gross yields, winter crops and legumes were faring particularly badly in both 1315 and 1316. Thus, the net yields of wheat, the single most important grain for human consumption, stood at about 43 and 57 per cent below average, while those for rye fell to around 38 and 68 per cent below their normal levels. The legume net yields were approximately 66 and 63 per cent below average. The net yields of spring grains were somewhat higher, but still

[5] Wrigley, 'Some Reflections on Corn Yields'.

Table 3.3. Year-to-year fluctuations in sheer quantities of crops sown on English demesnes, 1313–1319 (expressed as a ratio of each year to the previous year)

Year	Wheat	Rye and maslin	Winter barley	Legumes	Barley	Dredge	Oats	Total	No. of demesnes
1313–1314	1.01	1.03	1.02	0.96	1.03	1.03	1.04	1.03	88
1314–1315	0.96	1.01	0.97	0.98	0.98	1.02	1.01	0.98	83
1315–1316	1.08	0.87	0.96	1.25	0.95	0.84	0.97	1.00	101
1316–1317	0.98	1.01	0.89	0.61	1.05	1.03	0.98	0.97	81
1317–1318	0.97	1.07	0.83	1.22	0.97	0.97	0.85	0.92	86
1318–1319	1.02	0.97	1.07	1.00	0.98	1.02	1.12	1.06	92

Source: Slavin, Manorial accounts database (see Appendix 8).

Note: As the total number of demesnes for which seven consecutive accounts survive is comparatively low (I was unable to locate more than sixty-five demesnes), I have augmented the sample by calculating the ratio between the totals of two sets of demesnes for which consecutive accounts survive. For instance, I was able to identify 101 demesnes that had accounts for both 1314–1315 and 1315–1316. This figure appears in the right-hand column *no. of demesnes*, within the '1315–1316' row (given that here, I have calculated the ratio between the total amounts of crops sown in the course of 1315–1316 to those sown during 1314–1315)

low enough to ensure a catastrophic shortage. The situation was considerably better in 1317, with the exception of winter barley (55 per cent below average), dredge (29 per cent below average), and oats (44 per cent below average). The respective composite ratios of harvest to seed stood at about 1.68, 1.32, and 2.08. While there is no doubt that the abysmally low net yields reflect the failed harvests of the same years (namely, the gross yields), we still have to account for why their relative levels stood well below those of gross yields. It is precisely here where we enter into the realm of anthropogenic factors, in which human choices, decisions, and behaviour, and not Nature, are paramount players.

To understand the gap between the gross and net yields, it is essential to understand why such large proportions of already bad harvests were invested in seeding. A close analysis reveals that the *absolute* quantities of sown quarters were practically identical both during the famine and in 'normal' years (Table 3.3). In other words, manorial officials retained the same sheer quantities of crops in famine and non-famine years to be used as seed corn. At first, this choice appears as paradoxical and even irrational: why would they doom such large amounts of perfectly edible grain to rot in the fields under the incessant rain, instead of storing them in barns or releasing them onto local markets? There are two possible explanations to that. The first possibility is that without any ability to foresee the end of the bad weather, and, hoping that the rain would stop sooner rather later, the lords and their demesne officials were ready to take a risk and 'run business as usual', as far as the absolute quantities of seed corn are concerned. The second (and by far more likely) explanation is that unchanging sheer quantities of sown crops reflect the agricultural conservatism of lords rather than a managerial strategy of manorial officials. It should be borne in mind that in the pre-Black Death era, year-to-year variations in sheer quantities of sown crops were hardly noticeable. In other words, the lords traditionally expected each demesne to sow a specific quantity of crops, and as a result, sown quantities remained almost constant over long periods of time. It is true that there were some visible long-term changes, but those are measured in long runs rather than annual terms or even quinquennial means. It was manorial officials that were responsible for carrying out their lords' wishes by sowing the prescribed quantities of crops. Any significant deviation from the traditional quantities could carry serious consequences for manorial officials. At the very least, they would have to account for their daring changes to local auditors, before whom the accounts were rendered. In other words, the hands of the manorial officials were completely tied, and they were at no liberty to tailor some managerial decisions to the immediate reality and thus provoke the ire of their lords.

Lords, on the other hand, showed no signs of flexibility in face of the dire weather conditions of 1314–1317. For them, it was 'business as usual', or, to be more precise, 'business more profitable than usual'. They must have realized that the catastrophic shortage would lead to an unprecedented rise in market prices. Keeping the sheer levels of sown crops as they were in non-famine years would have served a two-fold purpose. First, it would have decreased the food supply available for sale (or other means of acquisition) even further, thus, driving prices even higher. As we shall see in the following chapter, the crop prices did indeed reach an unprecedented historical peak in 1315–1316 and 1316–1317. Second, it would assure that at least some of the sown seed would bear some profit when harvested. In the challenging circumstances of the inclement weather of 1314–1316, any positive yield, however meagre, would matter. Once the next year's harvest has been collected, the same strategy would be applied. This way, the producers could profit from the crop shortage and speculative prices for two consecutive years. That lords loved the years of dearth and high prices, while the years of abundance and low prices were sad news to them, is reflected in several literary sources. Thus, in the Middle English poem *Wynnere and Wastoure*, composed in the 1350s, the Waster assures the Winner that once the latter becomes rich, he would be driven into despair and suicidal thoughts by abundant harvests and low prices.[6] Such attitude is also found in Shakespeare's *Macbeth*, where Porter has aptly remarked that 'a farmer... hanged himself on the expectation of plenty'.[7] If this interpretation of the entrepreneurial behaviour of crop producers is correct, then there can be little doubt that a FAD scenario was being replaced with a FED scenario, through purely anthropogenic factors.

Nor did the sown acreage change. The seeding rates, measured by number of bushels sown on one acre of arable land, remained constant over the pre-Black Death period, including the famine period. On average, the per-acre seeding rates stood at 2.75 bushels for wheat, maslin, and rye, 3.15 bushels for winter barley, 3.25 bushels for beans, 2.75 bushels for peas and vetches, 4.25 bushels for barley and dredge, and 4.5 for oats. There were some local and regional variations. Thus, some Berkshire and Oxfordshire demesnes of the Winchester Bishopric had thicker rates, especially for spring grains, averaging around 6 bushels per acre.[8] Similarly high rates can be observed on the Kentish demesnes of Canterbury Cathedral Priory. On the other hand, some Somerset demesnes reported unusually thin seeding rates, ranging between 1.5 and 1.9 bushels per acre for winter grains.[9] Similarly, Hinderclay and Sapiston, both in the Blackbourn Hundred (Suffolk), and Newton, in west Norfolk, report low seeding rates of 2.2–2.8 bushels for oats.[10] It should be noted that exceedingly low seeding rates would encourage weed growth and competition with grains, which, in turn, could result in meagre harvests if the grains became choked by weeds.[11] Some weeds, such as ryegrass (*Lolium rigidum* Gaudin and *Lolium*

[6] *Wynnere and Wastoure*, ed. by Ginsberg, lines 368–74. I have used an online annotated edition, published by the *TEAMS Middle English Text Series*.

[7] Shakespeare, *Macbeth*, Act II, Scene 3.

[8] The demesnes in question are Harwell (Berkshire), Witney, and Adderbury (both in Oxfordshire).

[9] The demesne in question are Taunton, Nailsbourne, Poundisford, and Rimpton.

[10] UCL, Bacon Rolls 445–47 (Hinderclay); Cambridge University Library (henceforth, CUP), Ch(H) 4/4/30 (Newton); East Suffolk Record Office (henceforth, ESufRO), HD 1538/338/15–17 (Sapiston).

[11] Campbell, *Seigniorial Agriculture*, p. 310.

multiflorum L.), wild or animated oat (*Avena sterilis* L.), and various types of Bromus grasses, are particularly persistent.[12] This could especially be an issue in the climatic context of the famine years; as some research has shown, high precipitation levels stimulate excessive weed growth.[13]

Changes in Annual Gross Crop Receipt (AGCR) Disposal

Obviously, increasing the proportion of harvest used in seeding would change the overall structure and patterns of the disposal of annual crop stocks by demesne officials. In the course of each agricultural year, demesnes would acquire the annual gross crop receipt (*AGCR*), made up of the harvest (accounting for about 90–95 per cent of the *AGCR* in non-famine years), as well as occasional purchases, rent-payments, and, in the case of appropriated churches, tithes.[14] Each year, the *AGCR* would be disposed in the following manner: about a quarter of the *AGCR* would be invested in seeding. The remainder of the *AGCR* would be sold; consumed domestically by local *famuli* (hired labourers) and, especially during the harvest period, fed to working animals; sent to lord's headquarters for local consumption; carried over to the next year (namely, stored in local barns); and after threshing in granaries, baked or malted and subsequently brewed into ale.

As Table 3.4 shows, there were some clear changes in the patterns of the *AGCR* disposal. At first, these changes appear to have been not too dramatic, but a closer analysis of the account rolls reveals that they were significant enough to alter the existing managerial structures, which, in turn, reduced the availability of the food supply further. This is especially apparent in the balance between the share of purchased and sold grain. In normal years, crop purchases rarely accounted for more than 6 per cent of the total *AGCR*. During the famine years, the share of purchased crops rose by almost 10 per cent. Conversely, there was a pronounced decrease in the relative proportion of *AGCR* sales, from 32 to about 22 per cent (that is, decline by some 32 per cent). In other words, the demesnes managed to buy more and sell less. Some demesnes cut down the sale levels in a more drastic manner. Thus, Heckley (Hampshire) sold only 2.8 quarters of crops out of the total 113.1 quarters in 1315–1316 (2 per cent), while in

[12] Cussans and others, 'Survey of the Incidence of the *Bromus* Species'.

[13] Santín-Montanyá, Zambrana-Quesada, and Tenorio-Pasamón, 'Weed Management in Cereals', pp. 142–43.

[14] Slavin, *Bread and Ale*, pp. 69–75.

Table 3.4. Changing patterns of Annual Gross Crop Receipt (AGCR) disposal, 1301–1325

	Incomings			***Outgoings***					
Years	**Harvest issue**	**Other**	**Total**	**Seeding**	**Sales**	***Famuli***	**Working animals**	**Other**	**Total**
1301–1325 (excl. 1315–1317)	92%	8%	100%	24%	32%	10%	7%	27%	100%
1315–1316	85%	15%	100%	38%	21%	11%	9%	21%	100%
1316–1317	83%	17%	100%	38%	23%	12%	15%	12%	100%
1317–1318	90%	10%	100%	31%	28%	11%	11%	19%	100%

Source: Slavin, Manorial accounts database (see Appendix 8).

1300–1301 the same demesne marketed 106.4 quarters out of the total 382.8 quarters (28 per cent).[15] The decline in the relative share of *AGCR* sales, coupled with the abysmally low yields of 1315 and 1316, compounded the deficit of crops available for purchase at local hubs. The sheer quantities of crops sold by the demesnes in the course of 1315–1316 and 1316–1317 stood, respectively, at 33 and 47 per cent below their pre-famine levels. In the following year, however, the sheer sale levels rose to 15 per cent below their pre-famine years, notwithstanding the mediocre harvest of that year. This has to do with the fact that contrary to the spring crops, wheat performed quite well that year and, being the single most commercialized crop, determined the sale levels.

Unlike the proportions of the *AGCR* sales, the relative share of crops allocated to local *famuli* hardly changed, and if it did, then it changed in the *famuli*'s favour (Tables 3.4 and 3.5). It was, again, the absolute quantities that were affected, in conjunction with the decline in food supply and availability. The total quantities were some 12 per cent below the non-famine-year levels in 1315–1316 and about 22 per cent below the non-famine years during the second year of the famine. In order to understand the dynamics of crop liveries

15 QCA, Rolls 185 and 200.

Table 3.5. Annual crop allowances (mostly of barley) allocated to *famuli*, 1301–1325

(a) c. 1301–1325 and 1315–1316 (105 demesnes)				
Years	**Total *AGCR* (in quarters)**	***Famuli* allocation (in quarters)**	**As % of *AGCR***	**As % of 1301–1325**
1301–1325	31,730	3,302	10%	100%
1315–1316	28,144	2,979	11%	90%

(b) c. 1301–1325, 1315–1316 and 1316–1317 (64 demesnes)				
Years	**Total *AGCR* (in quarters)**	***Famuli* allocation (in quarters)**	**As % of *AGCR***	**As % of 1301–1325**
1301–1325	20,243	2,112	10%	100%
1315–1316	16,652	1,806	11%	86%
1316–1317	14,310	1,651	12%	78%

Source: Slavin, Manorial accounts database (see Appendix 8).

Notes: The table is based on two sets of demesnes accounts: (a) 105 demesnes, for which two consecutive accounts survive (c. 1301–1325 (excluding the famine years) and 1315–1316) and (b) sixty-four demesnes, for which three consecutive accounts survive (c. 1301–1325 (excluding the famine years), 1315–1316, and 1316–1317).

during the famine years, it is essential to appreciate the nature of late medieval working manorial personnel and their wage structure. In the pre-Black Death period, the core of manorial labour was done by various local labourers, the majority of whom were 'stipendiary' (that is, working for wages) rather than 'service' workers (working for rent concession).[16] These *famuli/famule* performed various agricultural tasks, including ploughing, carting, winnowing, threshing, stacking, harrowing, milking, herding animals and poultry, milling, carpeting, gardening, protecting harvests, supervising their fellow-workers, scaring birds, and cooking meals for other labourers. The overwhelming majority of labour consisted of what Claridge and Langdon have recently called 'first-tier' workers, consisting of adults (about 85–90 per cent, of whom the overwhelming majority seem to have been men), while the 'second-tier' workers, namely junior workers (referred to as 'boys', or *garciones*) performing mostly non-physical tasks (such as scaring birds and catching vermin), accounted for 10–15 per cent of all

[16] Postan, *The Famulus*; Farmer, 'The *Famuli* in the Later Middle Ages', pp. 207–36; Slavin, *Bread and Ale*, pp. 92–103.

the manorial personnel.[17] As Claridge and Langdon have meticulously shown in their study, the nature of each labour task, as well as the age and gender of each worker, commanded different wages, expressed in the number of weeks that each labourer had to work for 1 quarter of grain.[18] For instance, supervisors were found at the top of the *famuli* hierarchy, working, on average, between eight and ten weeks for 1 quarter of grain; most workers had to work between ten and twelve weeks, to earn the same amount. The number of weeks for female labourers would vary between twelve and sixteen; young workers would have to labour between sixteen and twenty-four weeks for 1 quarter of crop.[19] Also, one has to bear in mind geographic variations. For instance, the highest grain wages were found within the Home Counties around London, where most labourers would have to work for ten weeks to earn 1 quarter of grain. This undoubtedly reflects the high level of commercialization of that region, where the food demand of the growing population of the capital (and other regional capitals, such as Winchester and Colchester) meant higher demand for labour.[20] Conversely, the least commercialized areas of the west and the north saw the highest frequency of low-paid workers, reflecting their young age.[21] With those particulars in mind, we can return to the famine years.

Surprisingly, the individual stipends of *famuli* did not change at all during the famine years, despite the catastrophic dearth of grain, the rising prices, and, consequently, falling real wages. A manorial supervisor would still receive his 1 quarter of grain (mostly barley) for ten weeks of work, while a shepherd or cowherd would have to work the same twelve weeks to earn the same amount of grain. In other words, it appears that the levels of payment were firmly protected by custom, as reflected in some late-thirteenth century agricultural treatises (such as the *Seneschaucy* and the *Husbandry*), much to the labourers' benefit.[22] What did change, however, was the overall number of *famuli*. Because

[17] Deriving from the manorial accounts database (see Appendix 11). On 'first' and 'second-tier' manorial workers, see Claridge and Langdon, 'Composition of Famuli Labour', pp. 195–210.

[18] Claridge and Langdon, 'Composition of Famuli Labour', pp. 193–94.

[19] Those calculations derive from the manorial accounts database, and they fully agree with those in Claridge and Langdon, 'Composition of Famuli Labour'.

[20] Campbell, Keene, Galloway, and Murphy, *A Medieval Capital*; Claridge and Langdon, 'Composition of Famuli Labour'.

[21] Claridge and Langdon, 'Composition of Famuli Labour'.

[22] *Seneschaucy*, Chs 66–69, in *Walter of Henley*, ed. by Oschinsky; *Husbandry*, Ch. 13, in *Walter of Henley*, ed. by Oschinsky, p. 425.

Table 3.6. Number of *famuli* employed on English demesnes, 1301–1325

Years	A. *c.* 1301–1325 and 1315–1316 (86 demesnes)	B. *c.* 1301–1325, 1315–1316, and 1316–1317 (56 demesnes)	As % of 1301–1325
1301–1325	809	487	100%
1315–1316	707	425	87%
1316–1317		392	80%

Source: Slavin, Manorial accounts database (see Appendix 8).

Notes: Campbell, Bruce M. S., *Three Centuries of English Crop Yields, 1211–1491* The table is based on two sets of demesnes accounts: (a) eighty-six demesnes, for which two consecutive accounts survive (*c.* 1301–1325 (excluding the famine years) and 1315–1316) and (b) fifty-six demesnes, for which three consecutive accounts survive (*c.* 1301–1325 (excluding the famine years), 1315–1316 and 1316–1317).

of the reduced crop resources available for labourers' allocation and because of the customary nature of grain stipend, manorial officers could not employ the same number of *famuli* as in the pre-famine years, and, hence, the most obvious solution was to cut their number. As Table 3.6 shows, the manorial officials had to cut the *famuli* personnel by 13 per cent during the first year of the famine and then by an additional 7 per cent during the second year. On some demesnes, those cuts were even more drastic. Thus, at West Wycombe (Buckinghamshire), the number of *famuli* fell from around fifteen in the pre-famine years to eleven and ten in 1315–1316 and 1316–1317, respectively.[23] At Overton (Hampshire), the figure declined from ten to four in 1316–1317.[24] In some instances, manorial officials preferred to hire 'multi-duty' labourers to cover for more than one task. For instance, at Feering (Essex), a herder looking after both cows and sheep was hired in 1315–1316 and 1316–1317.[25] On the same manor, a man was hired to load horse-carts with grain and to be a domestic servant in 1315–1316.[26] Although 'multi-duty' workers were not unheard of in the pre-famine years, their number rose considerably during the crisis years. In other words, the impact of the crisis on manorial employment cannot be overlooked. Those lucky *famuli* that were kept undoubtedly profited

[23] HantsALS, 11M59/B1/58–69, HantsALS, 11M59/B1/70, membr. 25^{r}, and HantsALS, 11M59/B1/71, membr. 14^{r}.

[24] HantsALS, 11M59/B1/70, membr. 12^{r} and HantsALS, 11M59/B1/71, membr. 9^{v}.

[25] WAM 25632 and TNA, SC 6/841/5.

[26] WAM 25632.

a great deal from the situation because of being protected by custom, which fixed their wages in kind. Working for grain, rather than cash, has never been as beneficial as during the famine years, because it relieved the *famuli*, at least in part, from securing scarce and expensive grain for cash. As we shall see in the next chapter, crop prices rose to unprecedented levels, in a totally uncontrollable manner. On the other hand, those workers that were unfortunate to be laid off, or simply not able to secure employment, found themselves in the unenviable position of being deprived of the possibility to trade their labour for the most basic kind of food, which became increasingly dear and scarce. This seems to be another reflection of the decline in the entitlement to food resources, and it is likely that this led to social tensions among peasants. As we shall see in Chapter 8, the levels of crimes (especially related to food) rose a great deal during the famine years, and, in many cases, these were directed towards fellow-villagers rather than lords. This is hardly surprising, given that theft or other crimes can be regarded as survival strategies, in addition to being an outlet for social tensions.[27] Also, it is likely that the lack of employment opportunity may have strengthened the family and kinship ties, which served as an alterative channel of food security. In any pre-Industrial and autarkic society, which depend almost solely on agriculture, kinship ties were arguably the single most important social institution, while family was the most basic and dominant framework of labour organization. The predominance of family-oriented labour and production meant that *famuli* were by no means the representation of agricultural labour in late medieval England. Rather, they were a group of privileged workers, who were lucky to have secured an additional source of income as employees of lords, as well as beneficed clergy and some wealthier tenants. Although it is impossible to estimate their proportion within the total agricultural labour, it is unlikely that they accounted for more than 3–5 per cent of the rural population of England.[28] Notwithstanding their insignificant number, however, the disruption in access to food of some *famuli* still reflects a largely FED situation.

[27] Ó Gráda, *Famine: A Short History*, pp. 52–55; Hendrix (with Brinkman), 'Food Insecurity and Conflict Dynamics'; Scudder, *Global Threats, Global Futures*, p. 199.

[28] This is true only if we assume nine *famuli* per demesne, plus the under-recording of *famuli* working for beneficed clergy and substantial tenants. In all likelihood, however, the figures could not have been any higher than 5 per cent, since it is unlikely that every demesne could afford (or needed) to hire such number of *famuli*. Recently, Claridge and Langdon have estimated the total manorial labour force to have been about one hundred five thousand around 1300. Claridge and Langdon, 'Composition of Famuli Labour', pp. 212–13.

Table 3.7. Annual oat fodder distribution among demesne horses and oxen (in quarters per animal)

Years	Quarters per horse	Quarters per ox
c. 1301–1325 (147 demesnes)	2.4	0.3
1315–1316 (137 demesnes)	2.3	0.3
1316–1317 (124 demesnes)	2.1	0.3
1317–1318 (118 demesnes)	2.3	0.3

Source: Slavin, Manorial accounts database (see Appendix 8).

Notes: The '*c.* 1301–1325' row excludes 1315–1316, 1316–1317, and 1317–1318. I did my best to make sure that all four samples (*c.* 1301–1325, 1315–1316, 1316–1317, and 1317–1318) overlap in terms of manorial coverage.

Another component of the *AGCR* disposal was the feeding of livestock, and primarily working animals, namely horses and oxen. Draught animals were as instrumental as *famuli*, and their role in agricultural work was as vital as ever during the famine years, when every bushel mattered. As Table 3.4 demonstrates, manorial officials did their best to ensure the sufficient feeding of draught animals by augmenting the relative share of the *AGCR*. Thus, in non-famine years, about 7 per cent of the *AGCR* went to horse and ox fodder, made up almost exclusively of oats (or, in some instances dredge) and occasionally of legumes. This figure rose to 9 per cent during the first year of the famine and to 15 per cent during the second year of the famine, owing undoubtedly to the fall in oat yields and hay deficiency in those years. In 1317–1318, the share of draught-animal fodder within the *AGCR* was still higher than usual (10 per cent), due to the failed oat harvest of 1317, but it returned to the pre-famine figures in 1318–1319. But did it mean that the per-mouth figures remained, more or less, the same? As Table 3.7 shows, there was a slight decline in oat fodder distribution among horses, from 2.4 in pre-famine years to 2.3 and 2.1 quarters per animal in 1315–1316 and 1316–1317, respectively. In other words, the decline was quite insignificant. As far as oxen go, there was no change in their fodder allocation, standing at 0.3 quarters of untreated oats per beast a year. Obviously, these are average figures, and the situation would vary from place to place, depending on both environmental factors and local customs. Thus, in Kent and Somerset, two grassy counties, very few demesnes distributed more than 1.2 quarters per horse in 'normal' years, while during the famine years the figure fell even further. On the other hand, in Oxfordshire, Wiltshire, and parts of west Hampshire, where grass was relatively scarce, very generous allowances of oats were offered. This is reflected in the accounts of the Winchester Bishopric manors, where an average horse and ox would have received the overwhelming 5 and 0.6 quarters per mouth, respectively, namely twice as much as elsewhere, around 1300. In 1315–1316, the respective figures still stood at 4.2 and 0.5 quarters per animal.[29]

Given the dire straits of the famine years, it is quite remarkable that manorial officials managed to assure a steady supply of oat fodder to draught animals. It is possible that the relative stability in fodder allocation levels reflect the fact that reeves' hands were tied by custom, and any visible change would make them accountable before auditors (just as in the case of seeding, discussed above). It is equally plausible, however, that it derived from their fear that underfeeding horses would have depressed grain production even further. Feeding horses with oats was especially crucial because horses require more protein than oxen to retain strength and be in good health. Modern animal nutritionists recommend that the protein proportion within the equine diet should reach about 9.4 per cent for an adult working horse, 10.4 per cent for a progressively pregnant mare, and 11.7 per cent for a young horse.[30] Indeed, despite the low crop yields and the increased demand for oats by humans, manorial officials did not decrease the amount of oats allocated to horses and oxen during the famine years. The concern for the well-being of draught animals is reflected in some entries, according to which the officials had to distribute more oats than usual because of the catastrophic dearth of hay. For instance, the demesne of Byfleet (Surrey) allocated very little oats to horses around 1300, relying primarily on grass-based fodder. The challenging situation of the famine years prompted local officials to revise this policy and augment the oat portion of equine fodder. In 1315–1316, an average Byfleet horse received 3.4 quarters of oats, compared with just 0.3 quarters in 1309–1310. To avoid confusion and possible reproval from auditors, William de Rotthyng, local sergeant, made sure to state in his account that the unusually high quantities of oats were fed to horses 'because they had no hay this year' (*quia non habuerunt fenum hoc anno*).[31] Similarly, the 1315–1316 account for Waltham St Lawrence (Berkshire) reports that 3 bushels of oats were given to three stots 'because there was no hay' (*quia non fuit ibi fenum*).[32]

[29] Unfortunately, no 1316–1317 accounts survive for the Winchester Priory demesnes.

[30] See University of Minnesota, 'Nutritional Needs of Horses'.

[31] TNA, SC 6/1011/5.

[32] HantsALS, 11M59/B1/71, membr. 22^{r}.

Although there is no doubt that the tendency to keep the fodder distribution at its pre-famine levels was beneficial for the health of working animals, it might have equally been devastating for human well-being. Unlike *famuli*, draught animals could not be dismissed, and their number remained more or less steady during the famine years (with the exception of those manors that were ravaged by the bovine murrain of 1315–1316). In order to feed the steady number of draught animals with oats, manorial officials had to augment the relative proportions of animal fodder within the *AGCR*. This, in turn, meant that even fewer demesne crops were now available for purchase, either at farm gates or at local markets, and the potential availability of food for human consumption shrank even further. This may have been yet another hidden mechanism dragging the crop prices up and creating the widespread grain market chaos.

In addition to seeding, selling, paying the *famuli*, and feeding draught animals, the remainder of the annual crop receipt was expended on a series of other matters, all included under the 'other' column in Table 3.4. As the table indicates, the proportions of those 'other' expenditures declined from 27 per cent to 21 and 12 per cent, respectively, in the course of the two famine years. In 1317–1318, when the crisis was more or less over, the figure still stood at 20 per cent, reflecting the failed oat harvest. Here, manorial officials seemingly had freedom to manoeuvre, unlike within the seeding and *famuli* wage sectors, which were both protected by custom. Given that the 'other' column comprises a variety of expenditure sectors, a careful analysis of the account rolls is required to appreciate the situation and avoid generalization.

One sector, which had to be drastically cut, was fodder for non-working animals, in sharp contrast with working animals, whose fodder allocations remained largely unchanged during the crisis years. In particular, the proportions of legume fodder allocated to porkers were cut down to 60–80 per cent below their pre-famine levels, on account of the abysmally low legume yields in 1315 and 1316. Similarly, in most cases, manorial officials ceased distributing crops, mostly of lower quality, among sick sheep, otherwise a commonplace practice in normal years. It should be noted that in both cases, the amounts in question were too meagre to have a pronounced impact on the *AGCR* disposal. Indeed, few demesnes spent more than a few quarters of legumes on an entire pig-herd, often consisting of thirty mature porkers or more, and even fewer demesnes would allocate more than a quarter or two to occasionally sick animals. Moreover, the physiological peculiarity and omnivorous diet of swine meant that they would still produce *some* lard and pork, despite these cuts in legume fodder.

Similarly, one finds very few references to food distribution. Generosity of alms giving has never been the virtue of manorial officials: even in non-famine years, they would distribute very modest amounts of grain among local paupers. The hardships of the agrarian crisis, however, forced them to almost stop giving away any grains altogether. In the same vein, there is much evidence on reducing the amounts of legumes and oats boiled as breakfast pottage for the *famuli*, which they would have received in addition to their grain wages, in non-famine years. Similarly, many demesnes decreased the proportion of wheat and rye receipt to be baked into bread and distributed among local *famuli* and other labourers. The relative proportion of barley, dredge and occasionally oats converted into malt and ale was reduced, too. Although it is tempting to blame the lords and their officials of depriving local workers of their most basic food staples, it should be borne in mind that malting and brewing sizable amounts of winter grains would not have been the best strategy during the famine years. Malting and brewing result in the loss of some 70 per cent of calorific value of the raw grain, while serving those grains in the form of pottage would imply virtually no calorific loss at all.

While some *famuli* and other manorial labourers did lose access to their regular food supply, the provisioning of lords' headquarters, another *AGCR* element, went on as usual. It should be noted that although around 1315 the majority of English lords managed their demesnes directly, not every single seignorial household was provisioned by his/her demesnes. In fact, the vast majority of lay lords preferred to rely on the market, rather than on the demesne, to ensure their food supply. It was primarily conventual institutions (that is, monasteries, nunneries, Cathedral priories, hospitals, and, in rare cases, university colleges) that were provisioned, on a steady basis, by their demesnes with food supplies.[33] As far as conventual estates are concerned, it is essential to distinguish between 'procuring' demesnes, namely those shipping a part of their annual *AGCR* to the lord's headquarters, and 'non-procuring' demesnes, which disposed their annual *AGCR* locally, without transporting it to their lords' granaries. For instance, in 1315, the Canterbury Cathedral Priory held sixty demesnes (of which fifty-seven were exploited directly). Yet, only twenty-two demesnes supplied the monks and their household with crops. Similar situations prevailed on other conventual estates. In most cases, the procuring demesnes tended to be located in a reasonable proximity to the headquarters. Because of this distinction, I shall confine the following calculations to the procuring demesnes only. Whenever possible, I have filled up some gaps, based on external evidence and educated extrapolations, but in some instances, the surviving evidence is so patchy that such an exercise was impossible.

[33] This dichotomy has been dealt with in Slavin, *Bread and Ale*, pp. 15–25.

Table 3.8. Proportion of the *AGCR* Sent to Lords' Households by Procuring Demesnes

		Total *AGCR*			Sent to households			As % of total *AGCR*		
		c. 1301–1325	1315–1316	1316–1317	*c.* 1301–1325	1315–1316	1316–1317	*c.* 1301–1325	1315–1316	1316–1317
Estate	**Demesne**	**In quarters**			**In quarters**			**As %**		
Winchester Cathedral Priory	18 demesnes	9,184	8,545		1,561	940		17%	11%	
Canterbury Cathedral Priory	22 demesnes	6,351	4,497	4,632	953	1,034	1,297	15%	23%	28%
Westminster Abbey	8 demesnes	2,841	2,240	1,882	597	448	376	21%	20%	20%
Durham Cathedral Priory	5 demesnes	2,345		1,347	727		310	31%		23%
	Billingham only	322	268	299	109	107	117	34%	40%	39%
Bolton Priory	Bolton and 7 home farms	3,668	2,155	2,157	1,210	431	604	33%	20%	28%
Ixworth Priory	Sapiston	147	121	127	68	19	34	46%	16%	27%
God's House, Southampton	Cosham and Werrore	163	131		55	48		34%	37%	
	Heckley	383	113	58	42	42	2	11%	37%	4%
Ely Cathedral Priory	Newton	711	792		135	230		19%	29%	
Osney Abbey	Little Tew	116		46	61		4	53%		8%

Source: Manorial accounts database (see Appendix 8).

As Table 3.8 demonstrates, the situation varied from estate to estate. Thus, the procuring demesnes of the Winchester Cathedral Priory and Durham Cathedral Priory (with the exception of the demesne of Billingham) had to cut the relative proportion of crops shipped to the headquarters of their lords by some 6–8 per cent. In some poorer houses, such as Ixworth Priory in Suffolk and Oseney Abbey in Oxfordshire (both Augustinian houses), the reductions were even more drastic. As a result, those houses had to resort to local grain markets to satisfy their dietary needs. Conversely, the Kentish demesnes of Canterbury Cathedral Priory, and the Cambridgeshire demesne of Newton, belonging to the Ely Cathedral Priory, managed to augment the relative share of the *AGCR* transported to the monks' headquarters. The harvest failures on Kentish and East Anglian demesnes were not nearly as bad as elsewhere in the country, and this could well be a decisive factor here. Overall, there was little relative change on the procuring demesnes of Westminster Abbey, which dispatched about 20 per cent of their *AGCR* in both famine and non-famine years in the pre-Black Death era. Whatever the individual situation was, virtually every conventual household had to spend more cash on crops, in absolute terms, during the famine years.

Although there are plenty of references to grain hoarding during the famine years, very few demesnes carried over more than a few bushels to the next accounting year. As we shall see later, most officials opted to withhold crop supplies until the late spring/early summer, when the prices would reach their annual peak, and then release them on local markets. Therefore, it is hardly surprising that, with some exceptions, there are virtually no references to grain carry-overs. In fact, out of the total 42,659 quarters of crops, making up the total 1315–1316 *AGCR* on 140 demesnes, only 137 quarters (that is, 0.32 per cent) were carried over to the next accounting year.[34]

Net Yields: The Peasant Experience

What about the peasant sector? As we have already seen in the previous chapter, there are ways to quantify the harvest failures within the peasant sector. This is largely thanks to the surviving tithe receipts; they shed much light on peasant husbandry, which accounted for some 80 per cent of all English agricultural sector. The biggest limitation of tithe receipts, however, is their absolute, rather than relative, nature. In other words, they record approximately one-tenth of sheer quantities of crops harvested by villagers, but, unlike demesne accounts, they do not indicate their yields. Any estimates of yields on peasant holdings,

[34] Thus, at Ickham (Kent), 36 quarters of barley (out of the total 488.8 quarters of barley and 954.1 quarters of all crops) were carried over in the course of the same year: CCA, DCc/BR/Ickham/20.

as attempted in Chapter 2, are speculative and potentially inaccurate. To make things even worse, however, tithe receipts do not specify what proportion of the harvest was invested in seeding. It is unclear if peasants mimicked the seignorial policy of keeping the absolute levels of seed quantities constant or if they showed more flexibility in face of the disaster by sowing smaller quantities of seed than usual and, thus, retained some seed for consumption. Although several economic models regarding the 'rational behaviour' of the 'prudent peasant' have been offered, these are mere hypotheses, which cannot be tested because of either insufficient reliance upon or the non-existence of necessary data.[35]

A potential clue to this puzzle might be found in the comparison between harvest levels on the demesne and on tenant holdings within a same manor. As seen in Chapter 2 (Figure 2.1), there was a high degree of synchronicity in movement between the demesne yields and tithe receipts during the famine years. To achieve such synchronicity, a high degree of similarity in managerial decisions across the two sectors was required. Although, as scarce evidence indicates, the relative production levels on tenancy holdings were some 15–20 per cent lower than on the demesne, we still cannot be fully certain that the output levels within the tenancy sector were always lower than on the demesne during the famine years. Instead, we might somewhat 'optimistically' propose that the harvests fell across the two sectors in a similar manner. If that was indeed the case, then it would not be too outlandish to suggest that the net yields within the peasant sector may have been similar to those within the demesne. In other words, once a certain percentage of harvest was invested in seed, an average peasant (just as an average lord) may have been left with the respective composite net yields of approximately 1.68, 1.32, and 2.08 for 1315, 1316, and 1317, respectively (see Table 3.2). These figures, however crudely estimated, are alarmingly low and, in the cases of 1315 and 1316, definitely fall short of the minimal subsistence threshold. But does it mean that once the seeding was over peasants could rely on those meagre quantities of grain to get by during the famine? It should be borne in mind that the 'seeding deduction', which converts 'gross yields' into 'net yields' was not the only deduction to which the agricultural producers were subject. One has to account for tithes too.

So far, all the yields discussed in this study were inclusive of tithes (Tables 2.3 and 3.2). As we have seen, the demesne accounts almost invariably recorded harvests and yields *after* the tithe deduction, thus under-recording the total ara-

[35] McCloskey, 'English Open Fields', and 'The Prudent Peasant'; Richardson, 'The Prudent Village'; Popkin, *The Rational Peasant*.

ble production levels. When calculating both gross and net yields, I have always reinstated the tithe part of the harvest, dividing the total *recorded* harvest by the factor of 0.9. Apart from revealing the true harvest levels (and hence, true yields), the tithe-inclusive yields also determine the true levels of food availability. Tithes, paid to God through the local clergy (sometimes resident) or religious institutions (monasteries, bishoprics, colleges) which had appropriated the parishes, would not have been consumed or hoarded by these tithe owners in non-famine years. Rather, as tithe accounts indicate, the vast majority — usually between 90 and 100 per cent — of tithes were sold by the same tithe owners. After all, parish priests usually had their own parcel of arable land, known as 'glebe', to sustain themselves with food, while tithes should be regarded as a source of surplus income rather than food supply. As Robert Swanson has stated, tithes 'only had value if sold', while long-term storing could mean decay and waste, as indeed indicated in some contemporary sources.[36] Collectively, tithes represented about 20 per cent of the total seignorial income and about 8 per cent of the national harvest around 1300, and, as such, they were prominent in grain trade in late medieval England, supplying local communities, both rural and urban, with food.[37] Indeed, there is much evidence that institutional tithe owners (monasteries, colleges, bishoprics) were selling the lion's share of their annual tithe receipts. Hence, the tithes would theoretically be transferred from peasant holdings to local markets and, thus, would be available for re-acquisition by the same peasants. This was indeed an omnipresent phenomenon in non-famine years. But was such a scenario possible during the crisis of 1315–1317?

As rectory accounts indicate, during the famine institutional tithe owners conducted business as usual, selling the vast majority of tithe receipts. One major difference, however, was in the sheer quantities of tithed crops reaching local markets. As expected, these were much reduced quantities, reflecting the harvest failures. The catastrophic deficit of crops ensured that prices rose to an unprecedented level. As far as demesne accounts can tell, lords were purchasing crops for trebled or even quadrupled prices, but very few, if any, peasants could afford such luxury. If anything, it is possible that peasants could afford to purchase some lower-quality grains in the course of the first few months after the harvest, but as the manorial accounts suggests, lords did their best to withhold

[36] Swanson, 'A Universal Levy', p. 91.

[37] Britnell, 'La commercializzazione dei ceriali in Inghilterra', p. 639; Campbell, 'The Agrarian Problem', pp. 18–19; Livingstone, 'The *Nonae*', II, 250–51; Kowaleski, 'The Grain Trade', pp. 30–31, 37–38.

their grain surplus as much as possible for the prices to go up towards the end of an agricultural year.[38] In other words, the dire circumstances of the famine years would not allow most peasants to purchase tithed crops from local parish priests, vicars, or rectors, and this implied that the available food resources for peasant consumption (including tithes) dwindled further, intensifying the classic FED scenario.

Of course, we may assume some evasion of tithe payment during the famine years. Indeed, there is direct evidence of dishonest payment or withholding of tithes by parishioners in normal years, when in some cases the latter ensured that the tithe sheaves contained the worst quality grain.[39] At times, parishioners were caught and fined for dishonest payment of tithes, as indicated in some court rolls.[40] But at the same time, there was moral awareness that tithes were to be paid faithfully and honestly, even in bad years. This notion is exemplified in a number of late medieval vernacular sermons, all surviving in post-Black Death renditions, but likely reflecting earlier periods. Thus, a mid-fifteenth-century compendium, known as *Jacob's Well*, contains two curious *exempla* (anecdotal stories) related to tithing.[41] The first *exemplum* tells of a proud 'earl' who went to Hell because of false tithes in good years, while the other one relates the story of a poor knight who was rewarded with the Kingdom of Heaven because he gave full tithes even when his vineyard crop failed.[42] Elsewhere in the same compendium, the author instructs 'thou shalt pay the tenth sheaf…or the tenth part…even if all that grows is not worth 6d. And thou shalt reckon neither cost, nor expense [of production]'.[43] Likewise, there is no evidence that some parish priests excused their parishioners from paying tithes on account of the harvest failure and omnipresent poverty. After all, tithes were, at least in theory,

[38] See below, Chapter 4, pp. 146–49.

[39] Dodds, 'Tithe and Agrarian Output', pp. 46–51.

[40] For instance, on 16 September 1312, Nicholas Spaule of Heacham (Norfolk) was fined one shilling for paying dishonest tithes during the harvest of that year (NRO, Lestr/DA6).

[41] Although the anonymous author of *Jacob's Well* attributes the first *exemplum* to Helinand of Froidmont (*c.* 1160–1237) and Peter Venerable (*c.* 1092–1156), and the second one to Caesarius of Heisterbach (*c.* 1180–1240), the *exempla* in question are most likely taken (just as the majority of all the *exempla* of *Jacob's Well*) from a Latin manuscript of Arnold of Liège's *Alphabetum Narrationum* (*c.* 1307). See Gregg, 'The Narrative Exempla of *Jacob's Well*', p. 261.

[42] Salisbury Cathedral Library, MS 103, fols 17^{v}–18^{r} (summarized in *A Repertorium of Middle English Prose Sermons* (*RMEPS*), ed. by O' Mara and Paul, no. 2287. The relevant parts of the sermons are printed in *Jacob's Well*, ed. by Brandeis, pp. 46–47. For other *exempla* dealing with tithe withholding, see Tubach, *Index Exemplorum*, nos 1928 and 4867.

[43] *Jacob's Well*, ed. by Brandeis, p. 38.

Table 3.9. Estimated peasant holding yields, per seed, net of seed and net of tithe (representing the net surplus available for consumption by a hypothetical peasant household after the deduction of seed corn and tithe)

Crop	**1315**	**1316**	**1317**
Wheat	1.57	1.16	2.43
Rye	1.71	0.83	2.39
Winter barley	1.31	0.77	1.13
Legumes	0.59	0.65	1.73
Barley	2.02	1.62	2.02
Dredge	1.56	1.16	1.63
Oats	1.09	0.97	0.97
Composite	1.40	1.09	1.76

Source: Manorial accounts database (see Appendix 8); Table 3.2.

Note: All figures are ratios between the present year's harvest net of seed corn and tithes and the previous year's seed. For instance, the gross yields of wheat in 1315 was 2.88:1.00, but after the deduction of seed corn and tithes, the figure declined to 1.57:1.00. In other words, a hypothetical tenant household would have been left with a consumable surplus barely above what had been sown in the previous year.

due to God, and not priests, and thus the latter simply had no authority to exempt their parishioners from this duty.[44] Thus, we might argue that although tithe evasion may have indeed existed during the famine years, its extent was by no means ubiquitous and certainly not significant enough to alter the suggested estimates on Table 3.9. The table offers speculative, but not unrealistic, estimates of peasant crop yields, after the deduction of both seed and tithe. It appears that after losing a fair share of their harvest to seed and tithe, the peasants were left with very little. Furthermore, it is likely that some additional share of the harvest (perhaps some further 10 per cent) was given to peasant draught animals. Unlike the demesne, where the ox was still the most predominant working animal (with the exception of the London hinterland and East Anglia), it was the horse that peasants favoured. As we have seen in the previous chapter, horses are much more oat-dependent than oxen, and therefore some share of harvests was expended on feeding horses, especially given the widespread deficit of hay.

[44] I am thankful to Prof. Robert Swanson of the University of Birmingham for his suggestions regarding tithe payments.

How sufficient were these amounts in relation to dietary requirements of English peasants *c.* 1315? As we have seen in Chapter 1, England was a poor country on the eve of the famine, with about 60–65 per cent of her population living on less than a virgate (thirty acres or so), and perhaps as much as half of her population holding ten acres of land or less.[45] According to the most recent estimate, as much as about 41 per cent of England's population lived below the poverty line *c.* 1300.[46] Getting by on very little or no land, these people were undoubtedly the most vulnerable to any subsistence crisis, let alone a famine of disastrous proportions. This point will be dealt with in a much greater length later in this study. It has been suggested that the 'minimal' holding to support one rustic family throughout a year, without resorting to alternative food channels, would have been eighteen acres, but this estimate seems unrealistically pessimistic.[47] It appears, however, that as few as twelve acres could be sufficient to feed a peasant family without worrying about acquiring additional food resources from outside their holdings of land in non-crisis years (see Table 3.10). This is conditional upon the fact that an average commoner would have derived no more than 70 per cent of his calorific intake from crop-based products, and it precludes any possibility of crop surplus to be sold for cash or exchanged for other goods. To appreciate this point better, it is essential to consider an alternative measure of productivity: yields per area unit (here, measured in bushels per sown acre).

While yields per seed provide direct evidence on *total food supply*, yields per area unit indicates the *relative availability*, or *access to food supply*. This distinction is instrumental, given the high degree of landholding inequality in England *c.* 1300. According to one recent estimate, lords held and directly controlled 20 per cent of the arable land, despite constituting only 2 per cent of the total population, while, conversely, 37 per cent of the population, consisting of cottagers, landless peasants, and non-agricultural rustics, held no more than 5 per cent of all arable land.[48] In other words, *relative* output levels and *relative* access to food were determined by the size of each household's land and, consequently, by yields per area unit. This was as apparent as ever during the Great Famine years. Table 3.10 estimates an approximate calorific value of harvested crops during the famine years. Around *c.* 1300, an average arable acre would yield about 11.2 bushels of crops, gross of tithe and seed. This is a composite

[45] Broadberry and others, *British Economic Growth*, pp. 317–18; Dyer, *Standards of Living in the Later Middle Ages*, pp. 118–27.

[46] Broadberry and others, *British Economic Growth*, pp. 317–18, 321.

[47] Kitsikopoulos, 'Standards of Living and Capital Formation'.

[48] Broadberry and others, *British Economic Growth*, pp. 317–18.

Table 3.10. Yields per acre, calorific output per sown acre (net of tithe and seed), and its capacity to feed a peasant family *c.* 1300 and 1315–1317

	***c.* 1300**	**1315–1316**	**1316–1317**	**1317–1318**
Bushels per sown acre, net of tithe and seed	7.75	5.44	4.38	6.44
Estimated processed kilocalories per sown acre	287,368	221,324	188,702	246,006
Per diem kcal per household, if 15 arable acres (= 10 sown acres)	7,889	6,076	5,181	6,754
Per diem kcal per family member, if 15 arable acres (= 10 sown acres)	1,856	1,430	1,219	1,589
Per diem kcal per household, if 10 arable acres (= 6.7 sown acres)	5,262	4,053	3,455	4,505
Per diem kcal per family member, if 10 arable acres (= 6.7 sown acres)	1,238	954	813	1,060
Per diem kcal per household, if 5 arable acres (= 3.3 sown acres)	2,627	2,023	1,725	2,249
Per diem kcal per family member, if 5 arable acres (= 3.3 sown acres)	618	476	406	529

Source: Manorial accounts database; Appendix 3.

Notes: (a) I have assumed the sown portion of arable acreage to be two thirds of the total arable acreage. This is a likely over-generalization because of some regional variations, but it is unlikely to be too removed from reality; (b) I took an average rural family to consist of 4.25 members, which closely agrees with some estimates (see, Moore, '"Quot homines?" The population of Domesday England'.); (c) I have assumed that crop-based products contributed, on average, 70 per cent to the calorific intake of a commoner; (d) I have assumed that an average person would consume about 2000 kcal per diem (1,400 of which would have derived from crop-based products); (e) Spoilage rates are assumed to have been 10 per cent *c.* 1300, 1316–1317, and 1317–1318 and 15 per cent in 1315–1316; (f) the table is based on a thinner sample of accounts than the tables dealing with per-seed yields because of comparatively smaller number of demesnes with two consecutive accounts, necessary for such exercise. For 1315–1316, 1316–1317, and 1317–1318, 95, 89, and 85 accounts were consulted, respectively. For detailed calculations, see Appendix 3.

yield, taking into account each crop's individual yields and respective contribution to the total arable. Although this figure derives from manorial accounts related to the demesne sector, the high coefficient in production levels between the demesnes and tenancy, discussed at length in Chapter 2, allows us to speculate that this figure may be reflective of both sectors. After the deduction of tithe (10 per cent) and seed (a further 24 per cent), the same peasant is left with some 7.75 bushels of crops to be disposed as he pleased. This composite amount would yield approximately 287,400 kilocalories in processed form (see Appendix 3). During the three famine years, the calorific output of each acre declined considerably.

As Table 3.10 indicates, after the deduction of tithe and seed, an average landed peasant household was left with abysmally low surpluses. These were in the area of 5.44 bushels per acre in 1315, 4.38 bushels in 1316, and 6.44 bushels in 1317. Those figures were so low that those holding fifteen acres or less could barely support their families in 1315–1316 with their own produce, while in the following year their holdings were no longer sufficient, and they would have needed to secure about 13 per cent of deficient calories at local markets. These figures appear even more terrifying for smaller holders: holders of ten-acre holdings would secure no more than about 68 and 58 per cent of the required calories from their holdings in 1315–1316 and 1316–1317, respectively. Smaller holders bore even a higher burden, and it seems that cottagers and landless rustics were doomed to starve. Furthermore, Table 3.10 may well overestimate the total calorific supply because it assumes, in a speculative manner, spoilage rates of 15 per cent. In reality, however, the spoilage rates may have been even higher: unlike lords, who had means to invest a great deal of cash in storage facility repairs during the famine, very few peasants could afford that. Moreover, as we shall see in Chapter 7, the continual rain and lack of solar irradiance may have reduced the calorific content of harvested grains even further by corrupting their kernel contents.

It is true that peasants holding less than ten acres of land could not be self-sufficient in non-famine years either, but at least they could seek and rely upon alternative sources of food. The single most important channel was local markets, both 'formal' (chartered) and informal (within parish or at the demesne gates), whose food commodities were affordable for smallholders.[49] In the case of landless and needy rustics, we may add several additional channels and institutions, including the 'lord's table' (manorial repasts during harvest time), local churches, hospitals, and almshouses (offering poor relief in both coin and kind), and a network of kinsmen, friends, and fellow parishioners, whose binding ties were essential for the food security of the needy.[50] But as we shall see later, these

[49] Britnell, 'The Proliferation of Markets in England'; Dyer, 'Did Peasants Need Markets and Towns?', pp. 25–47; 'The Hidden Trade of the Middle Ages'; 'England's Economy in the Fifteenth Century', p. 209.

[50] Dyer, 'Changes in Diet in the Late Middle Ages', 'Did the Peasants Really Starve in Medieval England?', pp. 53–72, and 'Poverty and its Relief in Late Medieval England'; Slavin, *Bread and Ale*, pp. 179–87; Hanawalt, *The Ties That Bound*, pp. 55, 57, 63, and 254; Rubin, *Charity and Community in Medieval Cambridge*, pp. 144, 152–53, and 160.

channels and institutions became increasingly unreliable and failing during the famine years. Thus, smallholders could not rely on the market because of supply disruptions and abnormally high prices. Some wage-earners ran the risk of not being hired by local manorial officials. Although some ecclesiastical institutions continued providing support for the needy, their abilities depended on their socio-economic status, and even great religious houses could cater only for a limited number of paupers. Finally, community support on parish levels ceased to be reliable: because of the dire circumstances of the famine, there was an increasing tendency of selfish and anti-social behaviour, as reflected in the spike in crime levels, discussed in Chapter 8.[51]

Obviously, the late medieval English population was not fed only by bread. Although grain-based products contributed about 70 per cent to the total calorific intake, the remaining 30 per cent derived from pastoral and horticultural products. With all the potential danger of inaccuracy, it can be estimated that dairy products contributed a further 12 per cent, to which we might add an additional 8 per cent or so deriving from meat (including poultry) and eggs, and the remainder derived from fish and garden produce.[52] As late medieval lay subsidy rolls (aka, tax assessments), recording taxable property of tenants, indicate, peasant households held non-negligible numbers of animals and were actively involved in the pastoral sector of agriculture.[53] Even though swine (and particularly piglets) seems to have been chronically under-recorded in the lay subsidy rolls, it is clear that those having at least one sow could expect a fair litter of ten piglets a year, rendering some 193,80 kcal, which would have supplied an average household with 530 kcal on a daily basis.[54] Cows and sheep provided additional milk and occasionally meat. A preliminary analysis of all the surviving tax assessments indicates that an average English family, not exempt from taxations, held at least two heads of cattle, six sheep, and one sow, with some regions exhibiting even higher figures.[55] In other words, non-farinaceous products were instrumental in providing peasants with additional calorific intake,

[51] Hanawalt, *Crime and Conflict in English Communities*, pp. 251–52 and 266.

[52] Slavin and DeWitte, 'Between Famine and Death', pp. 58–60. Somewhat different estimates are provided in Broadberry and others, *British Economic Growth,* p. 289.

[53] Postan, 'Village Livestock in the Thirteenth Century'; Slavin, 'Peasant Livestock', pp. 7–15.

[54] Calorific figures derive from Harvey, *Living and Dying in England*, pp. 53 and 64–65; also, the 2002 version of McCance and Widdowson's *The Composition of Foods*.

[55] These are preliminary results, deriving from all the surviving lay subsidy rolls for individual parishes, compiled between 1225 and 1332. The Suffolk portion of those rolls is analysed in Slavin, 'Peasant Livestock'.

but their capacity to relieve the food shortage and save the situation was, at the best, limited. In order to take up the slack for the unavailable grain calories, it was essential to expand the pastoral sector, which could not be achieved, for four reasons.

First, an overnight shift from arable farming to pastoral husbandry would have been a very expensive enterprise, demanding colossal capital investment, unavailable to most peasants on account of their poverty. For instance, if a peasant household lived on ten acres of land and derived about 950 kcal from crops in 1315–1316, then, in order to make up for the dearth of grain, it would require him to purchase either four sows to produce forty piglets a year, or about thirty-three ewes or five cows to be milked and produce sufficient cheese.[56] Needless to say, the numbers would double and treble in the case of holders of smaller tracts of land. To make things even worse, murrain prevailed among sheep (and, to a lesser degree cattle) during the famine years. Second, even if some smallholders managed to acquire *some* livestock, then it was well below what was required, and its produce was consumed too fast. As John Trokelowe reports, 'even flesh of animals began to be deficient, and eggs and other dairy products began to disappear too. One could hardly find capons or geese; sheep were lacking, because of their murrain; and pigs could not be fattened, because of such dearth'.[57] Third, the expansion of animals would have required additional space, which could not be easily secured, despite the existence of common grazing rights and access to demesne pasture, which were undoubtedly enjoyed by most peasants. Finally, the torrential rain and freezing winters would have made the management and upkeep of a large number of animals particularly challenging. One can imagine the pitiful sight of a sheep flock roaming in a mudded pasture field, shaking from cold rain and strong wind. To that one may, perhaps, add a no less pitiful sight of a hungry shepherd supervising the flock and fancying it turning into delicate mutton.

[56] Milk extraction rates are assumed to be 48 per cent, as suggested in Broadberry and others, *British Economic Growth*, p. 285.

[57] *Johannis de Trokelowe Annales*, ed. *by Riley*, p. 92. The translation is mine.

Chapter 4

Declining Food Security: Food Markets and Trade

Famines and Markets

The mutual impact and interaction between famines and markets has long been a topic of both philosophical and scholarly interest. In the classic *Wealth of Nations* (1776), Adam Smith contended that food supplies are determined by the efficiency of food markets. He was careful to distinguish between 'dearth' and 'famine', the former being a physical deficiency of grain, created by harvest shortfalls, while the latter was brought about by violent governmental interference, through the misallocation of scarce resources. It is free markets and an efficient network of international grain trade that can save a population and make sure that dearth does not turn into famine. Free trade minimizes the damage done by harvest failures and bad governments by encouraging grain merchants, who enjoy light or no tariffs, to import crops from famine-free lands into starving countries. This ensures a steady replenishment of food supplies in hunger-struck regions, which keeps prices down, makes official markets function properly, encourages people to purchase foodstuffs, and, in this way, relieves or eliminates starvation and suffering. Furthermore, it forces grain hoarders to release their stored supplies into local markets, because reasonable prices make grain storing unprofitable and uncompetitive. According to Adam Smith, such a model is bound to work, as long as each region sticks to its comparative advantage and contributes to market integration.[1]

[1] Smith, *Wealth of Nations*, Bk 1:1–3.

Contrary to Smith's optimistic model, Amartya Sen's study of the Bengali famine of 1942–1943 reveals that rice prices rose, not because of harvest failures but because of panic purchases, food hoarding, and speculation. It was a failed market performance that disrupted the food supply and created a catastrophic crisis of food entitlement.[2] Similar conclusions have been reached regarding late twentieth-century 'Third World' famines, including the Bangladeshi famine (1974–1975), the Sudanese, Kenyan, and Ethiopian famines of 1983–1985, and the food crisis of 1999–2000 in Ethiopia. Here, scholars were able to show the omnipresent failure of markets, which manifested not only in their inability to cater sufficient amounts of food but also in a rising coefficient of variation of regional prices.[3] On the other hand, during some European pre-Industrial food crises, such as the French famines of 1693–1694 and 1708–1710, the Great Irish Famine of 1845–1852, and the Great Finnish Famine of 1866–1868, thoroughly studied by Cormac Ó Gráda, markets tended to function 'normally', and there is not much evidence for their disintegration or segmentation.[4]

In other words, famine can create two possible market scenarios, neither of which would preclude a food crisis. For sure, one should not accept the link between famines and market failure as granted. Each famine exhibits different market behaviour, depending on local contexts and circumstances particular to each event. The complexity of market functionality in famine contexts has been analysed by Ó Gráda in his magisterial monograph on the history of famine.[5]

Which of the two models does the Great European Famine fit? As I have demonstrated elsewhere, the disruption of grain supply had a negative impact on grain markets during the famine years. In particular, the omnipresent shortage caused: (1) aggressive and uncontrollable short-term soar in crop prices; (2) market segmentation (defined by an increased coefficient of variation in prices between different crop-trading hubs, namely, a degree of deviation from an average price exhibited by different grain markets); (3) the rise of 'preferential trade'; (4) uneven and unsteady chronologies of crop supply and distribution. Taken together, these four criteria together undoubtedly exacerbated the

[2] Sen, *Poverty and Famines*.

[3] Olsson, 'On the Causes of Famine'; Ravallion, *Markets and Famines*, pp. 111–13; Drèze and Sen, *Hunger and Public Action*, pp. 144–55 Ó Gráda, 'Markets and Famines', p. 156.

[4] Ó Gráda, 'Markets and Famines', pp. 143–66.

[5] Ó Gráda, *Famine: A Short History*, pp. 129–58.

crisis, transforming it from a grain shortage into the Great Famine.[6] However, to reach a better understanding of the supply side of the crisis, it is necessary to account for other factors as well. The current chapter revisits the grain market and grain trade during the famine years in light of additional evidence and conceptual frameworks, unexplored in my previous work on the crisis.

Price Movements, 1315–1317

To appreciate the price behaviour of crops during the Great Famine years, it is instructive to place it into a wider context of price movement in the late medieval period. In the long run, the accounting years of 1315–1316 and 1316–1317 (namely 29 September 1315 to 28 September 1316 and 29 September 1316 to 28 September 1317) mark the single highest point in the price history of crops before the Price Revolution of the sixteenth century (Figure 4.1). There were indeed several additional spikes marking exceptionally dear years (1294–1295, 1321–1322, 1350/1–1351/2, 1369–1370, 1390–1391, 1438–1439, and 1482–1483), stemming from failed harvests, but none of those spikes come close to those of 1315–1316 and 1316–1317. To make things even worse, 1315–1316 and 1316–1317 were the years when real wages (namely, nominal wages divided by Consumer Price Index, or CPI) reached their lowest level in the course of the late Middle Ages. As we shall see later in this chapter, at no other point were real wages so low. This fact was yet another reflection of the overwhelming poverty of the populace, compounding the widespread starvation and suffering during the famine years.

Having established the unique place of the 1315–1317 figures in a much wider context of price history, let us now reconstruct price behaviour during the famine years in a detailed manner. It should be noted that despite the relative wealth of accounting material from the famine years, deciphering regional variations and seasonal trends in grain prices is more than a frustrating task. Although manorial accounts always record the amounts and prices of crops, they omit, in most cases, the seasonality of transactions. For the two famine years (1315–1316 and 1316–1317), of a total of 282 accounts recording 2115 transactions, only 76 accounts documenting 278 transactions specify the seasonality of crop sales and purchases. In other words, only about 13 per cent of all recorded transactions are dated. Thus, it is uncertain how representative this sample is of general trends in seasonal price movement. Nevertheless, the dated

[6] Slavin, 'Market Failure during the Great Famine in England and Wales'.

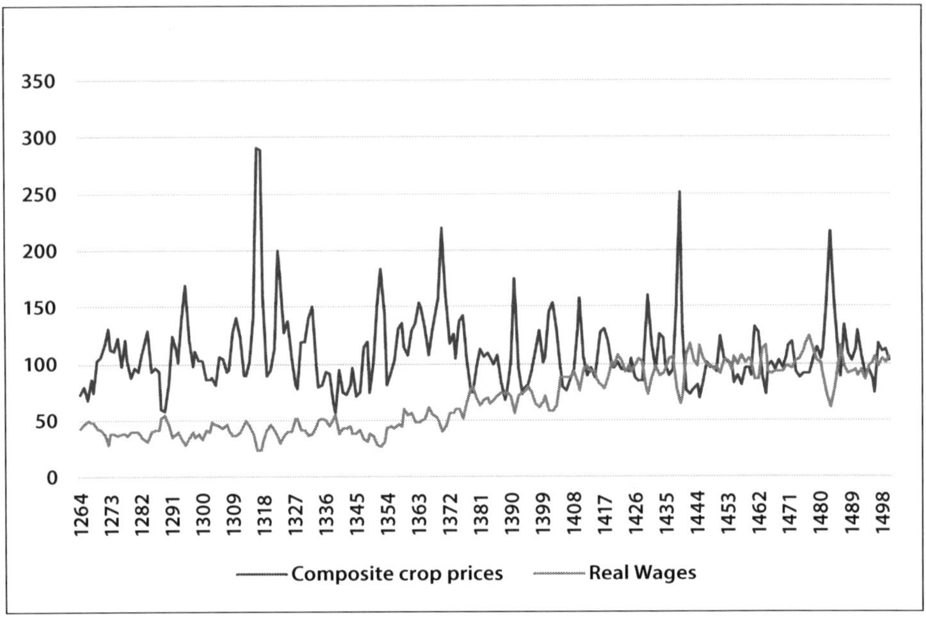

Figure 4.1. Indexed crop prices and real wages of urban workers in England, 1264–1500 (1451–1475 = 100). *Sources*: Munro, 'Revisions of Phelps Brown'; Farmer, 'Prices and Wages', in *The Agrarian History*, II, and in *The Agrarian History*, III.

Note: The respective figures indexed on 1451–1475 (1451–1475 = 100), appearing on the vertical axis, are as follows: (1) composite crop prices: 100 = 3s 8d per 1 quarter of composite grains; and (2) real wages: 100 = the average of the nominal wages (NW) (1451–1475 = 100 = 6d daily), deflated by the average of the Consumer Price Index (CPI) (1451–1475 = 100).

transactions still shed much light on different aspects of seasonal price fluctuations during the famine years.

As Figure 4.2 indicates, the price movement during the famine years consisted of a steep and cumulative rise between the aftermath of harvest and the eve of the next harvest. As we have seen, the harvest of 1315, which commenced around 1 August, after almost a year of virtually incessant rain and periodic frost, turned out to be very bad. To repeat, the composite *gross* crop yields (that is, gross of seed and tithe) stood at 25 per cent below average. The composite *net* crop yields (net of seed and gross of tithe) were considerably lower, standing at 38 per cent below average. In other words, the allocation of excessively high proportions of the harvest in seed corn depressed the net yields and thus the available food supply and, consequently, caused an aggressive rise in crop prices.

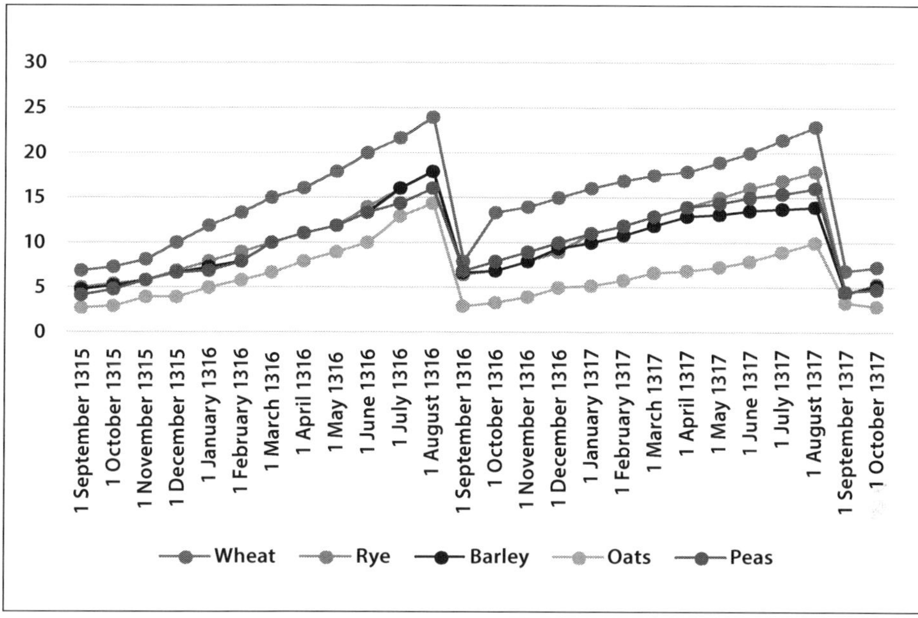

Figure 4.2. Crop price movement on demesnes, September 1315–October 1317 (in shillings sterling per 1 quarter). *Source*: Slavin, Manorial account database (see Appendix 8).

Note: The 'first day' of each month here signifies a period *around* the first day of month (±fifteen days) and not necessarily the actual first day of month. Thus, '1 February 1316' covers the period of 16 January 1316 to 15 February 1316. The figures appearing on the vertical axis are approximate monthly prices in shillings per 1 quarter of respective grains.

By September 1315 crop prices were already well above their normal level; on average, wheat was selling for as much as 7s a quarter (compared with about 5s a quarter in non-famine years between 1300 and 1349). But the worst was yet to come. By Christmas, a quarter of wheat was selling for 10s (in contrast with some 5s 6d in non-famine years). We can only imagine the gloomy mood of the Christmas celebration, when the general populace started feeling the first effects of famine. Around Easter (11 April 1316), the prices stood at 16s, while during the midsummer they reached their peak, standing at an overwhelming 24s per quarter (compared with the average of 6s 2d in 'normal' years). Other crops show similar trends. Between September 1315 and August 1316, a quarter of rye rose from 5s to 18s, the figures for barley soared from 4s 10d to 18s, those for oats rose from just 2s 9d to 14s 6d, and those for peas climbed from 4s 4d to 16s.

By mid-August 1316, in the middle of the harvest, wheat prices fell from 24s to 8s a quarter. This is quite puzzling, given that the harvest that year was disastrous, with the composite *gross* yields standing at 35 per cent below average, while the composite *net* yields stood at 50 per cent below normal. One possibility is that it took the producers some time to realize how bad the harvest was. An increasing influx of new crops, notwithstanding their meagre quantities, undoubtedly relieved the food shortage and perhaps may have created a sense of hope among both the producers and traders. But once the producers and traders were able to appreciate the extent of the disaster, around the time the manorial accounts were rendered at the end of September, the prices rose once more, at even a greater pace than in the previous year. The 'accounting panic' seemingly turned into a 'consumer panic' of better-off elements, who did their best to replenish their granaries and barns as fast as they could. Merely a month after the harvest, wheat prices rose to 13s per quarter. By Christmas they rose to 16s; around Easter they soared to 18s, while at the peak of the 'hungry gap' period in the early summer they climbed to 23s. Just as in the previous year, the seasonal trends in wheat prices mirrored those of other crops.

The harvest of 1317 was again worse than average (the composite *gross* yields were 14 per cent below average, while the composite *net* yields stood at 26 per cent below normal), but it was good enough to relieve the omnipresent sufferings and put an end, at least temporarily, to the deficiency of grains and legumes. In particular, wheat, maslin, and rye yields stood at about 13 per cent below average, and this fact ensured that the population had a sufficient supply of bread that year. The prices fell to 10s at the beginning of the harvest, and they fell further to around 8s shortly after the harvest was collected and the fear of famine was largely over. Between October 1317 and August 1318, wheat prices did not fluctuate much, rising, on average, from 8s to 10s. The harvest of 1318 was bountiful, and this fact drove wheat prices down to pre-famine levels. In the course of the 1318–1319 agricultural year, they stood at 4s 6d per quarter, and this figure did not change much in the course of that year.

Although no other part of the British Isles can boast such complete series of grain prices, there is some evidence that the situation was as dire in the marcher lordships in south and east Wales and the Lordship of Ireland. As the manorial court rolls from Dyffryn Clwyd (Denbighshire, in north-east Wales), studied by Matthew Stevens, indicate, the wheat prices were as high as in England (averaging at about 14s, 16s, and 9s per quarter in 1315–1316, 1316–1317, and 1317–1318), while the oat prices in 1315–1316 and 1316–1317 were even higher than their English counterparts (standing, respectively, at 12s and 9d a quarter), falling to 2s 8d in

1317–1318.[7] In Monmouthshire, as evidenced by the account rolls from Caerleon, Treigruk, Troy, and Usk, wheat, rye, oat, and barley prices were largely comparable with those in England.[8] The same situation prevailed in Glamorganshire (south Wales).[9] Irish evidence is even patchier because virtually all late medieval Irish chancery rolls and other state papers have been destroyed during the Great Fire of 1922. The famine-year rolls have been thankfully preserved in a calendared form and as their entries indicate, wheat was often selling for 20s per quarter, while the selling price of oats varied from 10s to 13s 4d a quarter.[10] Even *hastivel* (most likely, an early-ripening grain) was selling for exceedingly high prices — according to one entry, it was valued at 19s 5d a quarter at some point before the harvest of 1317.[11] To my best knowledge, no price data survives for Scotland from those years. There is much evidence on exceedingly high prices on the English side of the Scottish borders, as it is indicated in the accounts from Norham (North Northumberland) and several other places in that county.[12] Also, there is no doubt that Berwick-upon-Tweed and Carlisle, both under a disastrous dearth of food, experienced astronomically high prices. This, however, does not mean that the situation was necessarily similar in neighbouring Lothian and Galloway, which were spared English raids during the famine years.

Crop Yields and Prices

As Campbell and Ó Gráda have demonstrated, there was a great degree of correlation between harvest shortfalls and the subsequent year's crop prices, expressed as *price elasticity of demand* (namely, the responsiveness of the quantities of demand to changes in price). As they have estimated, the value of price elasticity to gross yields for the period 1268–1480 was −0.55 (in contrast with a much lower figure of −0.35 for 1750–1850) and −0.75 for net yields. This led them to conclude that in pre-Industrial England, harvests were an important determinant of price levels, in contrast with the Industrial Revolution era,

[7] Stevens, 'The Great Famine in Dyffryn Clwyd, 1315–1322', pp. 20–24.

[8] TNA, SC 6/921–24; SC 6/922/31; SC 6/923/6; SC 6/923/26–27; SC 6/925/15–16; SC 6/926/18–19; and SC 6/927/18, 22.

[9] TNA, SC 6/1202/8.

[10] For instance, NAI, KB 2/10, pp. 7 and 20.

[11] NAI, KB 2/9, pp. 46–47.

[12] DCA, Proctor of Norham accounts, 1314–1315, 1315–1316, and 1317–1321.

Table 4.1. Actual crop prices (indexed on 1270–1429) and 'expected' prices, deriving from Bouniatian's formula.

	Actual prices			*Expected prices*		
Crop	**1315–1316**	**1316–1317**	**1317–1318**	**1315–1316**	**1316–1317**	**1317–1318**
Wheat	2.55	2.94	1.41	2.41	3.58	1.10
Rye	2.23	2.90	1.23	2.17	6.18	1.13
Winter barley	2.85	3.09	1.74	3.43	5.24	3.58
Legumes	3.28	4.26	1.85	4.98	4.98	0.69
Barley	2.38	2.85	1.31	1.31	1.79	1.31
Dredge	2.46	2.76	1.40	1.21	1.97	1.35
Oats	2.61	2.61	1.55	1.74	2.33	2.25
Composite price	2.57	2.88	1.47	1.97	2.80	1.42

Source: Slavin, Manorial account database (see Appendix 8).

when crop failures were no longer a threat to food supply and access.[13] There is no doubt that the increased prices of 1315–1317 were caused by the disastrous harvests, but was this rise proportional to those harvest shortfalls? It should be borne in mind that the relationship between crop yields and crop prices does not imply inverse symmetry. In other words, a 25 per cent harvest shortfall does not mean that crop prices will rise by 25 per cent. In the case of harvest failures, prices always act much more aggressively, as it is indicated in Figure 4.2.

One way to measure the short-term responsiveness of prices to harvest is to use Bouniatian's formula, which establishes a proportionate ratio between indexed yields and next year's prices, by calculating an estimated price level on the basis of the quantity of goods.[14] The formula is as follows: $y = 0.757 \div (x - 0.13)^2$, where x is the indexed quantity of goods, that is, gross crop yields,

[13] Campbell and Ó Gráda, 'Harvest Shortfalls, Grain Prices, and Famines', pp. 875–77.

[14] Mentor Bouniatian (1877–1969) was an Armenian-French economist, based at the University of Paris. His contributions included studies on the relationship between goods and prices. Bouniatian's formula implies a widening negative gap between the level of goods supply and price level (see, Wrigley, 'Some Reflections on Corn Yields', p. 238).

Goods supply	1.0	0.9	0.8	0.7	0.6	0.5
Price	1.00	1.28	1.69	2.33	3.43	5.53

Table 4.2. Ratio between indexed next year's grain price and indexed current year's grain yield, during late medieval harvest shortfalls (all figures are indexed on 1270–1429).

	Net yields			*Ratio (price/yield)*		
Harvest year	**Wheat**	**Barley**	**Oats**	**Wheat**	**Barley**	**Oats**
1283	0.68	0.75	0.77	1.73	1.42	1.25
1293	0.60	0.91	0.86	2.33	1.39	1.10
1304	0.72	0.65	0.84	1.37	1.49	1.06
1315	0.57	0.81	0.63	4.47	2.94	4.14
1316	0.43	0.66	0.56	6.84	4.32	4.66
1317	0.87	0.81	0.56	1.61	1.61	2.81
1321	0.71	0.49	0.53	2.78	3.76	3.15
1349	0.62	0.53	0.30	1.50	1.76	3.62
1350	0.69	0.52	0.21	2.03	2.85	8.00
1351	0.63	0.75	0.78	2.74	2.23	2.29
1437	0.53	0.77	0.48	2.97	1.41	2.39
1438	0.41	0.88	1.37	6.04	1.90	0.93

Source: Slavin, Manorial account database; Campbell, *Three Centuries of English Crop Yields*; Munro, 'Revisions of Phelps Brown'; and Farmer, 'Prices and Wages', in *The Agrarian History*, II, 715–817.

in our case.[15] The calculations render mixed results (Table 4.1). On the one hand, the composite crop prices of 1316–1317 and 1317–1318 closely match Bouniatian's prediction. On the other hand, both the composite prices of 1315–1316 and, in many instances, individual crop prices did not match his model. This is especially apparent in the case of 1315–1316, in which spring grain prices soared much higher than the model predicts and matched the relative levels of winter grain prices, despite the fact that their yields fared much better than those of the latter. At the same time, the model predicted unrealistically high prices for winter crops and legumes in 1316–1317. These discrepancies imply some apparent flaws and limitations of Bouniatian's formula and potential dangers in applying it to historical data. Perhaps the biggest limitation of the equation is that it accounts for gross rather than net product. But even if a necessary adjustment is made to the formula, it is unlikely to render any 'universal law' of harvest-to-price ratio. Apart from the truism that bad har-

[15] This methodology has been utilized by Wrigley, 'Some Reflections on Corn Yields'. The formula is found on p. 238.

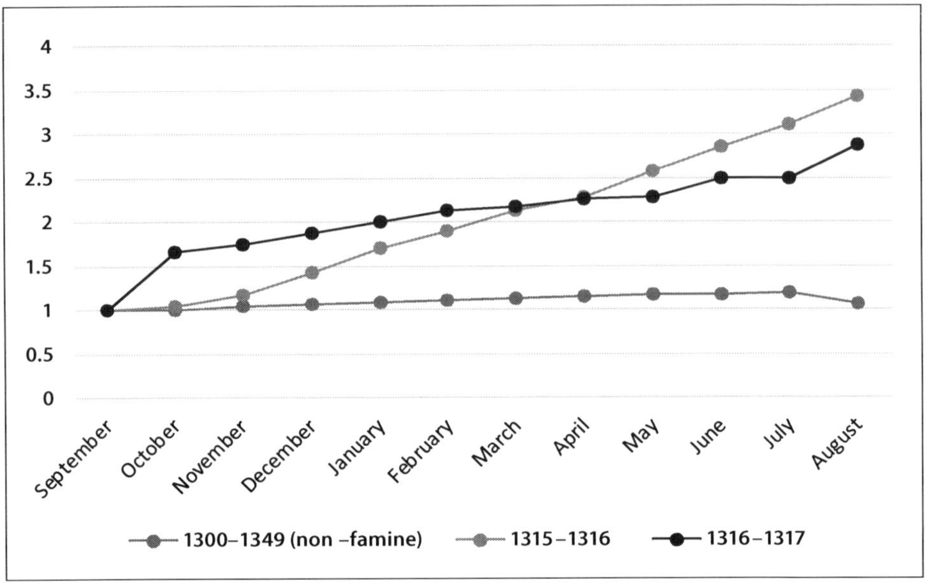

Figure 4.3. Seasonal wheat price movement (logged on September), 1315–1316 and non-famine years (1300–1349). *Source*: Manorial accounts database (see Appendix 8).

Note: The respective figures indexed on September (September = 1.0), appearing on the vertical axis, are as follows: (1) 1315–1316 = 7s per quarter; (2) 1316–1317 = 8s per quarter; and (3) 1300–1349 (non-famine years) = 5s per quarter.

vests and low net yields drag prices up (and vice versa), and this inverse correlation is always apparent (as shown by Campbell and Ó Gráda), it is difficult, if not impossible, to come up with any more precise formula that determines the *exact* ratio between *exact* levels of crop yields and *exact* levels of crop prices. The relation between the two variables has never been clear and straightforward and could vary from case to case, depending on a vast array of other factors. Such factors could include the relationship between net yields and population levels, carry-overs of stored crops, transportation costs, market behaviour, and availability of precious metals.

Perhaps an alternative method to appreciate the proportionality of the inverse relationship between yields and prices would be to compare the net yields against the ratio between the next year's prices and yields of individual crops (both indexed on the same baseline period) during the Great Famine and other late medieval harvest failures (Table 4.2). This method reveals two intriguing facts. First, it confirms the idea that there is no such thing as an 'expected price' or a 'magical formula' that can establish a precise relationship between yields and prices. For instance, during the Black Death years, the price-to-yield ratio (with the curious exception of oats in 1350–1351) was surprisingly low in relation to the abysmally low crop yields, comparable to those of the Great Famine years. This has undoubtedly to do with the human pandemic, which decimated at least half of England's population by late 1350, altered the land-to-labour ratio, and, hence, drastically decreased the sheer demand for foodstuffs.

Second, it appears that in bad years, failed grains had a 'dragging effect' (that is, the impact of price behaviour of one commodity on price behaviour of other commodities) on those crops that were faring comparatively better. For instance, the net crop yields of spring barley in the 1293 harvest were 9 per cent below average. Accordingly, the price-yield ratio was accordingly low (1.39) in the following year (1293–1294), in spite of the failure of the 1293 wheat harvest. This 'dragging effect' was even more apparent during the famine years, and it appears that compared to lesser harvest shortfalls and food crises, the crop prices during the famine years were indeed disproportionally high, as reflected in the excessively high price-yield ratios. In the 1315 harvest, the net crop yields of spring barley stood at 19 per cent below average; yet, the price-yield ratio in 1315–1316 was an overwhelming 2.94, instead of the expected 1.6 or so, as it was in 1317–1318, when the net barley yields were the same. A similar disproportionally high ratio can be seen within the oat sector. This can be explained by the growing demand for those crops that were available in fair quantities. This demand could be triggered by two factors: (1) food-hunger of the general populace, and especially the poorer strata and (2) grain market panic and speculation. The epic failure of winter grains and legumes may have created a widespread fear of total food scarcity, especially if harvests were to fail another time (which indeed happened). For some better-off elements, it could also mean a golden opportunity to acquire surplus quantities of crops, in order to hoard them until the prices would reach their seasonal maximum, during the 'hungry gap' period. Whatever was the case, the panic prompted local consumers to recruit whatever means they had and purchase the available quantities of barley and oats, either for practical consumption or speculative hoarding. As we shall see later in this chapter, there are plenty references to grain hoarding and speculation during the famine years.

Too Much even for a Famine?

The seasonal pace of the price rise during the Great Famine years was remarkable by both non-famine and famine standards. Between September 1315 and August 1316, the monthly price rise rates stood at almost 12 per cent for wheat, rye, barley, and legumes and an astonishing 16 per cent for oats.

This uncontrollable soar undoubtedly appeared shocking to contemporaries, who were used to rates of below 2 per cent per month in non-famine years (Figure 4.3). Since the comparison with non-famine standards is somewhat trivial, it would be instructive to consider the situations of other major famines in Europe. As Figure 4.4 shows, the Great Famine was indeed a disaster in its own league. Thus, during the troublesome year of 1596–1597, marking the peak of a major pan-Eurasian famine, in conjunction with a short-term weather anomaly, the monthly rates of wheat price rise in England amounted to 4.2 per cent. Similar figures are found for a more minor food shortage of 1621–1623.[16] Another intriguing comparator would be the food crisis of the 1840s, caused by widespread potato blight and a series of grain harvest failures all over northern Europe. Despite its seemingly pan-European parameters, it was only in Ireland that the crisis reached famine proportions. The Great Irish Famine of 1845–1852 killed some 10–15 per cent of the island's population and forced a further 10 per cent into emigration. Just as with the Great European Famine, the Great Irish Famine carried omnipresent signs of a FED crisis, initiated by environmental factors, and intensified by purely anthropogenic causes.[17] It was apparently the only other major western famine whose price behaviour can be compared in its aggressiveness to that of the early fourteenth-century catastrophe. Between October 1845 and September 1846, potato prices were rising at the rate of 14.7 per cent per month.[18] Intriguingly, Irish oat prices increased only at the monthly rate of 5 per cent, notwithstanding the disastrous starvation and the high demand for this grain as an alternative staple.[19] This was unlike the situation in 1315–1316, when oat prices rose even faster than the wheat ones, and when the 1315 harvest failure of winter grains and legumes dragged up the prices of spring crops a great deal.

Another aspect, which made the famine of 1315–1317 a truly remarkable phenomenon, is the amount of time it took prices to reach their peak (Table 4.3). During the Great Famine, it required merely eleven months, namely less than a full agricultural year, for the prices to reach their peak. In the case of

[16] Calculated from Poynder, *Database of England Monthly Grain Prices*.

[17] The economic aspect of *An Gorta Mór* is masterfully studied by Ó Gráda in his *Black '47 and Beyond* and *The Great Irish Famine*. See also Ó Gráda and Eiríksson, eds, *Ireland's Great Famine*; Paping, Vanhaute, and Ó Gráda, eds, *When the Potato Failed*.

[18] Calculated from Ó Gráda and Eiríksson, eds, *Ireland's Great Famine*, pp. 32–34 and Solar, 'The Crisis of the Late 1840s, p. 87.

[19] Ó Gráda, *Black '47 and Beyond*, p. 142.

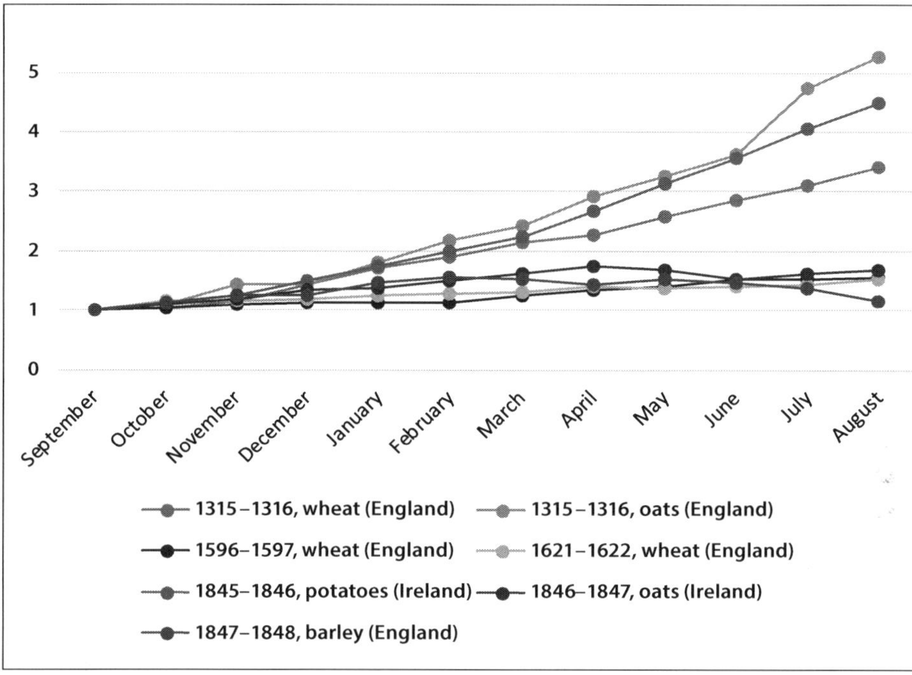

Figure 4.4. Seasonal crop price fluctuation during major European famines, 1315–1848. *Source*: Slavin, 'Market Failure during the Great Famine', p. 21

Note: The respective figures indexed on September (September = 1.0), appearing on the vertical axis, are as follows: (1) 1315–1316, wheat (England) = 7s per quarter; (2) 1315–1316, oats (England) = 2s 9d per quarter; (3) 1596–1597, wheat (England) = 38s per quarter; (4) 1621–1622, wheat (England) = 31s 7d per quarter; (5) 1845–1846, potatoes (Ireland) = 16d per hundredweight; and (6) 1845–1846, oats (Ireland) = 15s 10d per quarter.

other major pre-Industrial European food shortages, the process seems to have been much more gradual and slower. Thus, during the subsistence crisis of the mid-1550s in England, it took the prices thirty months (from September 1554 to March 1557) to reach their ceiling, from 10s to 40s per quarter of wheat. During the 1621–1623 crisis in England, the prices went up for seventeen months, from 30s to 60s per quarter, between September 1621 and January 1623. In the course of the French famine of the early 1690s, it took twenty-one months for a *setier* (156 litres) of wheat to skyrocket from 14 to 48 *livres Tournois*. During the 1815–1817 food crisis in Bavaria, 1 *scheffel* (= 223 litres) of wheat rose from 4900 to 13,600 Munich *denars* between September 1815 and August 1817, that is, in the course of twenty-three months. Finally, during the catastrophic potato famine in Ireland, it took twenty-two months for

Table 4.3. Price rise and fall speed in select pre-Industrial European famines

Famine event	Beginning	Peak	End	Commodity	Peak levels	Months of price rise	Months of price fall
Great Famine, England	Sept. 1315	Aug. 1316	Sept. 1318	Wheat	3.43	11	25
1554–1557, England	Sept. 1554	Mar. 1557	Sept. 1557	Wheat	4.00	30	6
1621–1623, England	Sept. 1621	Jan. 1623	Sept. 1623	Wheat	1.96	16	8
1692–1694, France	Sept. 1692	Jun. 1694	Nov. 1695	Wheat	3.43	21	17
1815–1817, Bavaria	Sept. 1815	Aug. 1817	Aug. 1818	Wheat	2.77	23	12
1845–1848, Ireland	Sept. 1845	Jul. 1847	Aug. 1849	Potatoes	8.63	22	13

Sources: Slavin, 'Market Failure during the Great Famine', p. 25.

Notes: '*Peak levels*' are indexed levels of prices logged on the first observed month.

potato prices to climb to their peak, from 1s 4d to an astonishing 11s 6d per hundredweight (50 kg).

At the same time, prices fell back very slowly during the Great Famine; in fact, it was not until September 1318, once the bountiful harvest of that year has been collected, that they finally returned to their pre-famine levels. In the cases of most other famines, however, prices tended to plummet at a much faster pace, as Table 4.3 suggests. There are several possible explanations to that. First, one has to account for better and cheaper storage facilities in the post-medieval period. Inevitably, releasing grain stores would have slowed down crop prices.[20] Second, in the case of the nineteenth-century crises, we have to consider the expansion of the international grain trade and the transportation revolution, which together decreased the transaction costs significantly and made food security a much less challenging affair than ever before.[21]

[20] Claridge and Langdon, 'Storage in Medieval England', pp. 1257 and 1260–61; Slavin, 'Market Failure during the Great Famine', pp. 42–46.

[21] Thus, in 1800, during the Napoleonic blockade, about 7 per cent of British population was fed by foreign grain. Around 1850, the figure rose to some 25 per cent. See Enrnæs, Persson, and Rich, 'Feeding the British'.

Nominal and Real Wages

Unlike prices, which are liable to a great deal of year-to-year volatility, depending on a previous year's harvest, nominal wages (that is, the numerical or face value of money paid to a labourer) hardly fluctuated from year to year. For instance, between 1310 and 1337, the daily nominal wages for an average unskilled urban worker stood at the unchanging rate of 2d per day, before falling to 1.75d in 1338–1339 and then to 1.5d in 1340–1351. The wages of agricultural workers tended to fluctuate more from year to year, but overall those fluctuations were insignificant when compared to annual price movements. During the period of 1310–1348, an agricultural worker would have received 5.2d for threshing or winnowing 1 quarter of grain, while his neighbour employed in reaping and binding of sheaves would have received 6.1d for working one acre of grain. During the famine years, urban wages remained unchanged, while agricultural ones declined in 1315–1316 from 4.3d to 4.1d for threshing/winnowing 1 quarter of grain and from 6.2d to 5.4d for reaping and binding one acre of grain. In the course of the second year of the famine, the respective figures rose to 5.1d and 7.7d, but this increase could not possibly match the remarkable speed of the grain price rise. In other words, nominal wages were lagging behind price levels, meaning that real wages were excessively depressed. At no other point in recorded English history was the gap between nominal and real wages, on the one hand, and crop prices, on the other, so wide, and it appears that that the real wages of labourers reached their lowest point in 1315–1316 and 1316–1317. As Table 4.4 shows, they were abysmally low, even by the standards of other extreme harvest shortfalls and food crises in England, except the disastrous year of 1597, marking the peak of the crisis of the late 1590s. Similarly, the real wages of the Great Famine years stand out, when compared to other major Eurasian famines, with the exception of 1847 in

Table 4.4. Indexed real wages of English labourers in the years following extreme harvest failures (indexed on 1451–1475 = 1.00).

Year	1316	1317	1438	1439	1440	1556	1557	1597	1598	1623	1624	1741	1801	1817
Mason labourer	0.23	0.23	0.79	0.65	0.72	0.45	0.43	0.29	0.35	0.34	0.37	0.56	0.33	0.52
Mason master	0.31	0.31	0.79	0.65	0.72	0.38	0.39	0.29	0.35	0.34	0.37	0.56	0.34	0.52

Source: Munro, 'Revisions of David Farmer's series of English "National Agrarian Wages"'.

Table 4.5. Indexed real wages of labourers in select pre-Industrial Eurasian famines (indexed on preceding ten years).

Place	Year	Urban	Agricultural
England	1316	0.48	0.51
England	1317	0.48	0.60
Northern Italy (Tuscany)	1329	0.66	
Sweden (Stockholm)	1509	0.58	
Turkey (Istanbul)	1607	0.73	
France (Paris)	1693	0.64	
France (Paris)	1694	0.77	
Ireland	1846	0.73	0.74
Ireland	1847	0.52	0.54
Ireland	1848	0.70	0.72

Source: Calculated from Munro, 'Revisions of David Farmer's series of English "National Agrarian Wages"'; Malanima, 'Wages, Productivity and Working Time'; Söderberg, 'Long-Term Trends in Real Wages'; Özmucur and Pamuk, 'Real Wages and Standards of Living in the Ottoman Empire; Hoffman, 'Paris 1380–1870'; Geary and Stark, 'Trends in Real Wages during the Industrial Revolution'.

Ireland, at the height of the Great Irish Famine (Table 4.5). This undoubtedly underlines the severity and extent of the Great Famine in the wider context of pre-Industrial food shortages and places it in a league of its own.

Although there is no doubt that the abysmally low real wages during the famine years reflect the severely depressed living standards of wage-earners, it must be stressed that they still provide only a partial and far from satisfactory measure of living standards of the general populace. As we have seen above, the agricultural *famuli*, whose wages have been included in the discussion, accounted for no more than 3–5 per cent (and likely less) of the entire rural population (see above, Chapter 3). Moreover, many of these *famuli* would have been hired as temporary workers, during the harvest seasons, when the demand for extra working hands was high. The vast majority of the English rural population around 1315 was either self-employed (performing their labour within the framework of their family) or unemployed. Wage labour was much more a feature of towns, where labourers would work for cash or cash and kind, either as short-term contractors or permanent members of guilds or confraternities.[22] In other words, real wages refer only to a small proportion of the total population.

[22] See. Epstein, *Wage Labor and Guilds in Medieval Europe*.

It can be argued that wage-earners were worse-off than their 'self-employed' counterparts, who had readier access to scarce food supplies. An alternative way to measure the living standards of general populace, based on landholding patterns, will be offered below, in Chapter 7.

Market Segmentation?

The rapid rise in crop prices went hand in hand with market segmentation. One aspect of market segmentation is the tendency of market prices to become more and more differentiated across various geographic loci. Roughly speaking, if the range of market prices is greater than 15 per cent across locations within the same region/country, then we may speak about a *segmented* rather than an *integrated* market. It should be noted that despite a significant process of commercialization that England underwent in the course of the thirteenth and early fourteenth centuries, the degree of integration of grain markets may have been somewhat overstated by some historians.[23] True, in many instances, some regional markets had nearly identical prices for the same grains, but this was by far not an omnipresent reality. Moreover, there is not enough sense of trans-regional market involvement or commercial collaboration and networking (other important traits of 'market integration') between different regions in England and indeed other parts of the British Isles. Therefore, 'market segmentation' may not be the best term to use here, since it implies a shift from integration to segmentation. Nevertheless, as the available evidence shows, grain markets became more segmented than they were in non-famine years.

As Table 4.6 indicates, the coefficient of variation for prices among the manors in the 1315–1316 sample was indeed higher than those in the non-famine sample. This rise is seen particularly with barley and oats, rising from about 0.16 and 0.13, to 0.28 and 0.30 respectively. Within the wheat sector, the increase in the coefficient of variation was much less pronounced. In all three cases, the variation in prices across markets fell back in the course of 1316–1317, despite the persistence of the famine. There are several examples when the differences in prices in two adjacent or nearly adjacent places were significant. For instance, in March 1316, 1 quarter of wheat was selling for 17s 3d at Kelvedon and 18s 8d at Feering, both in Essex and both belonging

[23] On late medieval markets, see, *inter alia*, Britnell, 'The Proliferation of Markets in England'. The money supply is surveyed in Mayhew, 'Silver Production and the Money Supply in England and Wales'. The 'market integration' is discussed in Galloway, 'One Market or Many?', pp. 23–42.

Table 4.6. The coefficient of variation (CV) across different markets in England (monthly terms) in non-famine and famine years.

Crop	**CV, 1280–1349 (non famine years)**	**CV, 1315–1316**	**CV, 1316–1317**
Wheat	0.125	0.152	0.115
Barley	0.157	0.279	0.149
Oats	0.131	0.300	0.178
n of wheat transactions	381	66	50
n of barley transactions	162	39	24
n of oat transactions	91	29	9
Total *n* transactions	634	134	83
Total *n* of accounts	87	52	44

Sources: Manorial accounts database (see Appendix 8).

Notes: Because of their fairly small transaction numbers, I have excluded other crops from this sample.

to Westminster Abbey.[24] The distance between the two manors is 1.8 miles. Similarly, in London, on 13 October 1316, the average price of 1 quarter of wheat was 15s 6d at Billingsgate Market and 17s at Pavement Market.[25]

One visible pattern in crop prices was a widening gap between country and town prices. This phenomenon is hardly surprising; the non-agricultural nature of urban occupations and the food dependence of towns on their surrounding hinterland implied that, almost as a rule of thumb, prices tended to be a bit higher in towns in normal years, and the gap between the two would become more apparent in years of dearth.[26] Thus, London prices were considerably higher than those in its hinterland; by mid-October 1316, the average price of wheat was 16s 5d a quarter, while the average 'manorial' price of the same grain was about 13s 4d.[27] On the other hand, on 18 July 1317, 1 quarter of wheat was selling for 18s at Exeter, which was considerably lower than the average 'manorial' prices of about 21s 6d a quarter nationally.[28] But then again, because of the

[24] WAM 25,632 and 25,803.

[25] Campbell and others, *A Medieval Capital*, p. 202.

[26] Campbell and others, *A Medieval Capital*, p. 202.

[27] See above, Figure 4.3.

[28] DALS (E), C5 G1 (Mayor Court Rolls transcripts, made by Ruth Easterling in the 1920s).

absence of comparable seasonal price data from elsewhere in the south-west, it is impossible to appreciate if that isolated reference reflects a regional trend, market segmentation, or an anomaly.

Although it is tempting to blame the variance in yields across the demesnes for the rise in the variation coefficient of prices, there was actually no clear correlation between the two variables. Thus, in January 1316 a quarter of wheat was selling for 15s 4d at Feering (Essex) and 10s at Monkton Deverill (Wiltshire), notwithstanding the fact that the respective 1315 gross per-seed yields for the same grain were 3.58 and 2.37.[29] Similarly, there is no clear sense of a correlation between the quantity of grain traded and the price of grain. Thus, in July 1317 the officials of Gamlingay (Cambridgeshire) sold 1 quarter of wheat for 20s, while demesne managers of Farleigh (Surrey) sold 4.75 quarters of the same crop at 14s per quarter.[30] But on the other hand, in February 1316 the officials of Berkeley (Gloucestershire) sold 1 quarter of wheat for 11s 6d, while those of Metham (Yorkshire) managed to sell 15 quarters of the same grain for 15s a quarter.[31] More similar examples could be added here.

Although in the vast majority of cases the prices obeyed the law of seasonal progression, there were a few instances when there was no clear correlation between prices and seasonality and when prices exhibited significant monthly volatility. Thus, at Cuxham (Oxfordshire) a quarter of wheat was selling for 20s in June 1316. Yet a month later it fell to 15s, notwithstanding the fact that the harvest was yet to be collected.[32] The following example is perhaps even more revealing. Between 13 January and 12 March 1316, the officials of Feering (Essex) sold wheat in four transactions. The price of 1 quarter was 15s 4d on 13 January; on 2 February, however, it fell to 12s, only to rise to 15s 6d on 14 February, and soar to 18s 8d on 12 March, which was well above the manorial average.[33] But then again, such instances were usually few, and they seem to reflect more the interactions between vendors and purchasers than sale seasons.

What do all these facts teach about market performance during the Great Famine years, in a wider comparative context of pre-Industrial famines? Although the connection between famines and market performance has been

[29] WAM 25,632 and LH 9764.

[30] MCA 5327 and MCA 4829.

[31] Berkeley Castle Muniments, GAR 291 (consulted on microfilm made available through the Gloucestershire Record Office).

[32] MCA 5839 and MCA 5840.

[33] WAM 25632.

Table 4.7. The spatial coefficient of variation in crop prices in major European famines (famine years in italics).

Place	Commodity	Years	Coefficient of Variance
England and Wales	Barley	1300–1349 (excl. 1315–1317)	0.157
		1315–1316	*0.279*
Scotland	Oats	1689–1695	0.260
		1696, 1698–1699	*0.238*
		1697, 1699–1703	0.236
France	Wheat	1690–1693	0.287
		1694–1695	*0.438*
		1696–1699	0.230
Bavaria	Rye	*1816–1817*	*0.311*
		1818–1827	0.158
Ireland	Potatoes	1840–1845	0.320
		1846–1848	*0.205*
		1849–1851	0.180
Finland	Rye	1859–1864	0.049
		1867–1868	*0.049*
		1870–1873	0.059

Sources: Manorial accounts database (see Appendix 8); Poynder, *Database of England Monthly Grain Prices, 1270–1955*; Poynder, *Database of French (Parisian) Monthly Grain Prices, 1549–1698*; Ó Gráda, 'Markets and Famines', pp. 158–59 Cullen, *Famine in Scotland*, p. 59.

extensively studied, there is no scholarly consensus on this issue, since different circumstances affected markets in different ways (Table 4.7). In some instances, such as the late seventeenth-century Scottish famine, the Great Irish Famine of 1845–1852, and the Great Finnish Famine of 1866–1868, markets functioned 'normally', responding quickly to emerging disequilibria of prices rather than segmenting.[34] This is despite the fact that the Finnish economy was still largely under-commercialized, and its grain markets were under-developed relative to other parts of Europe on the eve of the 1866–1868 famine.[35] Conversely, the coefficient of variation rose significantly in the case of the 1693–1694

[34] Ó Gráda, 'Markets and Famines', and *Eating People is Wrong*, pp. 92–129.

[35] Ó Gráda, 'Markets and Famines', p. 147.

famine in France and the 1816–1817 crisis in Bavaria. It should be noted that the degree of commercialization and market integration varied in each case. Thus, the *ancien régime* economy in the days of Louis XIV (1643–1715) was largely autarkic, marked by a pronounced lack of market integration between different regions.[36] Other factors aggravated famine in pre-Industrial Europe. Thus, during the potato blight in Ireland, the disastrous policies of the British Government and the rent-seeking tendencies of local lords are to be blamed. In the case of the late seventeenth-century French famine, on the other hand, market segmentation could be explained by a combination of warfare and a lack of poor relief, in addition to disastrous harvest failures. In the twentieth and twenty-first centuries, similar market segmentation can be clearly seen in some famines in the developing world. The omnipresent failure of markets, manifested not only in their inability to cater sufficient amounts of food but also in a rising coefficient of variation of regional prices, can be spotted in the Bengali famine of 1942–1943, the Bangladeshi famine of 1974–1975, and the 1983–1985 famines in Kenya, Ethiopia, and Sudan, as well as in the 1999–2000 food crisis in Ethiopia.[37] Thus, in Kenya, the coefficient of variation trebled from 0.15 to 0.45 between January and November 1984.[38] In other words, one should not take the link between famines and market failure for granted, since each famine exhibits different market behaviour, depending on local conditions particular to each event. In other words, 'famine' should not necessarily be equated with market segmentation, characterized by increasing differentiation in local crop prices. All the same, the case of the Great Famine, as well as the shortages of late seventeenth-century France and early nineteenth-century Germany, prove that in some instances, harvest failures can indeed have a devastating impact on market functionality.

The Rise of Preferential Trade?

Under these circumstances of market segmentation and unpredictability of prices, anyone could potentially be either a winner or loser. On the one hand, wealthy crop producers could abuse the instability and sell their produce to

[36] Ó Gráda and Chevet, 'Famine and Market in Ancien *Régime* France', pp. 714–18.

[37] Olsson, 'Causes of Famine'; Ravallion, *Markets and Famines,* pp. 111–13; Drèze and Sen, *Hunger and Public Action*, pp. 144–55; Ó Gráda, 'Markets and Famines', p. 156.

[38] Drèze and Sen, *Hunger and Public Action*, pp. 144 and 155; Ó Gráda, 'Markets and Famines', p. 156.

strangers for excessively high prices. But on the other hand, the same producers could become the victims of the similar arbitrary and speculative behaviour of other producers. For instance, in January 1317 the demesne officials of Haughley (Suffolk) managed to sell 6.38 quarters of wheat at 17s per quarter, just a bit over the 'national average' of 16s. Yet, the same officers significantly overpaid when buying 5.38 quarters of wheat at 20s per quarter around the same time.[39] Similarly, in October 1316 the authorities of Birdbrook (Essex) paid an astonishing sum of £10 16s for 12 quarters of wheat (at 18s per quarter), only to sell 20 quarters of the same grain for £16 (at 16s per quarter) two months later.[40] In both cases, it is unclear why the demesne officials failed to sell higher than they bought, but they undoubtedly illustrate the unpredictability of the market and the volatility of grain prices during the famine years.

One likely factor accounting for those unpredictable and volatile transactions is a shift from *generalized trust* to *particularized trust*. Roughly speaking, generalized trust is a tendency to engage in transactions with all persons indiscriminately, without considering the degree of familiarity and trustworthiness or their characteristics or affiliations. Generalized trust was undoubtedly suited better for grain transactions between strangers in non-crisis years. Particularized trust, on the other hand, denotes a scenario when transactions tended to be limited to a specific group of people who possess specific social and geographic attributes and affiliations; in many cases, both vendors and buyers shared those particular traits.[41] Almost invariably, particularized trust is based on a personal network of acquaintance, trust, reputation, and affinity. This, in turn, implies a large degree of commercial asymmetry allocating preferential terms of trade to 'insider' groups or individuals, and excluding, misinforming, or even defrauding strangers. It should be noted that such asymmetry of information between 'insiders' and 'outsiders' seems to have been commonplace in late medieval trade. Thus, while both local producers and consumers had a ready access to the information on local market prices, strangers or foreign merchants were zealously deprived of access to this kind of information by local authorities.[42] In other words, while local authorities could assist local grain merchants in obtaining the right quote, they did their best to keep for-

[39] TNA, SC 6/996/14.

[40] WAM 25423.

[41] Ogilvie, *Institutions and European Trade*, pp. 427–32.

[42] Britnell, 'Price-Setting in English Borough Markets', pp. 8–9; Davis, *Medieval Market Morality*, pp. 228–29.

eigners as ignorant as possible, to make them pay more and receive less. After all, the variability in price made the issue of information flow starker, and this fact had a knock-on effect on price variation. To avoid the undesired flow of information, serious punitive mechanisms were set up against price informers.[43] Can we, therefore, assert that the disastrous dearth of grain and a decline in price uniformity and price information across markets prompted both vendors and buyers to drift (perhaps in an undeliberate manner) to this mode of trade, which was more suitable for the crisis years?

Examples of such behaviour can indeed be spotted in those rare cases, when grain transactions recorded the names of trading parties. The remarkable paucity of such transactions may have to do with the fact that the vast majority of all seignorial crops were sold to local tenants.[44] Nevertheless, those few accounts specifying the names of purchasers shed much light on the emergence and predominance of particularized trust and preferential trade during the famine years. For instance, at an unspecified date in 1316–1317, the officials of Orpington (Kent), belonging to Canterbury Cathedral, sold 8 quarters of wheat for only 4s a quarter to the ministers of Queen Isabella, who happened to be in the region, as a form of gift-giving.[45] At no point during that year was a quarter of wheat selling so cheaply. A similar example is found in an account roll of expenditure for Mary of Woodstock's pilgrimage from her nunnery of Amesbury (Wiltshire) to Canterbury (April–June 1316).[46] During her voyage, Mary's entourage stopped at no less than twenty-two places, purchasing food for her household and fodder for her horses. As the account indicates, between late April and early June 1316, the price of 1 quarter of oats rose from about 6s 7d to 8s, which was considerably lower than the average market rise, which was from about 8s to 10s a quarter. All the more remarkable was the fact that urban prices were identical to the country ones: between 27 May and 1 June, a quarter of oats was selling for 8s both in Canterbury and Rochester and in the Kentish manors of Ospringe and Newington.[47] Moreover, the same price was offered in London on 3 June, despite the apparent gaps between city and country prices, as we have seen.[48]

[43] Britnell, 'Price-Setting in English Borough Markets', pp. 8–9.

[44] The Winchester Pipe Rolls are unique in this matter, since they, in great many instances, specified that the crops were sold to customary tenants. HantsALS 11M59/B1/71–72.

[45] CCA, DCc/Orpington 16.

[46] TNA, E 101/376/19.

[47] TNA, E 101/376/19.

[48] Campbell and others, *A Medieval Capital*, p. 202.

A similar example is found in the 1316–1317 account roll from Ketton (Durham), belonging to Durham Priory. In the course of the year, manorial authorities sold 2.12 quarters of wheat to William de Ros, 3rd Baron of Ros, charging just 8s a quarter.[49] William de Ros had strong ties with both Edward II and the Durham authorities. Although the account does not specify the seasonality of the transaction, it is known that at no point in that year were wheat prices so low. The financial loss stemming from this transaction was, at least, partially compensated when the Ketton authorities sold another quarter of wheat for 13s 3d to an unspecified buyer. It is unclear if selling grain so cheaply was a deliberate judgement of local manorial officers, on the instructions of manorial lords, or due to pressure from the royal ministers. It is certain, however, that charging royalty or high nobility (or their representatives) with high prices could have had negative repercussions, not only on local producers and traders but also on local manorial lords.

In some cases, social status could work equally to the disadvantage and exclusion of grain producers and traders. The following example is particularly revealing. The vill of Rippingale (Lincolnshire) consisted of several manors, two of which belonged to Sir William Inge (*c.* 1260–1322) and John Gobaud, respectively. According to the 1316–1317 account roll, the serjeant of Inge sold 30.5 quarters of crops to John Gobaud over the course of that year.[50] In each case, the transaction price seems to be above average. Thus, 1.88 quarters of wheat were sold for 18s a quarter around 31 October 1316, compared with the average 14s elsewhere. This may do with the social status of both the vendor and buyer. In the course of his political career, William Inge held several key administrative positions in the governments of Edward I and Edward II, and, at the time of the transaction, he was serving as the chief justice of the King's Bench.[51] John Gobaud (1300/1–1336), on the other hand, was merely a minor Lincolnshire landowner, sixteen years of age at the time of the transaction.[52] Despite the fact that both men held manors in the same vill, the gap in status and age between the two allowed Inge to reap high profits by selling his crops

[49] DCA, Ketton 1316–1317.

[50] Lincolnshire Record Office, Anc. 3/1.

[51] Brand, 'Inge, Sir William'.

[52] Waters, *Genealogical Memoirs of the Extinct Family of Chester of Chicheley*, I, 198. This John Gobaud is not be confused with Sir John Gobaud (also of Lincolnshire), creditor, landowner, sheriff of Lincoln, and commissioner of array: see Coss, *The Origins of the English Gentry*, p. 178; Gorski, *The Fourteenth-Century Sheriff*, p. 99.

well above their average market value. Interestingly enough, around the same time, Inge was also facing multiple allegations of corruption regarding illegal property acquisition and a forgery of a seal of the sheriff of Surrey.[53]

The following instance exemplifies the dynamics of preferential trade on both social and geographic levels. Towards the end of 1316, the officers of Preston (Kent) sold the combined volume of 23.07 quarters of barley to the officials of Ham, Mersham, and Little Chart, all in Kent. In each case, the Preston officers charged 8s a quarter. This price was below the 'manorial average', standing at about 9s 4d. Such behaviour might be explained by several social bonds shared by the four demesnes: they all belonged to the same lord, Canterbury Cathedral Priory; they all were run by manorial officials; and there had always been a strong degree of collaboration and coordination between the officials of these demesnes.[54] Around the same time, the same demesne officers of Preston sold additional 20 quarters of barley to a certain Robert Dod, charging 9s per quarter, namely one shilling higher than the price offered to their fellow officials of Ham, Mersham, and Little Chart. This price was around or a bit lower than the average market price. In other words, although Dod was a less 'preferential' customer than the three manorial officials, he still was not overcharged. The account roll does not give any details about the identity of the latter, but it is possible that he can be identified as Robert Dod, a baker from Sittingbourne (Kent), who was dead by June 1318.[55] It is sufficient to bear in mind that he was a member of the well-off Dod family, established in and around Faversham. If, however, the identification of Robert Dod as the Sittingbourne baker is correct, then the transaction between Robert Dod and the Preston officials may reveal some intriguing mechanisms of grain trade during the famine years. First, it underlines the importance of victuallers during the years of dearth, which, in turn, would deter the Preston officials from charging Dod above the average price. Second, it might suggest that Robert Dod, as a baker, purchased grain to produce and sell bread. Having purchased grain for average or slightly below-average price, he would most certainly reap a financial profit, since reselling grain in a processed form would imply a significant profit. As we shall see later, the famine years saw an increased number of bakers and bread regraters fined for speculating. In other words, Robert Dod (if identified correctly as the

[53] Brand, 'Inge, Sir William'.

[54] The Canterbury manorial tenants are described in DuBoulay, *The Lordship of Canterbury*, p. 165.

[55] TNA, C 241/112.

Sittingbourne baker) was not only a buyer and consumer of grain, but a middleman between producers and consumers. Does this, in turn, hint that market prices during the famine may have been higher than demesne-gate prices and that people like Robert Dod were the real winners of the catastrophe because of their ability to buy low and sell high? After all, the famine years led to a greater disparity between the market and demesne-gate crop prices — in contrast with non-famine years, when the gap between the two types of prices were certainly closer one to another.

What about the lower social echelons? Could they expect to benefit from the social networks of their kinsmen or fellow parishioners to get their badly needed grain cheaper? On the one hand, the spike in food-related crimes and baking and brewing fines, discussed at length later in this study, suggests that those ties and bonds had loosened a great deal during the famine years. But on the other hand, there is at least one example, which might be interpreted as personal ties between fellow parishioners as a decisive factor. At Cuxham (Oxfordshire), on 28 July 1316, in the middle of the harvest, a local villein named William Waldrugge managed to purchase 1 bushel of wheat for as low as 10d (that is, 6s 8d per quarter), 'because a keeper (of grain) thus conceded to him'.[56] Apart from the bonds of friendship or possibly familial ties between William Waldrugge and the granary keeper, which may explain such a low price, this transaction also implies that some demesne officials were ready to risk their lord's ire by selling his grain cheaply. The paucity of evidence, however, does not allow to establish with certainty that it was indeed personal bonds that accounted for Waldrugge's ability to acquire wheat so cheaply. Equally, one may interpret this transaction as an act of charity and poor relief conducted by a lord. As Christopher Dyer has most recently shown, some late medieval manorial lords indeed contributed, to a limited degree, to poor relief through the occasional distribution of alms and grain.[57] Whichever interpretation of the Cuxham transaction is correct, it still points out that local tenants may, at some instances, have been a preferential trade group too, whether enjoying personal ties with local manorial authorities or paternalism of their lords.

In other words, the scarcity of crops created a widespread phenomenon of preferential trade, based on personal networks, social capital, and status, on the

[56] 'Quia custos ei ita concessit' (MCA 5840). That William Waldrugge was a local villein is mentioned in Harvey, *Medieval Oxfordshire Village*, p. 31.

[57] Dyer, 'Poverty and its Relief', p. 47.

one hand, and discrimination towards 'outsiders' and weaker elements, on the other. The lords and their reeves always did their best to maximize their profits by selling as high as they could and buying as cheap as they could, but it was in the context of famine that they did it in an especially pronounced manner (through speculation and reliance on personal networks). Outsiders and buyers of lower social status, on the other hand, were clearly disadvantaged and, as such, were forced to pay higher prices for (very) scarce crop resources. This attitude reflects the collapse of generalized trust and a switch to a much more particularized sort of trust. In some cases, the profit was enormous, but in other cases, especially when dealing with vendors and buyers of high social standing, reputation, and prestige, it was ruinous.

Black Markets and Speculation in Raw Grain

The transaction prices found in the manorial accounts do not always agree with those reported by contemporary, near-contemporary, or later chroniclers and annalists. Of thirty-eight sampled chronicles narrating the famine, only eight recorded prices agree with the accounts' data.[58] But even in these few cases, the chroniclers reported only the highest figures (usually between 20s and 26s for quarter of wheat). The vast majority of the chroniclers, however, usually recorded much higher figures, with the most common price being 40s per quarter of wheat. One Leicestershire chronicler reported 44s a quarter.[59] Although some of these figures were given by late fourteenth- and fifteenth-century chronicles, who did not experience the famine, in some instances they were recorded by contemporary witnesses, who undoubtedly remembered the famine most vividly.[60] Such high figures were never recorded in the extant manorial accounts; in fact, the highest price found there is 26s 8d for a quar-

[58] Thus, *Vita Edwardi Secundi*, ed. by Childs, pp. 120–23; *Annales Londonienses*, ed. by Stubbs, p. 236; *Thomae Walsingham Historia Anglicana*, ed. by Riley, p. 145; *Chronique de Sepringham*, ed. by Glover, pp. 332–33; *Newnham Abbey Chronicle and Cartulary* (BoL, MS Top. Devon d. 5, fol. 5^{v}); *Chronicle of Gervase of Canterbury's Continuator* (Cambridge, Trinity College Library, MS R.5 41, fol. 113^{r}); *The Chronicle of St Mary's Abbey, York*, ed. by Craster and Thornton, p. 67; and *Castleford Chronicle*, ed. by Eckhardt, II, 1058.

[59] *Chronicon Henrici Knighton*, ed. by Lumby, p. 411.

[60] Thus, *Thorney Annals, 963–1412*, ed. by Hart, p. 33; *Annales Monasterii de Bermundeseia*, ed. by Luard, p. 470; *Annales Monasterii de Osenia*, ed. by Luard, p. 354; *Annales Prioratus de Wigornia*, ed. by Luard, p. 561; *Chronica Monasterii de Melsa*, ed. by Bond, p. 333; and *Johannis de Trokelowe Annales*, ed. by Riley, p. 92.

ter of wheat, reported at Hinderclay (Suffolk) and Long Sutton (Hampshire), most likely shortly before the harvest of 1316.[61] In many instances, it is the chroniclers' narrative, rather than the manorial reeves' figures, that are taken at face value by modern historians.[62] What accounts for this discrepancy between the two sources? One possibility, as I have argued elsewhere, is the tendency of the chroniclers to exaggerate the figures in order to build their narrative of famine to be as dramatic as possible.[63] But there is yet another possibility: the chroniclers' prices are, in fact, black market prices.

That famines give rise to black markets is a truism that needs no argumentation here. There is an abundance of historical examples illustrating it. One could think about the examples of the Soviet famine of 1918–1922, the Greek famine of 1941–1944, the Bengali famine of 1942–1943, China's 'Great Leap Forward' (1958–1961), the Eritrean-Ethiopian famine of 1983–1985 and the North Korean 'penal' famine of 1994–1998.[64] In August 2011 it became known that food aid from the United Nations, destined to relieve the starvation in Somalia, was stolen by local militants and sold on the black market.[65] Many more examples can be added. Roughly speaking, the essence of a black market is an 'underground economy' supplying goods (in many cases, illicit or deficient) and services illegally and for much higher prices than those on the official market.[66] In the case of famine, the most sought-after commodity on the black market is food, sold for considerably higher prices than on official markets, for two reasons. First, the deficiency of food supplies on the official market gives the black market an upper hand in this competition. Second, the costs of search and security of foodstuffs, which imply a great degree of

[61] UCL, Bacon Roll 446; HantsALS, DC/J1/14.

[62] Thus, Lucas, 'The Great European Famine of 1315, 1316 and 1317', pp. 343–77; Kershaw, 'The Great Famine and Agrarian Crisis in England', p. 8; Haines, *King Edward II*, p. 97; Slavin, 'Food Security, Safety, and Crises', p. 70. For a more balanced account, see Aberth, *From the Brink of the Apocalypse*, pp. 20–21. Aberth took 40 shillings as the highest point during the famine years.

[63] Slavin, 'Market Failure during the Great Famine', pp. 25–26.

[64] Ó Gráda, *Famine: Short History*, pp. 150 and 174; Clark, *Crime and Punishment in Soviet Officialdom*, pp. 10–12; Hionidou, 'Black Market, Hyperinflation and Hunger' and *Famine and Death in Occupied Greece*, pp. 82–108; Thaxton, *Catastrophe and Contention in Rural China*, pp. 160–62 and 170–76; Clay and Holcomb, *Politics and the Ethiopian Famine*, p. 98; Howard-Hassmann, 'State-Induced Famine and Penal Starvation'.

[65] CBC News, 'Somali Food Aid Stolen, Sold in Markets'.

[66] See Naylor, *Wages of Crime*.

time and risk, result in high transaction costs, which can be translated into high prices. This point will be discussed in the next chapter. Also, the black market is characterized by price segmentation, with each region, hub, or even dealer asking for a different price. For instance, during the Bengali famine, the official price of 1 maund of rice (83.3 lb or 37.3 kgs) was about 20 rupees in autumn 1943 (compared with 10 rupees in July 1942, on the eve of the famine). At the same time, its counterpart price on the black market of Calcutta was 40 rupees, while at Dacca and Faridpur it was selling for as much as 70–80 rupees.[67]

Excessively high prices are not the only indication of speculation, however. There is much evidence of the alteration of both quality and quantity of food during the famine years. Thus, a certain Ralph de Benton, keeper of king's stores at Berwick-upon-Tweed, was accused, perhaps falsely, of using false measures, taking bribes, and replacing good food with bad, while supplying local garrisons, at various points between 1312 and 1316.[68] In particular, he and his clerks were diverting the good foodstuffs from his stores and selling them to (starving) Newcastle by false measures, keeping one fifth to themselves.[69] Similarly, on 29 June 1316, Andrew Pilkeston of Horsham St Faith (Norfolk) was accused of using the 'small bushel' measure to sell grain and the 'large bushel' measure to buy grain.[70] In other words, Pilkeston would deliver less and acquire more, in relation to market price. At Sandal (West Yorkshire), on 10 April 1317, Matthew de Hyngindeheton was amerced of 2s for selling his neighbours flour with a dish of false measure.[71] Whether these allegations were true or false, they reflect that food speculation flourished during the famine years. A strong support in favour of this view is found in many manorial court rolls from the famine years, recording the cases of breaking the *Assize of Bread and Ale* (*Assisa Panis et Cervisie*).

[67] Ó Gráda, *Famine: Short History*, p. 174.

[68] TNA, SC 8/330/E970. This case is summarized, albeit not entirely accurately, in *Calendar of Documents Relating to Scotland*, vol. III: (*1307–57*), ed. by Bain, no. 522 (henceforth *CDRS*).

[69] *CDRS*, no. 427.

[70] NRO, NRS 19505.

[71] *Court Rolls of the Manor of Wakefield*, ed. by Lister, IV, 186.

Breaking of the Assize of Bread and Ale

The assize, enacted around 1254–1256, was a widespread and abiding law regulating, in a formulaic and tabular manner, the price, quality, and quantity of the basic foodstuffs.[72] Roughly speaking, according to the assize, the change in grain prices would affect the size and weight rather than the price of a loaf of bread. The relationship between the grain prices and size/weight was inverted: the higher the price of wheat, the smaller and lighter was a farthing (quarter-penny) loaf. As far as ale is concerned, the assize established that when 1 quarter of wheat, barley, and oats were selling for, respectively, 36d–40d, 20d–24d and 16d, then one penny should buy 2 gallons of ale in towns and 3 gallons of ale in the countryside. The *Statute of the Pillory* (*Judicium Pillorie*), enacted in 1266, established that when the price of raw barley increased by 6d, then the quantity of quarts of ale selling for 2d would have decreased by one. To compensate bakers for their initial expenses and costs related to salt, yeast, fuel, candles, sieves and, naturally, labour, the assize stipulated that for every 111 farthing (quarter-penny) loaves (totalling 27 3/4d), they should make a profit of 4d (that is, about 14 per cent of the price).[73] Neither the *Assisa* nor the *Judicium* spoke about the brewers' profits, but it has been estimated that for each quarter of barley, they could expect to earn somewhere between 3d and 6d (that is, about 6–12 per cent of the price around 1300).[74] The main purpose of the assize was to ensure a just profit for bakers and brewers, prevent frauds in food trade and legally protect consumers of all social strata. Although there is no doubt that the multitude of references to amercements related to breaking of the assize, as found in late medieval court rolls, reflects the fact that this quasi-statute was taken seriously by both urban and rural communities, not every amercement meant an offence. As several scholars have demonstrated, in many cases, those amercements were, in fact, licences paid for brewing, rather than actual cases of breaking the assize.[75] In some cases, manorial court rolls specified the nature of offence, such as the use of false measures and weights, and in such cases, it

[72] Davis, *Market Morality*, pp. 231–53; Seabourne, *Royal Regulation*, pp. 73–81.

[73] Davis, *Market Morality*, pp. 234–48.

[74] Davis, *Market Morality*, p. 242.

[75] Bennett, *Ale, Beer and Brewsters*, pp. 4–5, 100, and 'The Village Ale-Wife', p. 21; Britnell, *Growth and Decline in Colchester*, p. 89, and 'Morals, Laws and Ale in Medieval England', p. 28; Britton, *The Community of the Vill*, p. 25; Dyer, *Lords and Peasants*, p. 346; Davis, *Market Morality*, p. 247.

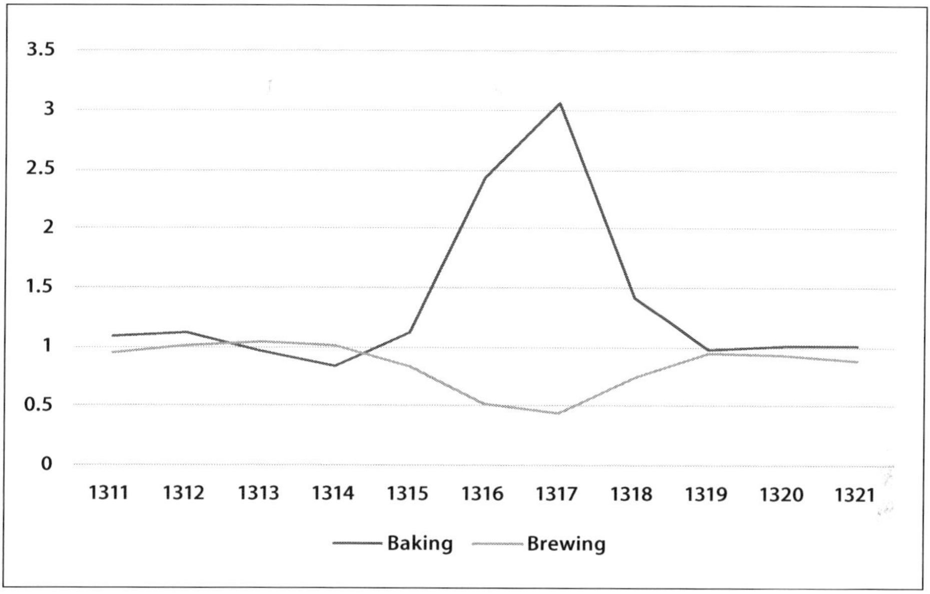

Figure 4.5. Number of baking- and brewing-related amercements for breaking the *Assize of Bread and Ale*, 1311–1321 (seventy-four manors), indexed on 1311–1314 (1.0 = 2027 amercements on all seventy-four manors in one year). *Source*: Slavin, Court rolls database (see Appendix 8).

Note: The manors are as follows: Combe, Hungerford, Wantage (Berkshire); Bledlow, Chesham Bois (Buckinghamshire); Balsham, Newton, Sutton-in-the-Isle (Cambridgeshire); Milbourne Stileham, Povington (Dorset); Dunton Waylett, Great Bromley, Great Waltham, Little Dunmow, Matching, Roding Marcie (Essex); Hawkesbury, Mitchinhampton (Gloucestershire); Anne Bec (in Monxton), Quarley, Southwick (Hampshire); Atterby, Grayingham, Harpswell, Helmswell, Snitterby (Lincolnshire); Isleworth, Ruislip (Middlesex); Aldeby, Bressingham, Burnham Thorpe, Coltishall, East Wretham, Gressenhall, Heacham, Hindolveston, Horsham St Faith, Kempstone, Kimpton (Weasenham), Langham, Lessingham, Marham, Sedgeford, Tivetshall, Wimbotsham, Wymondham (Norfolk); Brigstock (Northamptonshire); Cottisford, Great Horwood, Newington, Swyncombe (Oxfordshire); Tamworth (Staffordshire); Flixton, Great Blakenham, Hinderclay, Horham, Lakenheath, Redgrave, Rickinghall, South Elmham, Worlingworth (Suffolk); Streatham, Tooting Bec (Surrey); Cowbeech, Wartling (Sussex); Atherstone, Middleton (Warwickshire); Brixton Deverill, Chisenbury, Longbridge Deverill, Monkton Deverill, Ogbourne (Wiltshire); Halesowen (Worcestershire); and Dyffryn Clwyd (Denbighshire).

is possible to establish the ratio between the actual offences and licence fees.[76] It is less clear whether amercements for baking were actual fines or licence fees. On the one hand, court roll references to the breaking of the assize by bakers tend to be much less numerous compared to brewing-related amerce-

[76] For instance, the Redgrave courts of 20.7.1319 and 25.7.1321 (in Suffolk): UCL, Bacon Roll 13, membr. 27[r] and 39[r].

ments in the same rolls. On the other hand, there are several exceptional court roll series, such as that for the manor of Redgrave (Suffolk), which abound in such references, some of which might represent baking licence fees, rather than fines for breaking the assize. In the case of Redgrave, one way to distinguish between fees and fines is to consider the costs of each individual amercement: the higher were the costs, the more likely these were actual offences rather than licence fees. For instance, an amercement of 3d, especially if recurring on an annual basis, was unlikely to have been paid for an offence by the same baker. Conversely, amercements of 1s and higher were most likely to have stood for fines for the assize breakers. Thus, it is vital to distinguish between 'cheap' and 'expensive' amercements, in order to distinguish between licences and fines.

With all these caveats in mind, we may now turn to the court rolls to see if there were any visible changes in baking and brewing patterns during the famine years. In order to establish that, I have surveyed and analysed court rolls from seventy-four manors that survive with no or few gaps between 1311 and 1321 (Figure 4.5). The lion's share of the evidence comes from English manors, and unfortunately only one series survives for a Welsh manor of Dyffryn Clwyd (in Denbighshire, north-east Wales). Nevertheless, the Dyffryn Clwyd court rolls indicate that baking and brewing offences were an issue not limited to England but pertaining to other parts of the British Isles too — and certainly manorialized marcher lordships in Wales.

As Figure 4.5 clearly demonstrates, the annual number of baking-related amercements increased, while brewing-related amercements decreased, during the famine years. This, in turn, reveals several intriguing facts. To begin with, it reaffirms that the omnipresent grain shortage forced crop producers to cut down on malting and brewing and convert as much grain as possible into bread in order to save each calorie from being wasted. Indeed, brewing implies high extraction rates, meaning that a lot of calorific content gets wasted in the process.[77] The decline in both the supply and demand of ale within the peasant sector complements a similar trend within the demesne sector: as we have seen in the previous chapter, manorial officers reduced the proportion of grain harvest invested in malting and brewing. This temporary shift is reflected in a royal writ from 24 January 1317, in which Edward II stated that converting wheat and barley into ale instead of bread harms the poor masses.[78] Similarly, on 24

[77] See above, Chapter 3, p. 92.

[78] *Calendar of Close Rolls Preserved in the Public Record Office: Edward II, AD 1313–1318* (hereafter, *CCR, 1313–18*), p. 449.

August 1317, shortly before harvesting was completed, the London authorities forbade the malting of wheat, under pain of forfeiture.[79] This decree seems to have proliferated widely across Edward's dominions: on 9 May 1318, fifty-four individuals of Dyffryn Clwyd (Denbighshire, in north-east Wales) were fined for brewing wheat malt (most likely during the famine) despite the prohibition.[80] On the same manor, only thirty-four brewers were fined during the court of 19 April 1317, compared with the average of fifty-four brewers in the pre-famine years.[81]

In order to appreciate the spike in baking amercements more fully, however, it is necessary to view it in a wider context of the rules and stipulations of the *Assize of Bread*. According to the logic of the assize, the excessively high prices of the famine years would have produced an abnormally large number of loaves, with each being very small in size and light in weight. In theory, the 1317 price of wheat (16s a quarter) would have produced as many as 1200 meagre loaves of bread from a quarter of grain (compared with, say, 371 loaves of bread from a quarter of grain in 1313). The weight of each loaf would have been determined by the contents and quality of each type of bread: a loaf of 'wastel' bread (fine wheat bread) would have weighed about 0.43 lb Troy (in comparison with about 1.37 lb in 1313), a loaf of wholewheat bread may have been in the area of 1 lb Troy (in contrast with some 2.44 lb in 1313), and a loaf of whole-grain bread (otherwise known as 'all common crop bread', baked from a mixture of several crops) would have yielded 1.35 lb (compared with 3.25 lb in 1313).[82] The high prices of grain and the light weight of each farthing loaf meant that bakers were able *to bake more loaves*, and that they *had to sell more loaves*, in order to make their *usual* gains, as stipulated by the assize — with the exception of wastel bread. Under these circumstances, bakers found themselves in an especially disadvantageous situation, where the local authorities were encouraging them to bake lower-grade, black multigrain bread for their poor neighbours. According to the assize stipulation, it was impossible to make any significant profit from those low-quality breads when prices exceeded 8s per quarter.[83] The only way to increase their profits during the famine years was to bake and trade in bread contrary to the rules of the assize, and it seems that

[79] *Calendar of Letter-Books*, ed. by Sharpe, p. 77.

[80] Stevens, *Urban Assimilation in Post-Conquest Wales*, pp. 143–44; TNA, SC 2/216/1, m. 20.

[81] TNA, SC 2/215/71–76, 216/3.

[82] Davis, 'Baking for the Common Good'.

[83] Davis, *Market Morality*, pp. 241–48.

the harsh economic reality of the famine years could not have produced any alternative scenario. This, in turn, may explain the unusually high number of baking-related amercements. This reasoning, however, assumes that the same amounts of grain were available to bakers — which could not have been the case, given the low yields and declining quantities of grain available at local markets. Under these circumstances, it is apparent that there were fewer loaves, with each loaf weighing less than usual.

That some bakers were attempting to make illicit profits by baking smaller and lighter loaves is apparent in some court rolls entries. Indeed, as entries point out, the final products weighed less than they should have. Thus, John de Stonehurst of Wartling (Sussex), was amerced 4d 'pro pane suo invento minus ponderante quam pondere debuit' (because his bread was found to have weighed less than it should have).[84] But it is important to bear in mind that even if bakers did not opt to breach the assize, they would have undoubtedly faced technical difficulties to comply with its stipulations, in the context of the unusually high prices. As we have seen, the assize prescribed very clear ratio between the prices and the weight of each farthing loaf. At the same time, the astronomically high prices implied that the weight of each loaf would have been so unusually light that bakers were faced with the uncertainty of *how much* to bake in accordance with the assize. In particular, it was largely impossible to foresee the loaf weight at the end of the baking process, when bakers had to shape loaves of uncertain size out of dough.

Furthermore, the high number of amercements may reflect that local authorities wished to demonstrate to their communities that bakers were being monitored more closely than in non-famine years, and that the assize was being enforced rigidly. Even if they were *not found* guilty of flouting the normal margins of baking profits, they might have been *suspected* of doing so, and the authorities, distrustful on the verge of paranoia, imposed the amercements as a preventive measure to ensure that mis-baking was discouraged and the community was aware of that. All three interpretations of the high number of amercements are equally plausible.

Intriguingly, the lists of amerced bakers, as found in the court rolls of the famine years, included some names that were not active in the baking business before or after the famine. For instance, at Redgrave, one witnesses the emergence of no less than four 'new bakers' who were active during the famine

[84] BL, Add Roll 31503, rot. 4b.

years, alongside some twelve existing bakers.[85] At least two of these individuals, Agnes Redlister and Robert son of Simon Ireland, were related to well-established bakers of the manor, and it is apparent that the reputation of their relatives may have played a role in securing their licences.[86] Once the crisis was over, these individuals disappear from the court rolls. This reflects the fact that more and more individuals thought to make profit during the year, either through applying for baking licences or through breaking the assize. Indeed, the role of bakers, as middlemen between producers and consumers, was most vital for the survival of vulnerable and poor communities during the famine.

In some instances, those amercements were insignificant; thus, in the case of the four new Redgrave bakers, they varied between 3d and 6d, and it is likely that these individuals were applying for baking licences rather than paying their fines. At the same time, however, there is an unusually high incidence of excessively high baking amercements. For instance, at Redgrave, Adam Pistor (a Latin form of the 'Baker' surname; sometimes he was referred to as Baxter) and John le Coghneyser were amerced the princely fines of 2s 6d and 2s, respectively, in the court of 8 March 1316. On 29 September 1316, Adam Pistor was amerced 2s.[87] These were by no means two isolated cases: the average baking amercement of those two courts stood at 11.3d and 9d, compared with the average of 7.6d for all other courts between 1310 and 1321. These high figures undoubtedly imply an increased tendency to break the assize. But even smaller fines were not always applications for baking and brewing licences. For instance, at Middleton (Warwickshire) on 6 November 1317, Alicia Mersiche was fined 6d for brewing and selling ale against the assize. The same roll has an exceptional entry, explaining that Alicia was fined 'because of the bad quality' (*debilitate*) of ale she produced.[88]

The remarkable spike in baking-related amercements and the proportion of expensive amercements indicate that the catastrophic dearth of basic foodstuffs, coupled with a reluctance to follow the precise rules of the assize, encouraged speculative trade in bread and that more and more bakers were willing to take a risk and be involved in those speculative activities during the

[85] Those were Reginald le Warren, Agnes Redlister, Julian Porter, and Robert, son of Simon Ireland.

[86] Agnes Redlister was related to Gilbert Redlister and Christine Redlister, both bakers, while Robert Ireland was related to John and Edmund Ireland.

[87] UCL, Bacon Roll 14, membr. 1^{v}; Bacon Roll 13, membr. 10^{v}.

[88] University of Nottingham Library, (UNL), Mi M131/11.

famine years. If this interpretation is true, then it reaffirms the view that the famine loosened the traditional bonds of trust and collaboration within local communities. The basic needs of physical survival in those highly challenging circumstances were now placed above those traditional social and cultural bonds and conventions.

That some bakers and brewers tended to break the assize during the famine years is reflected in various contemporary and nearly contemporary sources. Thus, in his writ to the sheriffs from 24 January 1317, Edward II stated that there were numerous complaints that brewers brew and sell ale too dearly and in contempt of the assize.[89] The author of the nearly contemporary Middle English poem *The Simonie, or Poem on the Evil Times of Edward II* (composed *c.* 1320–1325) bemoans that bakers and brewers make a 40s profit by either paying small amercements at courts or bribing manorial court officials — much to the harm and hunger of the poor people.[90]

Regrating, Forestalling, and Speculating in Bread and Ale

In addition to baking amercements, the court rolls abound in references to bread and ale regrating. Put simply, regrating (aka, 'gannocking') was the method of buying foodstuffs in order to resell them later for a higher price within the same marketplace or community. In many instances, bakers and food regraters were, in fact, the same people amerced both for baking and regrating. As some scholars have shown, food regraters were, as a rule, poorer elements of the society, possessing small capital and seeking minor profits. In theory, regraters, just as bakers and brewers, were allowed to make a small profit (usually one penny for each shilling, by obtaining thirteen loaves for the price of twelve from bakers), as long it did not exceed what was allowed by legislation. In many cases, however, they were seen on a par with forestallers and engrossers, and their social stigma implied they were often living, as James Davis has aptly put it, 'on the margin of respectability'. All the same, regraters provided an invaluable service to bakers by purchasing their excess stock at the end of business day, and, because of their usefulness, their speculative activities were allowed for

[89] *CCR, 1313–18*, p. 449.

[90] 'Thes bakers and this brewers beth so bolde in here yifte | That fore a litel mercyment or fore a symple gifte, | On may fore xij d. at a court do xlti schillingwerd schame | But how so ever hit falle, the pouer han al the grame [a]t mele'. *The Simonie*, in *Medieval English Political Writings*, ed. by Dean, lines 493–97.

a minimal amercement in normal times.[91] Furthermore, despite all the moral opprobrium and social stigma related to food speculation, there seems to have been much theological understanding about the role of middlemen in famine economies. For instance, according to Thomas Aquinas, whose philosophical influence and impact cannot be overstated, a grain merchant had a right to keep quiet about a forthcoming influx of grain to local markets and sell his merchandise dearly, as long as it was conducted within reasonable moral limits.[92] The early fifteenth-century treatise *Dives and Pauper* asserted that although many corn-mongers are guided by their selfish interest and avarice, during the times of famine, they are still instrumental in relieving food shortages, just as did Joseph during the Egyptian famine (Genesis 47).[93] According to *Dispute between a Good Man and the Devil*, written around the same time, traders had to make profit in order to cater for their own needs.[94] One thirteenth-century sermon stated that both vendors selling their merchandise dearly and customers buying little or cheaply are driven by the same greed.[95]

Did the harvest failures have any impact on regrating activities? A close analysis of the court rolls reveals that there was a pronounced rise in the number of regrating amercements during the famine years. Unfortunately, not every court roll specifies whether these amercements were related to bread or ale, and, as a result, only a part of all regrating amercements can be linked to either type of commodity with certainty. Thus, at Bressingham (Norfolk), nine and fourteen bread-regrating amercements were recorded in 1315–1316 and 1316–1377, respectively, in comparison with 3 in 1312–1313 and none in other years in the same decade.[96] Similarly, the Redgrave court rolls reported sixteen ale regraters in 1315–1316 and seven bread regraters in 1316–1317, in contrast with just one or two in non-famine years.[97] At Dyffryn Clwyd (Denbighshire, in north-east Wales), forestallers and regraters appear both in famine and non-famine years, but it appears that their numbers was exceptionally high during the crisis. During the court of 4 May 1316, eleven forestallers were fined, in contrast with

[91] Davis, *Market Morality*, pp. 231, 253–63, and 323–35. The quote is on p. 261.

[92] Davis, *Market Morality*, p. 63.

[93] Davis, *Market Morality*, p. 64.

[94] Davis, *Market Morality*, p. 64.

[95] Cambridge, Trinity College Library, MS B.14.52, fols 71^{r}–72^{r} (summarized in *RMEPS*, p. 143).

[96] NRO, WAR 4–9.

[97] UCL, Bacon Roll 11–13.

the average of 2.3 persons per court in non-famine years. Likewise, ten forestallers appear in the court of 9 May 1318, possibly reflecting the pre-1317 harvest situation.[98] Again, this spike does not necessarily reflect the rise in speculative trade: in all cases, most regraters and forestallers were fined at 3d per person, with some individuals amerced at 6d per person. Both in non-famine years and during the famine, the vast majority (about 70 per cent) of all recorded regraters were women, some of whom seem to have been never-married, single women, judging by their patronymic surnames.

Just as in the case of bakers, it is difficult sometimes to determine whether those insignificant amercements stand for licences or for offence fines. On the one hand, it is likely that the food shortage of the famine years encouraged more individuals to try their luck in food speculation, even at the risk of being socially stigmatized. Such tendency may have been driven not only by the thought of individual profits of regraters but also by the unprecedented demand for readily available food. Whatever profit these regraters were making or were expecting to make, they were still playing an instrumental role in securing food for a starving population during the famine years. After all, finding bread and ale at times of catastrophic food dearth was by no means an easy and risk-free task. That unprecedented demand for bread may have, in turn, reduced or temporarily eliminated the social stigma of regraters and encouraged them to engage in their activities. At the same time, it is possible that just as with bakers, regraters were being increasingly targeted by manorial authorities, who suspected them of speculative activities and wished to demonstrate to the community that opportunities of regrating and price-raising were being combatted against.

How can we explain the success of regraters in securing bread and ale, given the catastrophic dearth of both raw grain and processed foodstuffs? After all, to engage in speculative activities, it was necessary to possess some sort of capital, which would have been running low during the famine years. Moreover, as we have seen, regraters were primarily people of low socio-economic standing in non-famine years, a fact that makes their proliferation during the famine all the more surprising. One possibility is that these people were, in fact, related to bakers and brewers, which would have allowed them to secure foodstuffs in an easier manner. A prosopographic analysis of the court rolls reveals that in some instances this was indeed true. Thus, during the court of 29 September 1316 at Redgrave, four out of the total eight regraters shared the same surnames

[98] TNA, SC 2/215/74, membr. 18 and SC 2/216/1, membr. 20.

with bakers and brewers that were amerced in the same court.[99] Likewise, four out of the total ten regraters amerced at the court of 11 November 1316 at Bressingham can be proven to have been related to local bakers or brewers.[100] In other words, in some cases regraters were indeed related to local victuallers, but this familial association was by no means widespread and should, therefore, not be taken for granted. In order to determine through what channels those regraters who were not related to bakers and brewers managed to acquire bread, it is necessary to discern their socio-economic status in the court rolls and to see in what capacities they appeared in the same documents.

It appears that only two out of the ten regraters amerced at Bressingham on 11 November 1316 appear as such both before and after the famine years, on a regular basis. Contrary to the view that regraters are often associated with lower socio-economic elements, these two seem to have been better-off individuals, possessing both means and resources to acquire food during the troublesome years. One of them, John Joye, managed to acquire five and a half acres of land between May and November 1316 alone. On 14 April 1317, his brother William appears on the list of bread regraters, suggesting that in some cases, the regrating business indeed carried a family character.[101] The other notorious regrater, Guy Tebbe, added three and a half acres of land to his possessions between May 1313 and April 1317. By late 1316 he was wealthy enough to extend credit to his fellow-parishioner, Warin le Rede, and to be involved in various financial dealings with his neighbours, which included the unjust detaining of 2s from Alicia Fankys. In the course of his career, Guy Tebbe was regularly amerced for regrating.[102] It appears that the relationship between the Joyes and the Tebbes was far from harmonious — for instance, on 5 August 1316, William Joye was accused of unjustly detaining 1s from Guy Tebbe.[103] On 14 April 1317, William Joye reciprocated by suing Tebbe in a plea of debt.[104]

Other Bressingham regraters seem to have come from lower socio-economic backgrounds. None of them were amerced for regrating outside of the famine years, and two of them appear only in a single court (11 November 1316). One regrater, Margaret, daughter of Robert Rager, was most likely a spinster, judg-

[99] UCL, Bacon Roll 13.

[100] Catherine Trees, John Joye, Alicia Norman, and Alicia Bunting.

[101] NRO, WAR 7–9.

[102] NRO, WAR 9.

[103] NRO, WAR 7.

[104] NRO, WAR 9.

ing by her patronymic surname. It should be noted that this is the only proven example of a never-married woman appearing as a regrater in famine-year court rolls utilized in this sample. Although there was no evident correlation between marital status and socio-economic situation in pre-Black Death England, there is some evidence that never-married women were worse-off than widows.[105] Without the support of husbands and children able to provide labour and income, it is possible that never-married women fared worse than their married counterparts. There is no evidence that Margaret was either a brewster herself or that she was related to a baker or brewer. The same Margaret could well be an exceptional case of a better-off single woman who had some capital for regrating activities during the famine years.[106] Poor regraters are also mentioned in other manors' court rolls. Thus, at Newton (Cambridgeshire) on 11 November 1315, three women, described as 'paupers', were excused of paying fines for ale regrating.[107]

In Ireland, forestallers were sometimes identified with thieves and felons. For instance, on 1 December 1316 the justiciar's court at Waterford accused Jordan le Waleys and Thomas son of Nicholas of being 'common robbers, thieves and forestallers' and of stealing three sheep and one lamb from the sheepfold of one William le Tylor.[108] Similarly, at another assize, held at Kilkenny on 19 May 1317, a certain John Hardy was accused of robbing several people and of being 'a common forestaller and robber and thief in diverse parts of Leinster'.[109] It is unclear if 'forestaller' was a mere synonym of 'thief' and 'robber' or if the suspects in question were involved in forestalling, in addition to other offences. The marginal social position of these felons strengthens the impression that in some instances, food-related criminals came from the lowest social echelons. But more than anything, these cases demonstrate that forestalling was seen as a serious crime, punishable by law.[110] The identification of the felons with forestallers did not endear them to the members of jury: John Hardy was

105 Slavin, 'Peasant Livestock Husbandry', pp. 19–20.

106 Despite the fact that never-married women were, in some instances, worse-off than widows, there were some remarkable examples of never-married women who managed to accumulate much wealth during their lives. The life and career of Cecilia Penifader of Brigstock (Northamptonshire) (*c.* 1295–1344), who, by the end of her life, has acquired some seventy acres of land, is one such example. See Bennett, *A Medieval Life*.

107 CUL, EDC 7/11A.

108 NAI, KB 2/8, p. 34.

109 NAI, KB 2/8, p. 72.

110 Britnell, 'Forstal, Forsetalling and the Statute of Forestallers'.

found guilty and sentenced to be hanged. The omnipresent dearth and famine undoubtedly made the jurors unforgiving towards someone who could potentially be held responsible for the dire situation.

Royal Response to Assize Breaking and Food Regrating

As Buchanan Sharp has noted, as a response to the disastrous harvest, on 21 September 1316 the London authorities set the maximum prices of ale in London at 1.5d for the best (stronger) and 1d for the lesser (weaker), compared with the standard price of 1d and 0.75d, respectively, as set in 1305.[111] On 24 January 1317, Edward II instructed his sheriffs to proclaim and enforce this assize in all towns and ensure that a gallon of the better ale was being sold in towns for 1.5d and for 1d in the countryside. On 14 April 1317, it was declared that the names of assize-breaking brewers and lords of liberties were to be sent to the king for justice.[112] Finally, on 24 August 1317, towards the end of that year's harvest, London authorities forbade the malting of wheat, under pain of forfeiture, and a similar measure had been already enacted in Winchester sometime in 1316.[113] As we have seen, the royal prohibition was implemented as far as north-east Wales, as evidence from Dyffryn Clwyd (Denbighshire) demonstrates.[114] Brewing implies considerable calorific waste, and there is no doubt that the main aim of the royal proclamations was to prevent it.

Despite the widespread geographic proliferation of the wheat-brewing prohibition, its implementation could hardly be successful, for several reasons. First, the legislation was passed too late, when the famine had reached its peak and claimed a fair toll of its victims. Second, the proposed prices were unrealistic. By late January 1317, the average price of barley, the most common drinking grain, was 10s 10d a quarter, namely some 2.7 times its 1305 price (4s 7d). Yet, the proposed price of ale was only 1.5 times higher than in 1305. Also, the

[111] *Calendar of Letter-Books*, ed. by Sharpe, pp. 71–73. The bylaw specifies that the lesser ale cannot be sold for more than three farthings or one penny a gallon; the former price stands, most likely, for the surrounding countryside, while the latter refers to the city.

[112] Sharp, 'Royal Paternalism', p. 636.

[113] And not barley, as both Braid and Sharp have mistakenly asserted (Braid, 'Economic Behavior, Markets and Crises', p. 346; Sharp, 'Royal Paternalism', p. 636). *Calendar of Letter-Books*, ed. by Sharpe, p. 77. Sometime in 1316, after the Parliament of Lincoln, the burgesses of Winchester appealed to Edward II, asking him to outlaw brewers brewing from wheat malt (TNA, SC 8/3/106).

[114] Stevens, *Urban Assimilation*, pp. 143–44; TNA, SC 2/216/1, m. 20.

assize did not take into account the aggressive rise in monthly prices during the famine years: by July/August 1317, the price of 1 quarter of barley has reached its astonishing ceiling of 14s. In other words, the proposed scenario, with all its good intentions, would not have worked, for the pure reasons of the rules of market forces. Finally, the increased amount of ale regraters in 1317 and early 1318 court rolls confirms that the assize was implemented with only limited success. As we have seen, the unusually high number of ale amercements at Dyffryn Clwyd confirms that point.

It was in this context that the assize was proclaimed, and it is unlikely that its implementation was successful. The impression that the government had little or no power to enforce the assize and impose prices is confirmed by the fact that on 20 February 1316, Edward II and his council, under pressure from the Lincoln Parliament, had to withdraw their earlier ordinance on livestock and poultry prices, issued in January 1315 and proclaimed in March of the same year. Although this ordinance initially had nothing to do with the famine, it became increasingly burdensome and impractical after the failed harvest of 1315, which increased the demand for non-farinaceous products. Having withdrawn the ordinance, all the king could do was to admonish his subjects that they should 'buy and sell as cheaply as they could' and 'at reasonable prices'.[115] According to the contemporary perception, the 'just price' can be determined only by 'the fruitfulness of the harvest, not the will of men'.[116]

Crop Hoarding

If the potential of regraters to impact the market can be debated, there can be little doubt that hoarders managed to disrupt food trade in a most pronounced manner. The very essence of grain hoarding was the practice of withholding grain supplies in storage facilities and releasing them only once prices have reached their ceiling. In the context of the Great Famine, this was largely the practice of lords, who could afford the luxury of setting a part of their annual harvest aside and who had the appropriate storage facilities to do so.[117] The hoarding trends are reflected, first and foremost, in the seasonality of crop transactions. I have dealt with this issue in greater detail elsewhere, but it would be worthwhile to repeat some of my earlier conclusions and place them within a larger context of late medieval attitudes to hoarding and hoarders.

[115] Davis, *Market Morality*, pp. 224–25.

[116] Davis, *Market Morality*, pp. 224–25.

[117] On storage facilities of lords and peasants, see Chapter 5, pp. 186–90.

Table 4.8. Seasonal distribution of grain sales and purchases, September 1315–August 1317.

	Harvest	As percentage of annual transactions		Crop amounts (in quarters)	N transactions
		Autumn–Winter	Spring–Summer		
SALES BY LORDS	Wheat, 1315–1316	29.3%	70.7%	2,901	329
	Barley, 1315–13116	23.6%	76.4%	1,700	186
	Oats, 1315–1316	85.2%	14.8%	2,315	150
	Peas, 1315–1316	24.4%	75.6%	180	59
	Wheat 1316–1317	69.6%	30.4%	2,162	181
	Barley, 1316–1317	64.5%	35.5%	1,160	99
	Oats, 1316–1317	92.1%	7.9%	1,900	107
	Peas, 1316–1317	62.7%	37.3%	174	53
	Weighted Average, 1315–1316	46.1%	53.9%		
	Weighted Average, 1316–1317	61.7%	38.3%		
PURCHASES BY LORDS	Wheat, 1315–1316	88.5%	11.5%	290	66
	Barley, 1315–1316	59.7%	40.3%	367	80
	Oats, 1315–1316	66.2%	33.8%	604	86
	Peas, 1315–1316	30.5%	69.5%	76	20
	Wheat 1316–1317	58.6%	41.4%	132	40
	Barley, 1316–1317	76.4%	23.6%	299	55
	Oats, 1316–1317	77.1%	22.9%	401	53
	Peas, 1316–1317	67.1%	32.9%	321	9
	Weighted Average, 1315–1316	67.2%	32.8%		
	Weighted Average, 1316–1317	61.0%	39.0%		

Source: Manorial accounts database; Slavin, 'Market Failure during the Great Famine', p. 45.

Notes: The autumn–winter term = September, October, November, December, January, and February. The range of prices (per quarter) for 1315–1316 are: wheat = 7s–13s 4d, barley = 4s 10d–8s, oats = 2s 9d–8s, and peas = 4s 4d–8s. For 1316–1317 they are: wheat = 8s–17s, barley = 10s 8d–10s 10d, oats = 3s–6s, and peas = 7s–12s. The spring–summer term = March, April, May, June, July, and August. The range of price (per quarters) for 1315–1316 are: wheat = 15s–24s, barley = 10s–18s, oats = 6s 8d–14s 6d, and peas = 10s–16s. For 1316–1317 they are: wheat = 17s 4d–23s, barley = 12s–14s, oats = 6s 8d–10s, and peas = 13s–16s.

Table 4.8 arranges all the sales and purchases of grain by lords according to seasonal transactions. As the table indicates, the lords managed to make a bit of profit by selling late and buying early, at least during the first year of the famine. In other words, they were reluctant to release their crop resources shortly after the harvest, preferring, instead, to hoard them in their barns and granaries, waiting for prices to rise. Thus, during the autumn/winter of 1315, just over 1 quarter of the available wheat was sold, while the remainder of the hoarded wheat was not released until the late spring/early summer of 1316, when prices approached their annual ceiling and acute starvation was an omnipresent reality. The seasonality of barley sales shows similar trends. The figures for oats and peas were, on the other hand, somewhat different. While the vast majority of oats were sold by March, the share of peas sold in the second semester accounted for about 75 per cent of all sales. At the same time, however, the lords managed to purchase the vast majority of wheat, barley, and oats by the end of February. Peas, on the other hand, were chiefly purchased, in modest amounts, in the course of the second semester. As we have seen in the previous chapter, there were practically no instances when lords and their demesne managers chose to carry over their hoarded stocks to the next fiscal year, even despite the disastrous harvest of 1316, when the prices collapsed. Moreover, although crop producers may well have predicted the bad harvest as the result of the inclement weather, they could not have predicted *how bad* it would be. Therefore, withholding stored grains beyond the harvest would have been impractical and potentially unprofitable for the lords and their managers.

Altogether different were the trends during the second year of the famine. Here, both sales and purchases moved, more or less, at the same pace. Between two thirds and three quarters of all transactions took place before March 1317, while in the second semester there was very little left to be sold or purchased. This does not mean that lords became, all of a sudden, more sensitive to the needs of the starving populace; rather, it reflects the fact that lords did not possess full control over the timing of sales and purchases, standing helpless in front of Nature's inclemency. Indeed, there were instances when they were forced to sell their hoarded crops earlier than planned because of heavy rain and defective barns. Thus, a 1315–1316 account from Longbridge Deverill (Wiltshire) has a marginal gloss stating that all barley had to be sold before Martinmas (11 November), 'since it could not be kept longer [at the barn], on the account of debilitation and inundation'.[118] In other words, the officials of

[118] 'Quia longuis custodui non potuit propter debilitatem et hinundationem (*sic*)'. LH 10688.

Longbridge Deverill must have had a clear plan to withhold their lord's grain longer but were forced to release and sell it sooner and cheaper than planned, for fear of having the grain ruined by torrential rain. As we will see in the next chapter, the strong wind, rain, and storms wreaked much havoc upon demesne barns, some of which were undoubtedly leaky and generally in deplorable shape. It should be remembered that around the famine years, at least half of all demesne barns were made of timber and, hence, susceptible to strong winds and rain.[119] Although lords and their officials did their best to repair defective buildings to prevent rot and unwanted germination, this was not always achieved. There are numerous references to the sales of rotten crops. In some cases, spoiled crops were sold for reduced prices, while in other cases lords managed to sell them for the full price. Thus, according to a 1315–1316 account from Downton (Wiltshire), 6.75 quarters of barley were sold for just 2s 9 1/2d a quarter because they were rotten.[120] It is likely that two years of inclement weather, lasting from the late summer of 1314 to the harvest of 1316, proved to be too much for long-term storage. Although lords managed to withhold fairly large proportions of their crop resources in the course of the first famine year, the ceaseless flood did not permit them to repeat this policy in the course of the second year. This, in turn, may explain why about three quarters of all lords' crops were sold by March 1317.

In other words, lords generally managed to buy early and relatively cheaply and sell late and dearly. This undoubtedly reflects not only the deliberate hoarding policy of the lords but also their advantageous position within the trade system over other crop producers. Lords certainly held the advantage on account of their vastly greater resources and their ability to use their crop surpluses effectively in a time of scarcity. On the top of that, they also managed to augment their grain stocks by purchasing additional crops. Unfortunately, the accounts did not specify how they managed to achieve that. It was rare that the accounts specify what people were involved in grain transactions, and they did only when these were people of high socio-political standing, worthy of mentioning. But those lords represented the minority of crop producers, and there can be little doubt that a fair share of crops purchased by lords came from the tenancy sector. The disadvantageous position of dependant peasants (whether tenants or serfs), in the context of the famine, is reflected in a contemporary Middle English poem, *Song of the Husbandman*, composed, most likely, after the

[119] Brady, 'The Sacred Barn', Appendix 1; Slavin, *Bread and Ale*, p. 121.

[120] HantsALS, 11M59/B1/71.

famine (perhaps *c.* 1340) but possibly reflecting the Great Famine. Lamenting his hardships during grain shortage, the Husbandman states, *inter alia*, that in order for him to make his ends meet and pay his tax to the king, he has either 'to deposit a security-pledge, or sell [his] grain that is still green on grass' (Ich mot legge my wed wel yef I wolle | Other sulle mi corn on gras that is grene).[121] 'Green on grass' presumably means immature grain, still unharvested. One way to interpret this passage is that in some cases, lords compelled their tenants to sell their produce well in advance of the harvest, regardless of its success or failure, possibly because it may have given them rent money. The pressure to sell grain before harvest may have come from grain merchants, who were speculating. On the other hand, it is possible that in some cases, rent might have been paid in kind from unharvested grain. Both interpretations may explain why lords managed to purchase their grain shortly after the harvest. Later in the same poem, the Husbandman complains that in addition to the unharvested grain, he also had to sell his seed corn in order to seek money to pay the king's tax.[122] The taxation in question may refer to the lay subsidies during the famine years: to finance his political and military ambitions, Edward II was granted the twentieth (country taxation) and fifteenth (urban taxation) in 1315 and sixteenth (country taxation) and fifteenth (urban taxation) again in 1316.[123] In other words, the English population was taxed no less than twice during the famine years. The impact of taxation on famine will be dealt with in the detail it deserves in Chapter 6.

The poverty and misery of starving peasants was described in a sharp contrast with the greed and wastefulness of lords. As we have seen in Chapter 3, according to *Wynnere and Wastoure* (*c.* 1350s), Winner delights in bad harvests because of the high prices and easily made profits, while good years and low prices drive him into despair and suicidal thoughts.[124] In his *Le livre de bonnes mœurs* (*c.* 1400–1410, translated into English later in that century), Augustinian preacher Jacques Legrand scorned traders, who always desire dearth and famine, because they can make wicked profits at the expense of the

[121] *Song of the Husbandman*, in *Medieval English Political Writings*, ed. by Dean, lines 45–46. A fully searchable online version is available courtesy of the University of Rochester-based *TEAMS Middle English Texts Series*.

[122] *Song of the Husbandman*, in *Medieval English Political Writings*, ed. by Dean, line 63.

[123] Jenks, 'Lay Subsidies'. p. 36.

[124] *Wynnere and Wastoure*, ed. by Ginsberg, lines 368–74; see p. 82 n. 6.

hunger and suffering of poor people.[125] In *Handlyng Synne*, a devotional treatise by Robert Manning (composed between 1303 and 1338), rich hoarders waste their stocks of grain, by feeding them to mice, in the course of a year or two of storage and thus deprive the poor of their food. The inevitable fate of those hoarders is corporal torments in Hell.[126] According to one late thirteenth-century poem, grain hoarders shall be driven away from Heaven by angels, just as they drive away the suffering poor from their stocks.[127] A rich hoarder's eventual departure to Hell is a recurrent motif in numerous late medieval sermons and poems.[128]

Royal Response to Hoarding

Apart from moral literature, grain hoarding practices are well illustrated in a letter of Edward II to his bishops (23–24 April 1316).[129] According to the missive, even despite the bad weather and poor harvest, there should have been enough grain to feed the poor population. Yet, poor people, reduced into beggary, could not find food anywhere because:[130]

> many people in the kingdom withholding (*obtinentes*) grain in their granaries in immodest quantities, keep it back (*retinent*) and hitherto refuse to release it for sale, so that they could sell them more dearly later (*et venditioni adhuc exponere recusant, ut postmodum carius illa vendant*).

Later in that letter, Edward asks his bishops to admonish hoarders, both lay and religious, living in the towns of their diocese, to store no more grain than

[125] Davis, *Moral Economy*, p. 118. The original French text has been printed in Legrand, *Archiloge sophie*, p. 49; the earliest English version is BL, MS Harley 149, fols 238^{v}–239^{v}.

[126] *Robert of Brunne's 'Handlyng Synne, AD 1303'*, ed. by Furnivall, p. 176 (lines 5377–88) (of Covetousness, the Fifth Deadly Sin). An online edition is available courtesy of the University of Michigan, *Corpus of Middle English Verse*.

[127] *The Minor Poems of the Vernon Ms*, ed. by Horstmann, pp. 172–74.

[128] Davis, *Moral Economy*, pp. 118–19.

[129] The letter has been studied in Sharp, 'Royal Paternalism', pp. 634–35 and *Famine and Scarcity*, pp. 41–42. Sharp stated that the letter survives in two copies, the one in the register of Richard de Kellawe, bishop of Durham, and the other one in that of Roger Martival, bishop of Salisbury (both dated 24 April 1316). In fact, at least one more copy survives, in the register of Walter Reynolds, archbishop of Canterbury (dated 23 April 1316): LPL, Register Reynolds, fol. 73^{v}, printed in *Concilia Magnae Britanniae et Hibernia* ed. by Wilkins, p. 457.

[130] LPL, Register Reynolds, fol. 73^{v}.

is necessary to feed themselves and their households until the next harvest, and to release the rest for sale. In their refusal to do so, hoarders are responsible for the suffering and death of the populace.[131] Although there can be little doubt that this letter exposes Edward's moral concern for the well-being of his subjects during the crisis years, it is unlikely that these moral injunctions were ever taken seriously by hoarders, despite clear admonitions of the bishops. Thus, Roger Martival, bishop of Salisbury, is known to have approached four archdeacons and the dean of Salisbury and asked them to carry out the king's wishes, for the sake of the public well-being (*publicam utilitatem*).[132] Yet, the account rolls from Longbridge Deverill (owned by Abbot of Glastonbury) and Downton (belonging to the Bishop of Winchester), both in Wiltshire and both under Martival's jurisdiction, indicate hoarding activities.[133]

Despite the general reluctance of the lords to release their hoarded stocks for markets, there is also limited evidence that some of them were more considerate of the needs of the starving population. On 11 November 1315, Edward II issued a letter of protection for servants of Master William de Briston, who intended to ship grain from his parsonage barns at Patrington (Yorkshire) to Beverley and York, presumably for sale there.[134] The same person secured a similar letter of protection on 1 January 1317, at the height of the famine, to release more grain from his parsonage barns for sale at York and Kingston-upon-Hull.[135] Master William de Briston was Edward II's clerk, and his willingness to release his crop stocks for sales might reveal the king's efforts to promote his anti-hoarding encouragement among his close associates. The case of William de Briston, however, seems to have been an exception rather than a rule, and it appears that the vast majority of lords would release their manorial stocks to supply their headquarters, rather than local markets, with food. I was able to identify eleven letters of protection, issued by Edward II, between the Michaelmases of 1315 and 1317, all dealing with the transfer of demesne issue or tithes from manorial barns to their households or conven-

[131] LPL, Register Reynolds, fol. 73ᵛ.

[132] Wiltshire Record Office (henceforth, WiltRO), Register Martival, fol. 126ʳ. The English summary is given in *The Registers of Roger Martival*, vol. I: *The Register of Presentations and Institutions*, ed. by Edwards, pp. 13–14 (no. 40). Martival's response is noted in Sharp, 'Royal Paternalism', p. 635.

[133] LH, 10,688; HantsALS, 11M59/B1/71, membr. 8ʳ.

[134] *Calendar of Patent Rolls Preserved in the Public Record Office: Edward II, AD 1313–1317* (hereafter, *CPR, 1313–17*), p. 365.

[135] *CPR, 1313–17*, p. 577.

tual institutions. Thus, on 18 February 1316, Master Roger de Heselarton was granted a safe conduct to carry his tithe grain from Thorpe-by-Newark Church (Nottinghamshire) to his household in York.[136] Similarly, on 17 October 1316, the Abbot of St Mary's Abbey, York, was given a similar letter to carry the crop issue from his Lincolnshire and Yorkshire manors to his headquarters.[137]

Hoarding practices by better-off elements was nothing unique to the Great Famine. There are abundant examples of similar behaviour during other historical famines and food shortages. For instance, the potato blight in Ireland gave rise to hoarding and speculation.[138] During the Bengali famine of 1942–1943 and the Bangladeshi famine of 1974–1975, large amounts of rice were stored by both wealthy individuals and colonial authorities.[139] In each case, hoarding and arbitrary speculation contributed a great deal to widespread starvation and subsequent excess mortality. During the Sudanese famines of 1984–1985, 1998, 2005–2006, and 2011–2012, local banks appear to have been culpable of hoarding substantial amounts of grain and driving up prices.[140] Similarly, ubiquitous hoarding activities have been reported during the 2002 famine in Malawi.[141] As recently as in 2011–2012, food speculation appeared to be a paramount factor in the East African famine.[142]

Foreign Grain Imports and Illegal Domestic Exports

One other governmental mean of crisis relief, studied recently in a detailed manner by Buchanan Sharp, was ban on grain exports and encouragement of grain imports.[143] Here, Sharp has identified two royal export prohibitions during the famine years. The first was dated 1 September 1315 and the second was issued on 14 February 1317. Although both letters mention a dearth of foodstuffs in England, the reason behind issuing both writs was the ongoing warfare with Scotland. As the first letter states, no grain shall leave England, lest it ends

136 *CPR, 1313–17*, p. 395.

137 *CPR, 1313–17*, p. 555.

138 Ó Gráda, *Ireland Before and After the Famine*, pp. 101–02 and 120.

139 Sen, *Poverty and Famines*, pp. 52–85; Ravallion, *Markets and Famines*, p. 19; Ó Gráda, *Famine: A Short History*, pp. 168–71 and 175–78.

140 Africa Watch, 'Sudan, Nationwide Famine'.

141 Devereux, 'Why does Famine Persist in Africa?', p. 29, and 'The Malawi Famine of 2002'.

142 SWI, 'Food Speculation Blamed for East Africa Famine'.

143 Sharp, 'Royal Paternalism', pp. 638–44; Sharp, *Famine and Scarcity*, pp. 45–54.

up feeding the Scottish rebels.[144] The second letter, on the other hand, prohibits food exports because the king needs a 'great store of victuals' to sustain his subjects on the Scottish frontier and the Lordship of Ireland.[145] As we shall see later, the Anglo-Scottish war carried a strongly ecological character, characterized by the destruction of food resources. During the famine years, there were numerous complains about illegal exports of grain to Scotland, which will be discussed at a deserved length in the following chapter.[146] Conversely, Edward II, on several occasions, encouraged grain imports. By the end of 1315, he asked Louis X of France and John (Yann) III Duke of Brittany to allow Newcastle merchants to buy local grain and ship it safely to England. He also appealed with the same request to the seneschal of the county of Ponthieu, then part of Edward's foreign domains. On 16 March 1316, he offered protection and safe conduct to foreign grain merchants, apart from Scottish and Flemish, to import grain into his kingdom.[147] Furthermore, between the Michaelmases of 1315 and 1317 he issued at least twenty-seven safe conducts for English grain merchants, primarily from Berwick-upon-Tweed and Newcastle (but also, in one case, Melcombe Regis), departing for France, Gascony, Genoa, Sicily, and Castile, and — somewhat surprisingly — Ireland, to purchase and import grain.[148] Judging by the importance of Berwick-upon-Tweed and Newcastle in the context of Anglo-Scottish warfare, it is apparent that Edward's pleas to the king of France, the duke of Brittany, and the seneschal of Ponthieu, as well as his safe conducts, were aimed at supplying his northern garrisons and northern frontier towns rather than empty market stalls elsewhere. Indeed, most foreign grain imported during the famine years ended up in northern strongholds. Moreover, securing grain from the north French or Breton markets was by no means an easy task — there is omnipresent evidence of crop failure all over northern France, including Brittany, during the famine years, mirroring the situation in England.[149] Despite the proliferation of vineyards and dependence on wine as the principal alcoholic beverage, grains still played an enormous role

[144] *CCR, 1313–18*, pp. 308–09.

[145] *CCR, 1313–18*, p. 455.

[146] Sharp has himself briefly noted on those inquests: see, Sharp, 'Royal Paternalism', p. 638 n. 69.

[147] Sharp, 'Royal Paternalism', pp. 639–40.

[148] And not thirteen, as Sharp asserted (Sharp, 'Royal Paternalism', p. 640).

[149] As it is indicated in a chronicle from the Holy Cross Abbey at Quimperlé. See, *Histoire de l'abbey de Sainte Croix de Quimperlé*, ed. by le Duc and le Men, p. 292. See also, Jordan, *The Great Famine*, pp. 33–34.

as the basic staple foodstuff of an average rustic in any part of northern France. Even in the pastoral uplands of Brittany, rye was grown ubiquitously.[150] But even if *some* grain could be purchased on the Continent, it might never have reached one of the English ports because of both sea-storms and shipwrecks, and the ongoing piracy in the North and Irish Seas, discussed in a greater detail in the next chapter.

All the same, some foreign grain imports were successful in reaching their final destinations. But was their volume sufficient to relieve, at least partially, the hardships of the starving population? An analysis of port custom accounts may provide a partial answer. It should be noted outright that few port custom accounts survive for the pre-Black Death period, and this paucity is especially apparent in the famine years. In fact, only the ports of Exeter, Scarborough, and Whitby (the latter two are both in Yorkshire and were accounted for jointly) are endowed with famine-year accounts.[151] As these accounts indicate, the quantities of foreign crops reaching England during the famine years were alarmingly low. No grain at all reached the two Yorkshire ports; similarly, no crops arrived at Exeter between Michaelmas 1315 and 16 January 1317, while between the latter and 20 August 1317 only 343 quarters of crops (199 quarters of wheat, 10 quarters of rye, 4.5 quarters of legumes, 126.5 quarters of barley, and 3 quarters of oats) reached that port. Calorifically speaking, if we allow the lowest wastage possible (namely if most grains were baked into bread and only a little barley was brewed), these amounts may have rendered no more than 135 million kcal, enough to feed some 447 people between 16 January and 20 August 1317, assuming a daily 1400 kcal per capita deriving from crops. In other words, these meagre amounts arrived at the very height of the famine and were capable, at best, of feeding either some 10 per cent of Exeter's population or a couple of parishes in its immediate hinterland.[152]

It is unclear if the situation was different in other ports, for which famine-year accounts do not survive. On the one hand, those few accounts that do survive for non-famine years reveal that the grain quantities imported were always small when compared with those in Exeter in 1317. Thus, in the

[150] Comet, 'Technology and Agricultural Expansion', p. 17.

[151] The Exeter port accounts have been edited and printed in Kowaleski, ed., *Local Customs Accounts*, pp. 130–52. The Scarborough and Whitby account, running from 8 June 1315 to 12 April 1317, is TNA, 122/34/2.

[152] The population of Exeter is estimated to have been about 3700 *c.* 1290 (Campbell, 'Benchmarking Medieval Economic Development', p. 909), but it may have well grown to over 4000 on the eve of the famine.

course of 1322–1323, some 538 quarters of crops were anchored at King's Lynn, all between 26 April and 23 May, figures comparable with the situation in 1308–1309.[153] At Bristol, no grain arrived in 1302–1303, 1303–1304, 1304–1305, 1305–1306, and 1308–1309, while in 1306–1307 and 1307–1308 this port received 160 and 63 quarters of unspecified grain from Gascony and Burgundy, respectively.[154] In the course of the accounting year 1308–1309, 470 quarters of crops (60 quarters of wheat, 290 quarters of legumes, and 120 quarters of barley) were imported into Southampton.[155] Neither Boston in 1308–1309 nor Ipswich in 1322–1323 received any foreign grain.[156] On the other hand, at Hull, whose eight port accounts survive for the period between 1304–1305 and 1324–1325, the figures fluctuated a great deal from year to year, from zero to slightly over 10,000 quarters, amounting on average to 2514 quarters a year.[157] Although Hull may have been an outlier, compared to other ports in non-famine years, it is highly unlikely that the amounts of foreign grains reaching it in the famine years matched the non-famine figures.[158] This assumption is based on the fact that the quantities of famine crops imported into Exeter during the famine were considerably lower than those anchored there in normal years; as we have seen, the respective figures for 1315–1316 and 1316–1317 were zero and 343 quarters, respectively, compared with 1165 quarters a year for 1295–1325.[159]

But even if we assume, in a clearly speculative manner, that contrary to the evidence of 343 quarters imported to Exeter in 1316–1317 and zero quarters imported to Scarborough and Whitby, an average major port received, on average, 500 quarters of foreign grain during that year, these amounts would still be far from sufficient to take up the slack. Collectively, if eighteen ports received about 9000 quarters of grain, these amounts would have been good enough to

[153] TNA, E 122/93/17. See also Hybel, 'The Grain Trade', p. 231 (who provided only the financial value of the imported goods to King's Lynn).

[154] TNA, E 122/15/1.

[155] TNA, E 122/36/8.

[156] TNA, E 122/6/3; E 122/50/3.

[157] Hybel, 'The Grain Trade', p. 231.

[158] At Exeter, the quantity of imported crops for 1295–1325 averaged 1165 quarters (986 quarters for the period 1295–1349) (calculated from Kowaleski, 'The Grain Trade', pp. 12–14). Although the Exeter figures seem to have been lower than the Hull ones, the many gaps in the surviving Hull accounts may disguise the real average figures for that port.

[159] Kowaleski, 'The Grain Trade', pp. 12–14.

feed no more than some 7000 people a year, at best.[160] Furthermore, as we shall see later, the grains reaching Newcastle were destined almost exclusively either for royal garrisons or besieged borderland towns rather than the rural population. Bearing in mind all those caveats, we may confidently state that the total number of people fed by foreign grain during the famine years must have been no more than a few thousand a year — and probably far less. Naturally, such supply was supplementary rather than alternative, and even if it was aimed at compensating for the shortfall of a few months during the 'hungry gap' period preceding the harvest, it would not have changed the situation in any way. In other words, the royal encouragements of grain imports did not materialize, partially because of the adverse conditions of the high seas, partially because of the ongoing warfare, and partially because of the dearth of crops in other parts of northern Europe.

To make things even worse, some English grain was leaving the country, mostly illegally and in conjunction with the ongoing warfare with Scotland and the political tensions with the County of Flanders. In many instances, corn-mongers were granted their safe conducts to buy grains on the Continent; these safe conducts were given 'on the understanding that these grains will not end up in Scotland or Flanders'.[161] Edward II had many reasons to worry about the danger of English grain leaving the country. For instance, in November and December 1315, a handful of Yorkshire merchants were accused of shipping local grain to the Scottish enemy.[162] It is possible that some of those merchants were, in fact, the same corn-mongers of Scarborough and Filey, imprisoned by 6 November 1316 for shipping their grain to Scotland.[163] Similarly, on 3 October 1316, Edward II instructed his officials to investigate and imprison various merchants of Gloucestershire, Somerset, Dorset, Devon, and Cornwall who were known to have exported grain to Scotland.[164] It appears that this investigation process took longer than the king expected, as he had to reissue the same

[160] Boston, Bristol, Chester, Chichester, Exeter, Great Yarmouth, Hartlepool, Hull, Ipswich, London, King's Lynn, Melcombe Regis, Newcastle, Southampton, Sandwich, Scarborough, Whitby, and Winchelsea.

[161] For instance, *CCR, 1313–18*, p. 322 (9 February 1316); *CPR, 1313–17*, pp. 218–19 (7 February 1315); *CPR, 1313–17*, pp. 386–87 (6 February 1316). Many more examples can be cited here.

[162] *CPR, 1313–17*, pp. 419–20, 423–24, and 429.

[163] *CPR, 1313–17*, p. 598.

[164] *CPR, 1313–17*, p. 593.

writ on two further occasions (4 January and 16 April 1317).[165] During the revolt of Llywelyn Bren (28 January–18 March 1316), some Welsh grain merchants took advantage of the situation to sell crops to the rebel army. One such person was John Odyn, burgess of Cardiff, who was imprisoned at the Cardiff gaol for supplying Bren with victuals.[166]

Domestic Grain Trade

The very limited scale of foreign grain imports, coupled with the segmentation of markets, instances of hoarding, and some illegal crop exports, imply, at least in theory, that the domestic grain trade was badly affected and hardly functioned. Still, there is evidence that attempts were made to secure some grain through the means of interregional trade, whereby local corn-mongers departed to other counties, adjacent or distant, to purchase and bring back grain for trade. In addition to the twenty-seven safe conducts related to foreign imports, Edward II issued at least thirty-four additional safe conducts, letters of protection, licences, and writs of aid in reference to the domestic grain trade. The single geographic cluster consisted of northern corn-mongers from the towns of Berwick-upon-Tweed, Newcastle, Lancaster, York, Beverley, and Ravenser Odd travelling south of the Humber in attempts to find some grain for their markets. Again, this stresses the degree of food crisis in the war-torn North and the dire need to secure victuals for local urban populations, as well as king's garrisons stationed there. Despite the predominance of northern corn-mongers, there is also a fair number of grain traders from south of the Humber. In May 1316, Edward II issued two back-to-back letters of protection for three London merchants to travel to Cornwall and the Channel Islands to buy grain there for the capital.[167] Similarly, on 24 February 1317, William Soreweles, a citizen of London, received a similar letter authorizing him to travel to Norfolk to buy grain there.[168]

As stated above, in many instances the original aim was to provision households rather than local markets. Thus, sometime in January 1316, the abbot of Kirkstead Abbey (Yorkshire) sent his servants to procure grains in

[165] *CPR, 1313–17*, pp. 679 and 693–94.

[166] *CPR, 1313–17*, p. 545.

[167] *CPR, 1313–17*, pp. 458–59.

[168] *CPR, 1313–17*, p. 622.

Cambridgeshire and Huntingdonshire.[169] Around the same time, Malton Priory (North Yorkshire), prone to Scottish attacks, sent its officials to Lincolnshire to purchase crops there.[170] On 20 February 1316, William de Gretton, a Lincoln corn-monger, received a letter of protection to travel to Cambridgeshire and Huntingdonshire to purchase crops there for his household, on account of scarcity in his town.[171] Close to that date, two servants of Mitford Castle (Northumberland) were sent, at their own risk, to the Scottish Marches to buy grain for the household of Aymer de Valence, 2nd Earl of Pembroke.[172] Although the letters of protection and safe conducts indicate that those purchases were made for the provisioning of conventual and noble households, it appears that, in some instances at least, their aim extended beyond that. Unfortunately, with the exception of one document, the licences do not specify the amounts of grain to be purchased. The exception in question is a royal licence, issued on 20 February 1316 to Walter de Cakhou, burgess of Ravenser Odd (Yorkshire), to buy 40 quarters of wheat and 60 quarters of barley to provision his household.[173] Even assuming very generous food allowances for the households of 'urban oligarchs' — twice as much grain per capita as commoners — it is highly unlikely that Walter de Cakhou's household would require 100 quarters for the period of about half a year, until the harvest of 1317. This view is strengthened by our knowledge of grain consumption by some aristocratic households in pre-Black Death England. For instance, in the course of 1336–1337, the household of Dame Katherine de Norwich, a wealthy widow of a powerful Norfolk magnate and landlord, consumed, on average, 1 quarter of wheat and 1.5 quarters of malted barley a week.[174] Unfortunately nothing is known about Walter de Cakhou's career, status, or household. It is certain that he was not a nobleman; he seems to have been a better-off burgess, possibly engaged in grain trade. If that is true, then it is likely that his household consisted of a few members of his immediate family and possibly several servants and assistants. Even if we assume that the size of Walter de Cakhou's household matched that of Katherine de Norwich, which is unlikely, it could not have demanded more than 20 quarters of wheat and 30 quarters of barley in the remaining twenty

169 *CPR, 1313–17*, p. 382 (the safe conduct was issued on 2 February 1316).

170 *CPR, 1317–18*, p. 394.

171 *CPR, 1313–17*, p. 397.

172 *CPR, 1313–17*, p. 396.

173 *CPR, 1313–17*, pp. 400–01.

174 Calculated from *Household Accounts from Medieval England*, ed. Woolgar, I, 179–87.

weeks before the end of the 1316 harvest. In other words, in either scenario, it is more than plausible that Walter de Cakhou procured grain not only to provision his household but also to trade it in his town or elsewhere in Yorkshire.

This brings us back to the issue of corn-mongers and their place during the famine years. Although, according to Smithian principles, they would be instrumental in relieving famine by replenishing market stalls with grain in times of dearth, relocating crops from one hunger zone to another can hardly be seen as famine relief. In effect, such behaviour may have contributed to a further intensification of the famine, when a certain quantity of grain resources would have been 'subtracted' from one needy region, only to be resold for excessively high (and in many cases, speculative) prices in another needy region. For instance, purchasing 100 quarters of wheat for, say, 14s a quarter from the Norfolk demesnes next to Norwich only to cart this amount to York and sell it there for, say, 16s a quarter, would relieve the crisis in neither Norfolk nor Yorkshire. The citizens of Norwich would have been deprived of 100 quarters of wheat, while only some better-off citizens of York would have been in position to purchase the grain. In other words, those corn-mongers involved in trans-regional grain trade during the famine years managed to abuse their financial advantage over most commoners to make an obvious profit from the chaotic situation.

Salt Trade

Although grains represented the single most important article of consumption by late medieval English communities, the crisis of the famine years cannot be fully appreciated without considering yet another important food commodity: salt. Salt was a vital ingredient in the late medieval north European diet; even though it was rarely used as a flavour condiment, it was most widely utilized as a long-term food preservative, either for dry-salting or brine-curing of meat or fish (mostly herring). Approximately, 1 lb of salt would be good enough to preserve 15 lb of meat or 9.3 herring.[175] Furthermore, salt was a necessary ingredient in dairy production, without which neither cheese nor butter could be produced. Approximately 1 lb of salt would be used in making about 40 lb of cheese or 10 lb of butter.[176] In other words, the importance of salt in late medieval food production cannot be overstated.

[175] Bridbury, *England and the Salt Trade*, p. 3.

[176] White, 'Account of the Making of Cheshire Cheese', p. 289; Cruess, *Home and Farm Food Preservation*, p. 255.

The climatic anomaly of 1314–1316 had a profound impact on domestic salt production. As we have seen in Chapter 2, the torrential rain either hindered or made the evaporation of brines impossible, rendering some domestic salterns useless. Although there are no extant salt produce accounts from the famine years, the abnormally high prices of salt, which rose even faster than those of crops, hint that next to no salt was produced in England during the crisis years (Figure 4.6). The prices started rising already in the course of 1314–1315, concurrently with the rain, averaging 4s 11d per quarter (compared with the average of 3s 2d in 1310–1314). But in the course of the subsequent year, the figures reached to an unprecedented average of 13s 5d per quarter; at no other point in English history did relative salt prices rise so dramatically. In 1316–1317, the prices were excessively high too, with 1 quarter selling for 10s 10d. Although dropping to 4s 6d in 1317–1318, they remained above average until 1326, perhaps reflecting solar activity and its impact on saline production in England. Just as with crops, it is important to distinguish between official market (or 'demesne-gate') and black market prices. As some contemporary chronicles indicate, in 1315 1 quarter of salt could be selling for as much as 35s, while in the course of 1316, it stood at 30s until the Feast of St John (24 June), rising to 40s by the Assumption of Blessed Virgin Mary (15 August).[177] Although these figures could well be exaggerated, they may still be taken as a reflection of a dire dearth of salt, driving up prices above those paid at official market prices. It is likely that the unprecedented rise in salt prices, more aggressive than that in grain prices, can be explained by both the dysfunction of domestic salterns and the excessive demand for south European salt all over northern Europe, on account of omnipresent dearth.

The situation in the Lordship of Ireland was at least as dire as in England, as indicated in some entries in the justiciar of Ireland rolls. Here, 1 *crannock* (roughly, an equivalent of 1 quarter) of salt was selling for 10s in 1315–1316 and 1316–1317.[178] The crisis was undoubtedly exacerbated by the Bruce Campaign of 1315–1318, when much salt was demanded and consumed by the fighting armies. Just as in England, the omnipresent shortage of salt created a black market in Ireland. Thus, according to one entry, in February 1316, a certain Michael de Trym of Drogheda was accused of selling 1 *crannock* of salt to the royal army for 40s, despite Edward II's prohibition to sell salt for more than 10s a *crannock*.[179] This passage is of a great interest because it indicates a

[177] *Johannis de Trokelowe Annales*, ed. *by Riley*, p. 92.

[178] NAI, KB 2/7, p. 58; KB 2/8, pp. 89–90.

[179] NAI, KB 2/7, p. 58.

royal attempt to set a price ceiling for salt in Ireland; something which was not done, as far as the surviving evidence can tell, in England. It appears that just as Edward II's attempts to fix the livestock and poultry prices in January 1315 were doomed to failure, so was his Irish salt-price fixing. The economic reality was too dire to allow local traders to respect his prohibition to sell salt for no more than 10s a *crannock*. The omnipresent dearth and disastrous demand for salt is reflected in several entries in the justiciar of Ireland's rolls reporting salt thefts and attacks on salt merchants.[180] Food-related crimes in Ireland will be explored in a detail in Chapter 8.

Therefore, it was no longer possible to rely on domestic salt production, and there was an increasing need to secure foreign salt to take up the slack. Unfortunately, the inclement weather of 1314–1316 had a negative impact on salt production in other parts of northern Europe, Brittany first and foremost, but also Normandy and Lüneburg (Lower Saxony). The Breton salt trade was especially prominent in Cornwall and western Devon, owing to the linguistic and cultural proximity between Breton producers and Cornish consumers.[181] As one English chronicler lamented, in an exaggerated manner, 'no salt was produced in France' during the famine years.[182] The vast majority of foreign salt, however, seems to have been imported from Bourgneuf Bay, a coast situated on the frontiers of southern Brittany and Poitou and spanning twenty-two miles.[183] The Bourgneuf salterns were located outside the north European 'famine zone', and there is no evidence of any short-term climatic anomaly similar to that in north-western Europe, affecting local salt production. All the same, a recent palaeoclimatalogical study has revealed that around 1300, the Bay of Biscay waters started exhibiting similar oscillations to those found in northern Atlantic waters, although there was no significant decrease in sea surface salinity until *c.* 1400.[184] In any event, foreign salt did reach England during the famine years, but in more modest quantities when compared to non-famine years. The single most complete and reliable source of information in this matter is the custom accounts of Exeter port. As Table 4.9 demonstrates, the famine years represented the lowest point in the quantities of imported salt reaching the port in the course of the second decade of the fourteenth century.

[180] For instance, NAI, KB 2/8, pp. 89–90.

[181] Bridbury, *Salt Trade*, p. 121.

[182] 'The Chronicle of John Somer, OFM', ed. by Mooney and Catto, p. 272.

[183] Bridbury, *Salt Trade*, pp. 56–75.

[184] Yannick and others, 'High Frequency Environmental Changes'.

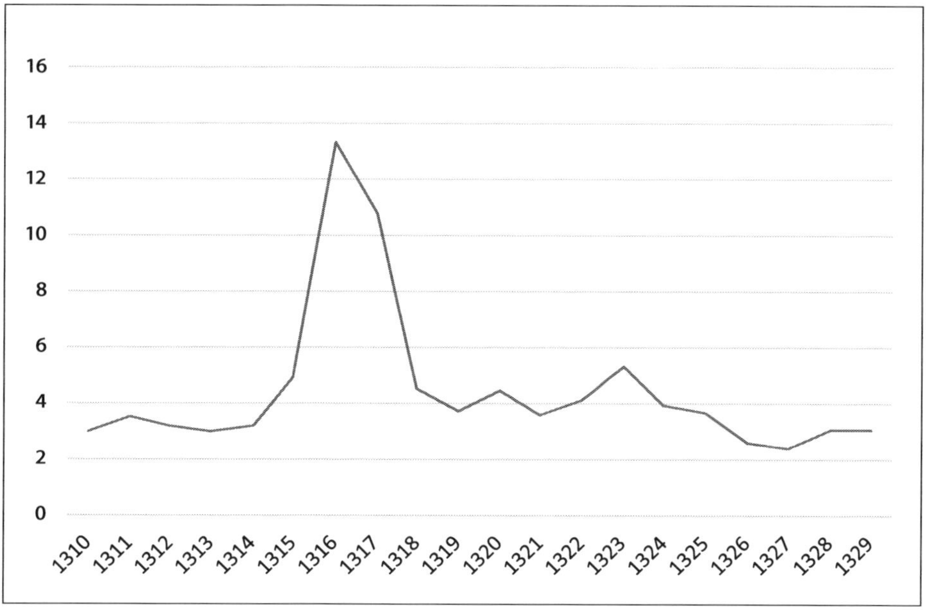

Figure 4.6. Salt prices in England, 1310–1329 (in shillings per quarter). *Source*: Munro, 'Revisions of Phelps Brown'; Farmer, 'Prices and Wages', in *The Agrarian History*, II, 715–817.

Note: 1 quarter of salt = 28 lb.

In 1315–1316, only 105 quarters of salt entered the port, while in the course of the following year the figures rose to just 483 quarters. This was a meagre sum, good enough to preserve some 44,000 lb of meat or some 27,200 herring, or to make either some 117,600 lb of cheese or about 29,400 lb of butter. It should be noted that in the pre-Black Death period, the total amounts of salt imports have never been impressive — thus in the 1320s, they do not seem to have exceeded several thousand quarters, at most.[185] This was in contrast with the post-Black Death era, when in the last quarter of the fourteenth century the figure rose to about 16,000 quarters a year.[186] To make things even worse, just as with grain, some salt may have been lost to storms, robberies, or confiscations. For instance, sometime between 8 July 1315 and 7 July 1316, John de Clyve, a Suffolk merchant, had his ship containing 800 quarters of 'gross salt' (namely, 'black' salt, evaporated in the sun rather than boiled), worth 800 marks (=

[185] Extrapolated from Bridbury, *Salt Trade*, p. 169.

[186] Bridbury, *Salt Trade*, p. 171.

Table 4.9. Quantities of foreign salt reaching Exeter port, 1312–1321.

Year	In quarters	In pounds
1312–1313	705	19,740
1315–1316	105	2,940
1316–1317	483	13,510
1317–1318	874	24,472
1318–1319	754	21,112
1319–1320	642	17,976
1320–1321	1,985	55,580

Source: Kowaleski, ed., *Local Customs Accounts*, pp. 130–52.

Note: Although Exeter port custom accounts begin in 1268, it was not until the 1312–1313 account that exact quantities of salt (mostly in quarters, but also in pipes or tons) are specified. 1 quarter of salt = 28 lb; one pipe = 2 quarters; and one ton = 5 quarters.

£536), confiscated for the 'king's use'.[187] This was truly a remarkable amount, and we can only speculate that it was seized by the king's officials either for provisioning his army in the northern counties and Scotland or for further sales at profit.

But what about that salt that did manage to reach the shores of England? An incomplete account for the port of Scarborough reveals a few interesting details. Between 10 July and 16 August 1315, thirty-four ships laden with salt, collectively worth about £57 3s (which, using the 1314–1315 prices, may be translated into approximately 231.3 quarters), entered Scarborough port. The largest share (fourteen ships, carrying salt worth £23 19s) was imported by Dieppe merchants. Three Dieppe merchants shared the same surname (Gardyn), suggesting that salt trade could be a family enterprise. Other merchants originated in Westkapelle (Zeeland), Fécamp (Normandy), Dunkirk (Flanders, on the Pais-de-Calais border), Bremerhaven (a Hanse town in Saxony), Brielle (south Holland) and an unspecified place in Lombardy. The ethnic composition of salt merchants reflects the wider political reality of north-western Europe in general and the foreign policy of England since 1303 in particular. In that year, Edward II's father, Edward I, signed the Treaty of Paris, ending the Anglo-French War over Gascony. According to this document, the then-prince Edward was to marry Philip IV's daughter, Isabella of France. In the same year,

[187] Bridbury, *Salt Trade*, p. 52; TNA, SC 8/98/4865.

Edward I issued the so-called *Carta Mercatoria* (The Market Charter), which granted extensive privileges to alien merchants.[188] The charter was temporarily called off in 1311 and the goods of most alien merchants were seized, under baronial pressure. Moreover, in 1313, German merchants were to be arrested because of their suspected involvement in piracy in the North Sea. In face of those adverse measures, the Hanse managed to convince Edward II in April 1314 to reconfirm its original privileges and exclude its merchants and their goods from those reprisals.[189] On the other hand, the commercial relationships with Flanders could not get any worse after 1315, when Edward II imposed an embargo on trade with the county, as a response for Count Robert III's refusal to expel Scottish merchants from his domain. When Louis X of France, together with William, Count of Hainault, Zeeland, and Holland, invaded Flanders in August 1315, Edward II wholeheartedly supported both allies. Such a geopolitical situation explains the strong presence of the Norman merchants, followed by their Dutch and German colleagues. It is somewhat perplexing to see two Dunkirk merchants entering Scarborough port, given that they were subjects of the Count of Flanders. It should be borne in mind that in the early fourteenth century the burgesses of Dunkirk tended to be unstable in their political allegiance, hesitating between King of France and Count of Flanders. To ensure their subordination and avoid a potential rebellion, in 1311 Robert III granted his citizens a degree of autonomy, which included extensive trade rights.[190] Because of all that, it is possible that Edward II did not consider Dunkirk traders as his enemy's subjects, and, as such, they were exempt from the 1315 trade embargo.

Conclusions

There can be little doubt that inclement weather and, consequently, failed harvests had a most negative impact on grain trade and crop markets during the Great Famine. The unprecedented lack of a steady supply of grain caused exceptionally aggressive price behaviour, by which market prices of crops trebled and quadrupled and the price variance across places became more pronounced, when compared to non-famine years. The rise of what may be seen as pref-

[188] Although the term is traditionally translated as 'the Charter of the Merchants', a more accurate translation, linguistically speaking, would be 'the Market Charter').

[189] Lloyd, *England and the German Hanse*, pp. 22–28.

[190] Demarcq, *Dunkerque 1 000 ans d'histoire*, pp. 11–15.

erential trade, as well as hoarding tendencies by better-off elements, emptied grain market stalls further. Trans-regional and international trade was being conducted on a very limited scale, and the volumes of crops shipped from one region to another were certainly less than enough to satisfy the needs of the starving population. All these facts made the task of securing food more challenging than ever before, and they undoubtedly encouraged the black market, where the prices seem to have been even higher than the 'official' ones. To make things even worse, however, nominal wages either remained at their pre-famine levels or rose very insignificantly, much slower than prices. This depressed real wages to an unprecedented point and reduced wage-earners into miserable poverty. The declining living standards of wage-earners, moreover, reflect those of self-employed tenants. In other words, while the producers and corn-mongers had little to sell, the commoners had little money to spend.

Under these circumstances, it is hardly surprising that individuals tended to act in a selfish manner to get by during those challenging times, when everything was scarce and dear. The instances of baking against the *Assize of Bread and Ale* surged, as well as the references to food regrating. Although the authorities, both local and royal, attempted to curtail those speculative activities, their measures seem to have worked with only a limited degree of success. Furthermore, it is unclear if some of those measures were practical or symbolic. Thus, the baking and regrating amercements might reflect that the authorities indeed attempted to exercise a more stringent control of food processing and distribution among their community members to prevent any potential wrongdoing. But they might equally be interpreted as a purchase of the right to break the assize.

In his recent article, which tackles some issues I raised in my 2014 paper on market failure, Phillipp Schofield came up with a different verdict regarding market performance during the famine years.[191] According to his interpretation, the seasonality of grain prices, mimicking exaggeratedly that of normal years, indicates that the market was functioning efficiently within the expectations of a FED model of famine, and hoarding was not bad enough to depress it. Had hoarding been as widespread as I have assumed, grain prices would have spiked right after the harvest rather than gone up evenly in the course of the year.[192] Here, Schofield based his assertion on Cormac Ó Gráda's study of European pre-Industrial grain markets, without considering the possibility that the situa-

[191] Schofield, 'Market, Economic Growth and Famine'; Slavin, 'Market Failure during the Great Famine'.

[192] Schofield, 'Market, Economic Growth and Famine', pp. 278–79.

tion may vary across different famines, given their different chronological, geographic, and cultural settings.[193] Moreover, as we have seen, in the wake of the second back-to-back harvest failure, the prices soared quickly between August and September 1316. Finally, even if hoarding within the seignorial sector, so well-documented in manorial accounts, is not indicative of the situation in general, there is nothing to suggest that better-off tenants with some crop surplus would not act in the same manner. In the same vein, Schofield suggested, relying on Buchanan Sharp's recent work, that the multitude of letters of protection and safe conduct issued by Edward II to grain merchants indicates that the market was functioning as usual.[194] However, as we have seen, bringing a certain amount of grain from region A to region B *may have* facilitated trade in region B, but it could also depress trade in region A. Moreover, the amounts of imported grain were never large enough to ensure a steady trade. Finally, it is important to distinguish between royal licences and their materialization: granting safe conduct or a letter of protection would not automatically mean locating, securing, and delivering grain to market stalls. All these facts tend to disagree with Schofield's interpretation of grain market behaviour during the famine. Rather, they strengthen the view that the Great European Famine, similar to the famines of developing countries, indeed exhibited the signs of what Amartya Sen called 'food availability decline' (FAD) and 'food entitlement decline' (FED). A clear manifestation of both FAD and FED was the disastrous disruption of grain supplies to local markets, coupled with the acute poverty of the masses. It was arguably the most decisive factor that intensified the extent of the famine and, consequently, aggravated the suffering of the disadvantaged. It is certainly possible that the crisis would not have been as acute had the grain trade and grain supply been more efficient. But it should be borne in mind that market inefficiency was not the only factor impacting the famine. When Sen wrote his monograph, he had in mind the Bengali famine of 1942–1943, caused by the disruption of rice supply because of the Japanese conquest of Burma, at the centre of the Pacific theatre of WWII. During the years of the Great Famine, the British Isles were unfortunate to suffer significantly from warfare, revolving primarily around the conflict between England and Scotland and affecting the grain trade a great deal. As we shall see in Chapter 6, warfare was yet another factor aggravating the food crisis and intensifying the famine.

[193] Ó Gráda, 'Markets and Famines in Pre-Industrial Europe'.

[194] Schofield, 'Market, Economic Growth and Famine', pp. 277–78; Sharp, 'Royal Paternalism'.

Chapter 5

Declining Food Security: Transportation and Storage

Speculators, Profiteers, and Hoarders: Some Second Thoughts

There can be little doubt that the black market, the breaking of the *Assize of Bread and Ale*, speculation, hoarding, and illegal exports all appeared as highly immoral and damaging activities in the eyes of contemporaries, as witnessed in royal proclamations, local court proceedings, and moralist literature. Excessive profiteering, especially during the times of crisis, was linked to usury and to avarice, one of the seven deadly sins, punishable by the tormenting flames of Hell. Hence, it is hardly surprising that in the eyes of contemporary authors, these morally illicit activities were seen as contributing factors to the famine and the hardships of the starving populace. While it is always vital to appreciate historical events and issues through the eyes of contemporary viewers, it may be also instructive to consider the situation from a modern economic perspective. In today's world, no economy can be efficient without commercial profit and interest, the two very backbones of free markets. According to Smithian philosophy, it is economic selfishness that promotes competition, prosperity, and market economy. As we have seen in the beginning of the previous chapter, Smith's solution to dearth and famines was a free market and liberal state with governmental non-interference. Grain merchants played a pivotal role in eliminating famine and minimizing dearth because of their ability to even out sales and consumption over the agricultural year and, thus, force hoarders to surrender their stocks to their own disadvantage.

Such a scenario, however, is possible in a liberal state, relying upon free trade and promoting economic security and property rights. An altogether different situation prevailed in early fourteenth-century England and, indeed, the rest of famine-hit northern Europe. In a society where the international grain trade was conducted on a limited scale and was stringently controlled by the government, while local production and sales were regulated by local institutions (rural and urban), famine relief was by no means a light challenge. Despite some initiatives to secure foreign grain, the idealized Smithian corn-monger travelling to far lands to secure food for the benefit of a starving populace and to the disadvantage of hoarders had everything, but necessary governmental or public support. Instead, local corn-mongers had to operate under very restrictive and challenging circumstances. Apart from the risk of being stigmatized as usurers and deadly sinners (and potentially being isolated from members of their communities), grain producers and corn-mongers had to face a series of additional challenges related to securing food supplies. All these challenges can be summarized as 'transaction costs', namely, supplying goods and services from Point A to Point B. These involved search and measurement costs, negotiation costs, transportation costs, protection costs, monitoring costs, and storage costs. According to Douglass North, in order to attain efficient economic performance, institutions need to regulate mechanisms to ensure low transaction costs.[1] The gloomy reality of the early fourteenth-century crisis could ensure everything but low transaction costs. As John Munro has convincingly shown, the sudden outbreak of warfare on a virtually pan-European level around 1290 raised transaction costs and caused a widespread depression of international trade.[2] Much the same can be said about the Great Famine, which was an omnipresent disaster in not only the British Isles but most of northern Europe. This implied that the catastrophe reduced the volume of trade and left local communities at the mercy of their own self-sufficiency. This, in turn, meant that without access to foreign stocks, corn-mongers would have had hard times finding domestic grain available for sale. Any attempt to secure grain would have involved a very high search and, therefore, opportunity cost.

Let us imagine a hypothetical scenario involving a corn-monger on his quest to secure grain for sale. The most immediate challenge that the corn-monger faced would be search costs. The omnipresent disruption in the grain supply and the rise of preferential trade undoubtedly decreased the value of market information and made his search most challenging. Corn-mongers would often face the risk of investing their time and resources in hunting goods, without any

[1] North, *Structure and Change in Economic History*.

[2] Munro, 'The "New Institutional Economics"'.

assurance that the same goods could actually be found. This would have been especially challenging in the context of increased transaction costs, because of both exogenous and endogenous factors. The torrential rain undoubtedly transformed roads into fields of mud and deep puddles, and this fact made carting conditions increasingly difficult. Cart-horses had to spend more energy, as they were struggling to pull carts through muddy soil and cold, damp, and windy weather. As we have seen in Chapter 3, an average manorial horse did not receive more oats than usual during the famine years, although there were some instances when the hay shortage prompted some demesne managers to increase the proportions of oat fodder. It is possible that some corn-mongers, too, had to allocate more fodder to their cart-horses to ensure that grain would be delivered from one location to another. Underfed, cold, and weak horses could severely delay the delivery of grain to local markets. Water transportation, the most obvious alternative to inland shipping, would command increased transportation costs, too, because of obvious related risks. The inclement weather increased the chances of delivery failure, due to potential shipwrecks and the loss of lives to the sea and inundated rivers. But it is also important to account for the ongoing problem of piracy, targeting commercial vessels, as we shall see later in this chapter. Because of all these difficulties, risks, and hazards, it is hardly surprising that the transportation costs (inland and especially maritime) went up during those years. In addition to search and transportation costs, however, we also have to account for protection costs. Years of dearth and famine always cause a pronounced spike in crime rates, especially food-related crimes, when more and more desperate and hungry people resorted to alternative channels of food security, often through illegal means. The Great Famine was by no means an exception, as we shall see in Chapter 8. To assure appropriate protection, corn-mongers may have had to hire additional personnel to guard their carts, stalls, and storage facilities from potential highway marauders, thieves, and pirates.

Grain hoarders are much more difficult to justify, especially given their omnipresent condemnation by both commoners and authorities. There is no question that they managed to make significant profit during the years of scarcity and famine — in sharp contrast with years of plenty, when good harvests could potentially ruin their investment, making stored grains an undemanded commodity. All the same, it is essential to consider the poor state of grain storage facilities during the famine years, when they were prone to physical damage and destruction by windy and rainy weather. Although impossible to quantify, the spoilage rates of stored crops were undoubtedly high during the crisis years. This may have been especially true in the case of peasants, who could not boast good storage facilities, on account of rural congestion and poverty. By contrast, better-off lords and merchants, provided, in effect, physical protection for meagre stocks of grain that were available in the context of the disastrous deficit. In other words, it was not only the *grains* themselves that were in high demand, but also *facilities* to store and protect them. Moreover, even the undoubtedly selfish tendency to hold crop stocks until the prices became abnormally high in the spring and summer was not entirely destructive. Releasing the grain earlier, when the prices were lower, could have potentially exhausted the stocks completely, leaving the population with a complete lack of food resources. It is true that at that point the prices were too high for the majority of population to afford any purchases, but at least it allowed those with *some* resources to secure *some* meagre amounts of food.

All these facts do not attempt to justify grain speculators and hoarders, but rather they explain the sad logic of their activities and, consequently, the price soar during the famine years. Corn-mongers saw themselves entitled to receive very high prices, not only because of their selfishness, deriving from their financial advantage over the vast majority of commoners, but because of all the risks, hazards, and insecurities involved in their commercial activities. Grain hoarders, at least in theory, not only deprived the starving masses from access to food but also served as custodians of grain stocks, assuring that there was at least *something* available during the disastrous dearth of the annual 'hungry gap' period of the late spring/early summer. This was not a moral economy by any means, but it was an expected outcome of the dire circumstances. Although the high prices were undoubtedly devastating for the well-being of most commoners, they were devastating not because of the self-interest of corn-mongers and grain hoarders, but because of the general levels of overpopulation and poverty. To put it simply: there was not enough money per capita to afford buying the meagre food resources that may have been available on local markets thanks to the corn-mongers' efforts. This, in turn, lends support to the idea that overpopulation and poverty played a pivotal role in transforming the crisis of 1315–1317 from a food shortage to a disastrous famine of unprecedented proportions. This point will be addressed in a due length in Chapter 7. Let us, meanwhile, return to the point of the logistics of crop supply, which is closely related to the idea of transaction costs.

Transportation Costs

The single best source of information for assessing the impact of the crisis on transportation (and consequently transaction) costs of grain and other foodstuffs is sheriffs' accounts, rendered in conjunction with the provisioning of royal armies. Sheriffs' accounts record, in meticulous detail, various expenses related to acquiring and transporting foodstuffs and other goods to royal garrisons. Relative transportation costs (measured in pence per ton-mille) would vary not only from year to year but also from place to place within the same year.[3] Normally, the ratio of land to river- to sea transport costs would have been in the area of 7:1.5:1 (that is, carting costs would have been seven times higher than sea transport costs, while the river transportation costs were about one-and-a-half times higher than sea transportation costs).[4] As some studies have clearly demonstrated, in pre-Industrial economies, high market prices tend to drag transportation costs higher, making interregional food transfers more expensive and challenging and, thus, exacerbating famine. This was, for instance, the case in eighteenth- and nineteenth-century Japan, in Scotland in the 1690s, and in Ireland in 1740–1741, as well as in the contemporary African lands.[5] The fall in transportation costs, with the introduction of faster means of transport during the Industrial Revolution, was one of the most paramount factors in decreasing vulnerability to food shortage and famine in industrialized countries.[6]

The abnormal weather of 1314–1316 had undoubtedly reduced travel and communication conditions to a deplorable state. Flooded roads were undoubtedly clogged with dirt in spring, summer, and autumn, in addition to being icy and slippery in winter. To make things even worse, sea storminess, coupled with the ongoing piracy, would have made maritime voyages unpredictable and risky. Because of those hazards, it is only natural to assume the costs of carting and shipping must have gone up during the famine years. As Table 5.1 demonstrates, this expectation is fulfilled only partially: while sea transport costs indeed rose significantly, standing at about 95 per cent above average during the famine years, the land transportation costs (at least as indicated in the 1315–1316 sheriff of Devon's account) did not change. Does it mean that the crisis had no impact on inland transportation costs? Unfortunately, no other surviving account from the famine years records land transport costs, and it is unclear to what extent the Devon account reflects reality elsewhere. On the one hand,

[3] Masschaele, 'Transport Costs'.

[4] These estimates disagree with those of Masschaele, 'Transport Costs', p. 273, suggesting the ratio of 8:4:1, but agree more closely with Campbell, Keene, Galloway, and Murphy, *A Medieval Capital*, pp. 193–98.

[5] Bassino, 'Market Integration and Famines'; Garenne, 'An Atypical Urban Famine', p. 191.

[6] Ó Gráda, *Famine: A Short History*, pp. 10 and 155–57.

it is possible that inland transportation costs were not affected nearly as badly as shipping costs. After all, carting was almost a universal skill among manorial tenants and, in many cases, town dwellers, and, thus, carting services would have been available in abundance. Sea faring and navigating, on the other hand, were relatively rare, high-qualified, and much-demanded skills. Also, seafaring became exceptionally dangerous during the famine years, not only because of the increased storminess but also because of ongoing piracy in Britain's waters. Various risks and hazards associated with seafaring made shipping a much less attractive enterprise, and crew members were able to command higher wages. For instance, the daily wages of a cargo-delivery supervisor, usually a sheriff's clerk staying on board for the duration of voyage, rose from about 8d to the overwhelming 18d.[7] Also, the dangers of storminess and piracy made journeys longer than usual, as we shall see later. All those facts undoubtedly raised the costs of transporting good by sea. Once the weather returned to normal and piracy activities declined, however, maritime transportation costs plummeted once more to their pre-1315 levels, as Table 5.1 indicates. Roads, on the other hand, may have been more secure travel routes, despite references to armed robbers, as we shall see in Chapter 8. At the same time, however, it is important to bear in mind that the sheriff of Devon's account does not indicate the feeding costs of horses, which undoubtedly went up during the famine years, on account of the rise in oat and hay prices, and, thus, it may only partially reflect the real costs of inland transportation. To appreciate the rise in maritime transportation costs, it is essential to consider the situation at sea.

Gone with the Sea: Storms and Pirates

During the troublesome decade of the 1310s, Britain's waters became an increasingly unsafe space for travel and trade because of both endogenous and exogenous factors. As we shall see in the next chapter, the natural environment of the British Isles was undergoing the process of militarization in the context of ongoing warfare. Although all battles were fought on ground, the English

[7] TNA, E 101/376/6, E 101/574/24; SoA, MS 120, fols 13–20 (famine years, in all instances giving the figure of 18d per diem); TNA, E 101/559/5 (the sheriff of Gloucestershire account, summer 1319, 6d per diem); E 101/556/14 (the sheriff of Essex account, summer 1327, 10d per diem); E 101/556/15 (the sheriff of Essex account, May 1334, 8d per diem). The figure of 8d per diem was also paid in the summers of 1333, 1334, and 1335, as indicated in the accounts of the sheriffs of Hampshire and Northamptonshire: TNA, E 101/561/9; E 101/577/8–9 and E 101/561/10. More examples can be cited.

Table 5.1. Transportation costs in the early fourteenth century, famine (1315–1317) and non-famine years (1319–1335), weighted according to the volume of grain in each transaction.

Type of transportation	Number of shipments	Quarters of grain	Equivalent in pounds	Transport-cost per ton-mile (weighted average, in d)
Famine years, inland transportation	1	480	176,320	1.11
Non-famine years, inland transportation	19	12,168	4,896,120	1.31
Famine years, maritime transportation	11	4,840	1,740,517	0.37
Non-famine years, maritime transportation	21	18,811	7,324,998	0.19

Source: Sheriffs' accounts (see Appendix 8); SAL, MS 120; TNA, E 101/14/40.

Note: For transport calculation details, see Appendix 4.

Channel, the North Sea, and the Irish Sea all became 'unofficial', yet integral, loci of environmental warfare.[8] To make things worse, the high winds and the torrential rain of 1314–1316 may have contributed to the climatic instability and storminess of the North Atlantic in the period of *c.* 1270–1430.[9] The increased storminess made maritime journeys challenging and risky, with the potential outcome of getting cast ashore in the best case, or wreckage and disappearance in the worst case. Although it is impossible to estimate the total number of ships wrecked in Britain's waters in the early fourteenth century, there is a good deal of written evidence about that, in particular in conjunction with the ongoing warfare. So far, I was able to identify fifty-seven inquests (mostly royal) related to shipwrecks due to storms, issued between Michaelmas 1313 and Michaelmas 1323 alone.[10] Of the fifty-seven inquests, seventeen (30

[8] See Chapter 6.

[9] See Chapter 1, pp. 13–14. Dawson and others, 'Greenland (GISP2) Ice Core'; Cage and Austin, 'Marine Climate Variability'.

[10] TNA, SC 8/2/99; SC 8/31/1522; SC 8/164/8196; SC 8/238/11866; SC 8/323/E592; SC 8/162/8093–96; SC 8/112/5576; C 66/146; KHLC, Fa/ZB/52; *CPR, 1313–17*, pp. 55, 67, 142, 143, 147, 149–50, 224, 231–32, 270, 313, 323, 501–02, 571, and 694; *Calendar of Patent Rolls Preserved in the Public Record Office: Edward II, AD 1317–1321* (hereafter *CPR, 1317–21*), pp. 28, 94, 169, 373, 403, 481, 483, 484–85, 538, 604–05, 606, and 609; *Calendar of Patent Rolls Preserved in the Public Record Office: Edward II, AD 1321–1324* (hereafter *CPR, 1321–24*), pp. 58, 158–59, 160, and 254; *CCR, 1313–18*, pp. 127, 168, 171, 227, 252–53, 291, 301–02, 326, 341, 453, and 488; *Calendar of Close Rolls Preserved in the Pub-*

per cent) can be dated to the famine years. These inquests deal with merchant vessels shipping grain either to be sold at local markets or to procure royal garrisons in north England or Ireland. It should be noted that all the writs were issued *post factum*, and in some cases the chronological gap between a shipwreck and an inquest was four of five years. Obviously, shipwrecks were not unique to the famine years.

Although the stormy weather may have increased the chances of shipwrecks, the difficulty to date the actual wrecks under royal investigation does not allow us to estimate the impact of sea-storms on the incidence of wrecks. What is certain, however, is that losing food to the sea had a much more devastating effect during the famine than in non-famine years, at least on a local level. Sometime in 1315–1317, three ships carrying collectively 300 quarters of wheat, 328 quarters of barley, 611 quarters of oats, and 800 quarters of salt from Dartmouth (Devon) to Berwick-upon-Tweed wrecked off the Devon coast, with one ship, the 'Cog St Marie' of Faversham, sinking. Most victuals were lost, except 60 quarters of wheat, which, although saved, became rotten and had no value.[11] At some point in late 1316 or early 1317, Edward II sent thirteen ships laden with various victuals at Sandwich, which were purchased from his chief grain supplier Antonio Pessagno, to be shipped to Newcastle and Berwick-upon-Tweed. In the course of their turbulent voyage, the ships lost 357 out of the total 1541 quarters of wheat, in addition to 35 out of 344 pipes of wine.[12] Although these entries indicate rather massive loses, one should not exaggerate the impact of the shipwrecks on the decline in food availability. There is no doubt that losing several hundreds of quarters of grain may have been devastating to some local communities, which, as a result of the wreck, was deprived of its food supply. But all the same, it is unlikely that the total quantities of wrecked grain were great to the point that they contributed to the disruption of food supply on a regional, let alone national, level.

In some cases of wrecking ashore, victuals and other merchandise would be salvaged by local coastal communities. Between Michaelmas 1313 and 1323, at least seventy-three instances of coastal scavenging or robberies have

lic Record Office: Edward II, AD 1318–23 (hereafter *CCR, 1318–23*), pp. 171, 297–98, 410, 660–61, and 713; and *Calendar of Inquisitions Miscellaneous (Chancery) preserved in the Public Record Office* (hereafter *CIM*), p. 397.

[11] TNA, SC 8/162/8093.

[12] *CCR, 1318–23*, pp. 660–61; and *CPR, 1317–21*, p. 28.

been reported. Of these, only seventeen could be dated on the basis of external evidence, of which nine can be dated to the famine period.[13] Thus, at some point in 1315, Laurence de Seynt Jak, a Gascon merchant and master of a ship called 'Our Lady of Fuenterrabia', petitioned Edward II regarding the loss of his merchandise, pillaged by the people of Rye, Winchelsea (both in Sussex), and Romney (Kent).[14] Around 1320, a Spanish merchant, Frauncois de Pymount, asked Edward II to compensate him for the loss of his merchandise, which included grain, but also wax, fur, hides, oil, and almonds, all pillaged by the people of the Isle of Wight, upon the wreck of his ship, most likely at some point between 1315 and 1317.[15] Between 18 and 23 May 1316, Berenger Blanc, Louis X's admiral, attacked three Genoese ships (called *dromonds*) carrying Spanish grain and other provisions for Berwick-upon-Tweed, in the Downs, off the Kentish coast, stretching roughly from Sandwich to Walmer. The *dromonds* contained at least 2750 quarters of wheat, 40 quarters of legumes, 34 quarters of wheat flour, and 36 quarters of salt. Berenger was successful in carrying the first two *dromonds* to Calais; the third ship, however, was brought back by storm to Sandwich, where the French pirates were immediately arrested and the *dromond*'s contents disappeared, presumably into the hands of profiteers.[16] On 13 November 1316, Thomas de Stotford, serjeant of the keeper of Berwick-upon-Tweed, petitioned Edward II to investigate the depredation of his ship carrying 85.5 quarters of grain (mostly rye), as well as 2.25 quarters of rye flour and 1 quarter of salt, by the people of Warkworth (Northumberland), near which the vessel wrecked.[17]

[13] TNA, SC 8/2/99, SC 8/134/6684, SC 8/164/8196, SC 8/323/E592, *CCR, 1313–18*, pp. 127, 302, 326, 341, 345, and 565; *CCR, 1318–23*, pp. 195–96, 201, 259, 284, 297–98, 398–99, 453, 565, and 591–92; *CPR, 1313–17*, pp. 23, 55, 67, 71, 140–41, 142, 143, 144, 147, 149, 224, 231–32, 246–47, 251, 261–62, 316, 318, 323, 324, 326, 409, 469, and 585; *CPR, 1317–21*, pp. 90–91, 94, 169, 186, 284, 285, 291, 293, 296, 302, 304, 306, 360, 364, 365–66, 373, 403, 472, 473, 478, 481, 484–85, 538–39, 546, 604, 605–06, and 609; *CPR, 1321–24*, pp. 147, 151, 158–59, 160, 254, 313, and 315; *CDRS*, no. 602; and *CIM*, nos. 211, 294, 397. The instances datable to the famine period are: *CCR, 1313–18*, pp. 326, 341, and 591–92; *CPR, 1313–17*, pp. 469 and 501–02; *CPR, 1317–21*, p. 285; and *CIM*, no. 294.

[14] TNA, SC 8/2/99.

[15] SC 8/323/E592.

[16] TNA, C 54/133 and C 66/146. The case has been calendared in *CCR, 1318–23*, p. 341, and *CPR, 1317–21*, pp. 501–02, and published, with a discussion, in Arnoux, 'Famine, blés marocains, marins génois et pirates français', and transl. into English with a discussion in Eddison, *Medieval Pirates*, pp. 94–95 and 173–75.

[17] *CPR, 1317–21*, p. 597.

The case of the three Genoese *dromonds* and the Berwick ship reflect the ongoing problem of piracy in the pre-Black Death era in general and in the famine years in particular. Piracy was nothing new; it flourished in the English Channel over the course of the thirteenth century and was often associated with the Cinque Ports (a commercial association of seven Kentish and Sussex towns), which supplied both ships and raiders.[18] But it was between *c.* 1280 and *c.* 1330, as Thomas Heebøll-Holm has shown, that piracy has reached a new level of intensity, serving as a tool at the hands of local ports, mariners, and merchants to compete with their rivals, both domestic and foreign. In this context, pirate attacks were not mere robberies, but well-organized and coordinated privateering.[19] In the context of the famine, coupled with military hostilities, food piracy can be seen as an alternative and illicit method of securing food; or, to use Amartya Sen's term, 'illegal transfers'.[20]

For the famine years, I was able to identify twenty-five recorded instances of piracy.[21] The pirate activities were ubiquitous. In the south, pirates were operating in the Channel and in the Celtic Sea and all the way from Cornwall to Kent, with the Isle of Wight, the Cinque Ports, and Kentish ports being the hottest spots. In the west, they were giving significant trouble in St George's Channel and in the Irish Sea. In the North Sea, piracy flourished along the coastal line of Essex, Suffolk, Norfolk, Lincolnshire, the East Riding of Yorkshire, Durham, and Northumberland, as well as along the eastern coast of Scotland. In the Irish Sea, it was reported around Anglesey and other parts of the north-western coast of Wales, Lancashire, Cumbria, and presumably Galloway. In most instances, pirates seem to have been privateers, acting under licence or letters of marque and attacking alien ships, and, in some cases, vessels of their compatriots. In other words, pirate enterprise respected neither boundaries nor ethnicities.

Eight out of twenty-five recorded maritime attacks were committed by Flemish pirates (and of course, 'Flemings' could be mistaken for any Dutch-speaking individual). This reflects, first and foremost, the political and military hostilities between Count Robert III of Flanders and King Edward II. It would

[18] Eddison, *Medieval Pirates*, pp. 69–85.

[19] Heebøll-Holm, *Ports, Piracy and Maritime War*.

[20] Sen, *Poverty and Famines*, p. 45.

[21] TNA, C 66/146; TNA, SC 8/112/5576; SC 8/162/8093–96; SC 8/238/11866; SC 8/31/1522; *CCR, 1313–18*, pp. 291, 315, 326, 341, 385, 455, 457–58, 466, 488, 563–64, 571, and 594–95; *CCR, 1318–23*, pp. 162, 171, 176, 192, 194, 257–58, 262, 394, 399, 401, 411, 414, 591–92, and 600–01; *CPR, 1313–17*, pp. 501–02, 571, and 678; *CPR, 1317–21*, p. 28; and *CIM*, no. 294.

not be an exaggeration to state that while the Scots were harassing the English on land, the Flemings did so in the sea. It should also be borne in mind that the Low Countries were devastated by the famine — in contrast with Scotland. The combination of the political hostilities and the dire famine prompted Count Robert III, in early 1316, to send a fleet of ships in order to seize victuals from enemies and ship them back to Flanders to provision the count's subjects.[22] As royal petitions reveal, most Flemish pirates came from either the *Zwin*, an inlet in the North Sea between Bruges and Sluys, or from Sluys proper. One particularly infamous gang was that of John Crabbe, an adventurer and pirate leader, who, in the course of his forty-year naval career (*c.* 1310–1352), changed allegiance on numerous occasions. In the given period, he considered himself loyal to both Count Robert III and Robert Bruce, the two sworn enemies of Edward II, and, in such capacity, he carried a series of naval attacks on merchant ships in the North Sea.[23] Scots, too, were spotted committing pirate activities in the North and Irish Seas. Thus, on 1 October 1315, a band of Scottish and Welsh pirates plundered and seized a ship called the *James of Caernarvon*, carrying it away from Holyhead to Scotland.[24] The leader of the gang was Thomas Dun, a Scottish equivalent of John Crabbe. Between 1315 and 1317, Dun's ships created terror in the Irish Sea, to the point that on 9 May 1317, Edward II gathered a crew of 140 men to lead a punitive expedition against Dun and his men. Thomas and his gang were at last captured near the Isle of Man on 2 July of the same year.[25] As the justiciar of Ireland's rolls reveal, Ireland was not immune from piracy either. Thus, on 14 November 1317, a certain John Daniel and his companions were accused of killing an unnamed Spanish merchant and taking his ship, laden with salt worth £100.[26]

The link between famine, war, and piracy is by no means a coincidental one. Neither was the Great Famine unique in this matter. Thus, during the 1626–1627 famine in China, when, because of a disastrous drought, the rice yields were half their normal level and the food prices soared a great deal, Chinese pirates targeted and pillaged rice junks off the Fujian coast (south-

[22] TNA, C 145/80/19 (calendared in *CIM*, no. 258).

[23] John Crabbe's career is surveyed in Lucas, 'John Crabbe', and Eddison, *Medieval Pirates*, pp. 99–105.

[24] *CDRS*, no. 451.

[25] *CDRS*, no. 549; Penman, *Robert the Bruce*, pp. 187–88. Dun was executed shortly after his capture.

[26] NAI, KB 2/12, pp. 53, 63–64.

east China).[27] During the harsh north European famine of the 1690s, which was coupled with the War of the Great Alliance (1688–1697), one witnesses increasing pirate activity in the North Atlantic, when French, English, and Scottish privateers attacked and robbed food-laden ships.[28] One more recent example is the ongoing piracy off the Somali coast, which commenced with the second phase of the Somali Civil War in 2005. Although their activities are best known as ransom piracy, Somali pirates are equally busy in targeting ships carrying food and fuel supplies. As such, it is important to distinguish between 'ransom piracy' and 'resource piracy' or 'food piracy', seen by some as a manifestation of 'moral economy'.[29] While 'moral economy' is undoubtedly a controversial term here, given that most of the looted resources seemingly end up under control of local warlords and tribal leaders rather than in the mouths of starving people, there is no doubt that food looting is an important feature of Somali piracy. What is clear is that the dire straits of food crises, at times coupled with military hostilities, encourages more and more desperate people to seek alternative and 'illicit' methods of securing food—Sen's 'illegal transfers'.[30]

Sea Transportation Costs: Speed and Spoilage

The inclement weather of 1314–1316 and the ongoing piracy created numerous obstacles and challenges to shipping grain by sea. First, they had a pronounced impact on the duration of voyages, whereby the average speed of sailing was considerably reduced. English cargo vessels could never boast a fast speed, especially when fully laden with grain: on average, the speed of one third to three thirds of a knot per hour was expected.[31] This was in contrast with balingers, barges, and specialized battle ships (occasionally built by the Crown), capable of achieving the speed of four knots.[32] But sailing has never been as challenging and slow as during the crisis years. The sea-storms, caused by the torrential rain and high winds, could impel cargo ships to change their

[27] Cheng, *War, Trade and Piracy in the China Seas (1622–1683)*, pp. 43–45.

[28] Cullen, *Famine in Scotland*, pp. 73–74.

[29] Samatar, Lindberg, and Mahayni, 'The Dialectics of Piracy in Somalia'; Klein, 'The Moral Economy of Somali Piracy'.

[30] Sen, *Poverty and Famines*, p. 45.

[31] Prestwich, *Armies and Warfare in the Middle Ages*, p. 273.

[32] Lambert, *Shipping the Medieval Military*, p. 199.

routes, by distancing themselves, as much as possible, from a coastline, to avoid crushing onto the land and wrecking. Conversely, periodical lack of wind (as reported in 1316–1317) would render ships unable to sail with a proper speed. Also, foggy conditions would force sailors to reduce the sailing speed. At times, unfavourable weather conditions would imply unplanned anchoring and harbouring. Finally, the danger of pirate vessels may force cargo ships to manoeuvre. These challenges are revealed in the Edward II's Wardrobe account book for 1316–1317. During that year, a fully laden cargo vessel would not exceed, on average, some 0.12 knots, compared with some 0.33 knots in non-famine years (Table 5.2). In some instances, the slowness of voyages was truly remarkable. For instance, it took 124 days for several ships to traverse from Boston to Berwick-upon-Tweed (via a stop in Newcastle-upon-Tyne), rendering an average speed of only 0.06 knots. In some instances, marginal glosses account for the unusually slow speed of delivery. Thus, a fleet of Sussex vessels sailing from Shoreham-by-Sea to Newcastle was severely delayed because they had to anchor at numerous ports, on the account of the 'fear of robbers' (*propter timorem praedonum*) and 'lack of wind on their voyage' (*defectum ventri in itinere suo*). As a result, it took the ships seventy-nine days to reach their destination on 25 November 1316.[33] Obviously, such setbacks were nothing new to the famine years; but again, the unusually the bad weather, and possibly the unusual volume of pirate ships, could have had some devastating impact on sailing speed and, consequently, the timing of food delivery.

Long sea-voyages could have some long-reaching implications. Apart from obvious delays in grain delivery to garrisons or to local markets, they also increased the risk of grain spoilage. Thus, one account notes the loss of 110.25 quarters of wheat carried onboard one Bristol ship bound for Skinburness (16 September 1316–10 February 1317) 'because of the sea-storm' (*per intemperiem maris submersarrum*).[34] By the time a Southampton ship bound for Newcastle has completed its perilous journey on 10 July 1316, it lost 16.88 quarters of oats (out of the total 315) to 'destruction and rot' (*in perditione et putrificatione*).[35] According to another account, of 561 quarters of grain purchased between May and approximately October 1315 around Dublin and Swords (County Dublin) for the provision of royal garrisons during the Irish campaign against Edward Bruce, as many as 331 quarters (that is, almost

[33] SAL, MS 120, fol. 82.

[34] SAL, MS 120, fol. 82.

[35] SAL, MS 120, fol. 73.

Table 5.2. Duration of sea-voyages and approximate speed of sea transportation.

Itinerary	Distance (nautical miles)	Day of departure	Day of arrival	Days at sea	Nautical miles per day	Knots per hour
Boston–Newcastle–Berwick-upon-Tweed	257	1.9.1315	31.10.1315	60	4.28	0.18
Southampton–Newcastle	409	11.7.1316	17.10.1316	99	4.13	0.17
Shoreham-by-Sea–Newcastle	359	8.9.1316	25.11.1316	79	4.54	0.19
Boston–Newcastle–Berwick-upon-Tweed	183	13.7.1316	24.11.1316	124	1.48	0.06
Weymouth–Ireland and Skinburness	533	18.7.1316	5.2.1317	203	2.63	0.11
Bristol–Skinburness	330	16.9.1316	10.2.1317	122	2.70	0.11
King's Lynn–Newcastle	165	1.9.1317	6.3.1317	187	0.88	0.04
King's Lynn–Berwick-upon-Tweed	257	April 1311	May 1311	44	5.84	0.24
King's Lynn–Perth	283	24.2.1311	5.4.1311	40	7.08	0.29
Boston–Newcastle	204	20.3.1311	24.4.1311	35	5.83	0.24
King's Lynn–Newcastle	165	Unspecified		24	6.88	0.29
King's Lynn–Newcastle	165	Unspecified		15	11.00	0.46
King's Lynn–Newcastle	165	Unspecified		20	8.25	0.34
King's Lynn–Newcastle	165	Unspecified		22	7.50	0.31
King's Lynn–Newcastle	165	Unspecified		18	9.17	0.38
King's Lynn–Newcastle	165	Unspecified		21	7.86	0.33
Bristol–Skinburness	330	12.5.1319	21.6.1319	40	8.25	0.34
Bridgwater–Skinburness	393	15.5.1319	14.8.1319	42	9.35	0.39
Maldon–Newcastle	275	11.6.1323	22.7.1323	43	6.40	0.27
London–Newcastle	315	10.5.1323	9.7.1323	61	5.16	0.22
Maldon–Newcastle	275	24.6.1327	1.8.1327	38	7.24	0.30
King's Lynn–Newcastle	165	11.6.1327	2.7.1327	21	7.86	0.33
King's Lynn–Newcastle	165	28.5.1333	21.6.1333	25	6.60	0.28
Maldon–Newcastle	275	5.5.1333	24.5.1333	20	13.75	0.57
Maldon–Newcastle	275	9.6.1333	1.7.1333	22	12.50	0.52
Bristol–Skinburness	330	May 1333	June 1333	48	6.88	0.29

Itinerary	Distance (nautical miles)	Day of departure	Day of arrival	Days at sea	Nautical miles per day	Knots per hour
Southampton–Newcastle	409	11.5.1333	27.6.1333	47	8.70	0.36
Sheerness–Newcastle	317	1.4.1333	31.7.1333	121	2.62	0.11
King's Lynn–Newcastle	165	7.5.1333	13.6.1333	37	4.46	0.19
King's Lynn–Berwick-upon-Tweed	219	12.3.1334	24.4.1334	43	5.09	0.21
Sheerness–Newcastle	409	29.5.1335	9.7.1335	42	9.74	0.41
Famine years					3.29	0.14
Non-famine years					7.67	0.32

Source: SAL, MS 120, fols 73–83; sheriffs' accounts (see Appendix 8).

Note: Sailing speed calculations included occasionally unplanned stops and anchoring.

60 per cent) were lost in the course of carting and shipping — truly remarkable figures.[36] Around the same time, John, son of Philip de Carrik, was put in charge of shipping 600 quarters of wheat from Dublin to Carrickfergus and Northburgh Castle (in Ulster). En route, the ship was diverted by a storm to Skinburness. A certain proportion of wheat was lost to the sea, and the rest was sold by John at Carlisle.[37] Even in those cases, when crops were either salvaged or merely dampened by rain or overboard water, they became, in many instances, rotten — especially when dunnage was deficient. Moreover, the exposure to marine humidity and lack of proper ventilation increases the moisture content in grains to over 17 per cent, causing mould, fermentation, sprouting, spoilage, and self-heating.[38] Again, grain spoilage through shipping was nothing unique to the famine years; indeed, this has been an ongoing issue in non-famine years, as sheriffs' accounts indicate. For instance, in early 1304, 1125 quarters of wheat and other victuals arrived in Berwick-upon-Tweed from the Isle of Wight and Southampton totally rotten.[39] The main difference, however, was that there were neither extraordinary weather events in 1304 nor any sign of grain shortage or starvation. The disastrous weather of 1314–1316

[36] TNA, E 101/14/40.

[37] *Historic and Municipal Documents of Ireland*, ed. by Gilbert, pp. 340–41.

[38] SKULD, 'Transportation of Wheat'.

[39] Prestwich, *War, Politics, and Finance Under Edward I*, p. 134.

may have placed grain at a much greater risk of spoilage, and large-scale losses would have had much more devastating implications on starving communities during the famine years than they would have on the same communities during 'normal' years.

River Transportation: Flooding, Blockage, and Obstacles

The abnormal weather of 1314–1316 also impacted river transportation and related costs. Although river transport system in pre-Black Death England was never as extensive as inland or maritime transport systems, it still played an important role in shipping goods in both commercialized and under-commercialized regions. In particular, in the London region, and the estuaries of the Humber, Great Ouse, the Trent, and the Severn, river transportation was crucial to delivering grain and other foodstuffs to local markets.[40] This view is perhaps a middle ground between the 'maximalist' position of Edwards and Hindle that medieval river navigation system was very extensive, and Langdon's 'minimalist' view that river transportation was secondary to other uses of water resources, such as watermills and fishing nets.[41] The importance of water navigation system, at least in the pre-Black Death period, is evidenced by a growing number of complaints regarding various deficiencies and obstacles in waterways from *c.* 1250 onwards.

Unfortunately, neither the 1316 account of the sheriff of Devon nor the wardrobe books for 1315–1316 and 1316–1317 provide any information on river transportation. Instead, we have to turn to three royal writs in form of special river commissions, that were issued at two separate points in 1316 and that indicate the growing inconvenience of shipping goods by river waterways. According to the first commission, issued at Lincoln on 20 February 1316 and directed to three Nottinghamshire officers, recent flooding damaged bridges and clogged the banks of rivers, streams, and dikes in that county. The officers were instructed to inspect the damage and have bridges repaired and blockage cleaned.[42] According to the second writ, issued on the same day, commercial

[40] Campbell and others, *Medieval Capital*, pp. 59–60, 194–98; see also, Caffyn, 'River Transport'.

[41] Langdon, 'Inland Water Transport'; Edwards and Hindle, 'The Transportation System of Medieval England and Wales'; Langdon and Claridge, 'Transport in Medieval England'. See also Jones, 'River Navigation in Medieval England'.

[42] TNA, C 66/144, membr. 3^{v} (calendared in *CPR, 1313–17*, p. 430). I am thankful to

vessels could no longer navigate freely the Trent waterway from the Humber to Nottingham Castle because its course had been diverted by weirs towards a watermill at Colwick (about 3.7 miles from Nottingham Castle) by its owner, William de Colwyk.[43] The problem of course diversion is also reflected in the third writ, issued on 27 May 1316. According to the commission, some landlords of Oxfordshire and Berkshire had restructured local weirs on the Thames waterway between Oxford and Wallingford to such a height that the banks on each side were badly flooded. Furthermore, they constructed certain obstacles called 'locks' (that is, flash-locks), attached to the same weirs, because of which vessels loaded with food supplies could not reach or return from Oxford.[44]

Of course, the issues of river clogging and course diversion were nothing new. In theory, the three royal commissions may have been issued in any given year, in either a period of plenty or shortage. But it is in the context of the crisis of 1315–1317 that they reflect, better than ever, the complex interaction between exogenous and endogenous/anthropogenic factors, which, when acting in tandem, create yet another aspect exacerbating the crisis and contributing to the transformation of shortage into famine. It was the torrential rain and accompanying storms that broke down bridges, flooded rivers, and clogged water reservoirs and dikes with mud and debris. This certainly made water transportation difficult. But the situation was made even more dire because of the decision making of local lords, who constructed various diversion mechanisms, such as weirs and locks, to facilitate transport across waterways to their manorial structures and to hinder shipping to other destinations. The complaints of the people in Nottinghamshire, Oxfordshire, and Berkshire that commercial vessels carrying victuals could not reach their destinations because of the obstacles should be seen and appreciated in the context of the widespread food scarcity and starvation. The deficient infrastructure of river navigation was yet another factor that contributed to the malfunctioning of the grain market and aggravated the food shortage, transforming it into famine.

late Prof. John Langdon, who advised me about this document and kindly supplied me with its digital image.

[43] TNA, C 66/144, membr. 3v (calendared in *CPR, 1313–17*, pp. 430–31). I am thankful to the late Prof. John Langdon, who advised me about this document and kindly supplied me with its digital image.

[44] TNA, C 66/145, membr. 16v (calendared in *CPR, 1313–17*, p. 501). I am thankful to the late Prof. John Langdon, who advised me about this document and kindly supplied me with its digital image.

Storage Costs: Demesne Barns

Transaction costs were determined not only by transportation efficiency but also by storage facilities. As it has been shown recently, the question of grain storage facilities and storage costs is of great importance to our understanding of the late medieval (and indeed, any pre-Industrial) economy.[45] It is essential to distinguish between two types of storage space: 'on-farm storage' and urban granaries. In this context, 'on-farm storage' means either demesne barns and granaries or peasant barns (or, in some instances, in-house compartments where grain stocks were deposited). Urban granaries were hired by corn-mongers or, in the case of purveyance, by royal officials to store grain stocks on a temporary basis before moving them to their final destination — namely, market stalls or royal household and garrisons.

Because of the chronic paucity of data on urban granaries during the famine years, this section will be dealing with demesne storage facilities. To ensure that harvested crops were stored in good condition, it was essential to closely supervise local storage facilities — barns in particular. Barns were large buildings, often more than 40 m long and 15 m high, in which unthreshed sheaves of harvested crops were kept in preparation for threshing. Because of their large size, barns were often susceptible to wind damage and needed constant maintenance. By contrast, granaries were small and compact edifices in which crops were stored after threshing, and, unlike barns, they did not require much attention. Each year, manorial officials had to spend a certain amount of money on maintenance and repairs of barns, which included expenditure on roofing, thatching, fixing of ridge-pieces in the walls or roof, daubing, plastering, and repairing floors, padlocks, and keys. Virtually every account roll gives a very detailed break-down of those expenditures. But apart from those forms of maintenance work, recorded into manorial accounts, there were other actions required to keep storage facilities in a good shape. First, it was vital to assure a low moisture content to prevent mildew and unwanted germination. Second, most of remaining crops were to be threshed at some point before March, when newly hatched insects would begin to attack cereals in the sheaf. Third, we have to account for other types of vermin (mostly mice and rats, but also additional types of rodents) as notorious consumers of stored crops. Unfortunately, it is impossible to quantify the scale of damage done by vermin to stored crops. Finally, once threshed, grains had to be moved from barns to granaries, often

[45] Claridge and Langdon, 'Storage in Medieval England', pp. 1242–65; Slavin, *Bread and Ale*, pp. 119–39.

adjacent to the former. Once there, it was essential to stir and ventilate crops twice or thrice a month to prevent spoilage.[46] During the famine years, the issue of storage supervision became more vital than ever before because of the inclement weather, which inflicted significant damage and spoilage and commanded higher costs of maintenance.

There are frequent references to strong winds and tempest, commencing in the summer of 1314. Some accounts speak about the 'great wind' (*magnus ventus*) breaking storage roofs and, in some instances, destroying entire barns. For instance, sometime in the course of 1314–1315, a barn of Deerhurst (Gloucestershire) was broken down 'because of the great wind'.[47] Similarly, the storm devastated a barn and a byre of Wield (Hampshire) in the same year.[48] It should be noted that demesne managers seem to have reacted fairly fast to the weather and, to avoid further damage, did their best to do at least some essential repairs between summer 1314 and Michaelmas 1315. In some instances, some structures were mended before the weather became catastrophic: thus, according to the 1313–1314 account from Kelvedon (Essex), a local barn was roofed 'before the arrival of a great wind' (*ante magnum ventum*, presumably in the summer of 1314).[49] Nevertheless, those preventive repairs were not entirely sufficient to ensure good storage conditions. As we have seen in the previous chapter, some manorial accounts contain references to the spoilage of stored crops, implying that some demesne barns were still leaky and generally in deplorable shape. It should be borne in mind that, around the famine time, at least half of all demesne barns were made of timber and, hence, susceptible to strong winds and flooding.[50]

It should be noted that the issue of weather is even more important for the temporary storage of grain in ricks, before they get transferred to barns, and later to granaries. As the author of the late thirteenth-century treatise *Seneschaucy* stated, it was up to manorial bailiffs to make sure that stacks were secured and covered.[51] The torrential rain and the strong winds were certainly a serious concern for local manorial authorities, who invested in sheltering crops in ricks during the crisis years. For instance, during the 1315 harvest at

[46] Slavin, *Bread and Ale*, pp. 123–24.

[47] WAM 8426.

[48] HantsALS, 11M59/B1/70, membr. 19^{v}.

[49] WAM 25800.

[50] Brady, 'The Sacred Barn', Appendix 1; Slavin, *Bread and Ale*, p. 121.

[51] *Seneschaucy*, Ch. 23, in *Walter of Henley*, ed. by Oschinsky, pp. 270–71.

Burnham Thorpe (Norfolk), 4d was paid to a worker for thatching a stack covering peas in ricks (*coopeiertura super tassum pisarum*).[52] Although the authorities were undoubtedly aware of all the hazards related to a prolonged exposure of recently harvested crops in ricks to the downpour and storms, covering crops with appropriate materials and bringing them in could take several days. For instance, at Werrore and Cosham (Hampshire), two customary tenants were hired for five days to place covers on harvest ricks.[53] Likewise, crops in ricks were prone to the consumption by sylvatic rodents (most notably, the common vole) and birds. Although there are no references from the famine years, some non-famine years' accounts contain occasional entries recording the hiring of servants (usually children or young people) to scare birds.[54]

Likewise, there are references to the disastrous state of demesne barns. As we have seen, a 1315–1316 account from Longbridge Deverill (Wiltshire) contains a marginal gloss stating that all barley could no longer be stored because of the debilitation and inundation of the barn.[55] The concern for the physical conditions of crop storage is reflected in the increased investments of manorial officials in various repair projects (Figure 5.1). Indeed, the sheer sums spent on building projects rose dramatically towards the end of 1313–1314, only to fall back in the course of the following three years and rise again in 1317–1318. The spike of 1313–1314 can be interpreted as the conscious decision of manorial officials to invest in 'preventive repairs' to avoid further tear and wear during the torrential rain. Conversely, the decline in expenditure during the famine years may reflect what John Langdon and his colleagues has identified as a period of building bust during the crisis years.[56] According to the same logic, the renewed rise of 1317–1318 could be deciphered as the end of the crisis and the return of building projects back to business. But it is essential to consider that the sheer figures do not always reflect the trends within the barn sector of the building projects. In fact, there seems to have been an inverse correlation between the absolute and relative figures. While the total building expenditure levels indeed

[52] NRO, WAL 63 269x1.

[53] QCA, Roll 246.

[54] Campbell, *Seigniorial Agriculture*, p. 415.

[55] 'Quia longuis custodui non potuit propter debilitatem et hinundationem (*sic*)'. LH 10688.

[56] Langdon, Walker, and Falconer, 'Boom and Bust'. Unfortunately, this otherwise excellent study omits the 1315–1316 Winchester Bishopric Pipe Roll (see p. 144).

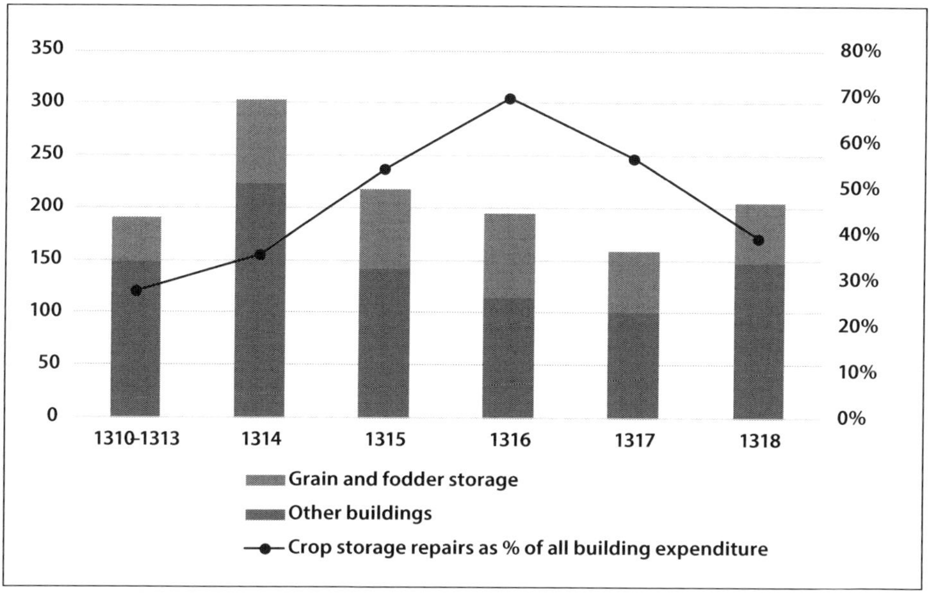

Figure 5.1. Annual expenditure on building (excluding mills) and barn repairs on 125 demesnes, 1310–1318 (in £ sterling). *Source*: Manorial accounts database (for the specific shelf-marks, see Appendix 8, Section 1).

Note: The figures, appearing on the left vertical axis, are annual expenditure on building (excluding mills) and barn repairs (in £ sterling), while the percentages, appearing on the right vertical axis, represent the relative share of barn repairs as a proportion of the total expenditure on all building repairs.

declined during the famine years, the proportion of money spent on crop storage repairs remained, in fact, high compared to the non-famine years. Thus, the combined figures stood at about £80 in 1313–1314, 1314–1315, and 1315–1316, compared with about £41 in 1310–1313 and £58 in 1316–1317 and 1317–1318. The figures reached their peak, unsurprisingly, in 1315–1316, when all expenditure on barns accounted for as much as 70 per cent of all building costs — compared with some 30 per cent in 1310–1313 and 40 per cent in 1317–1318. In other words, there is no doubt that the catastrophic weather implied considerable damage and, hence, considerable expenditure on storage facilities.

The inefficiency of grain storage still remains a challenge to the economies of some developing countries. For instance, during the 2007 floods in Uganda, numerous local granaries were destroyed. As a result, some better-off Ugandan farmers turned away from this traditional method of food storage and turned to more modernized food facilities.[57] Interestingly, in the post-Black Death era, one witnesses some pronounced improvements in barn construction in England, when more and more wooden facilities were substituted with stone-based storages.[58] This fact undoubtedly minimized the risk of food shortages and partially mitigated food shortages in the subsequent periods.

Conclusions

As this chapter aimed to show, the malfunctioning of grain markets, caused by the catastrophic disruption in supply, had to do not only with selfish behaviour of large-scale producers and middlemen but also with the rise in transaction costs. Although a much more complex and eclectic idea, here transaction costs are considered mainly through the prism of two main aspects: transportation and storage costs. In both instances, these costs rose during the crisis years because of the adverse combination of both exogenous and endogenous factors. It is true that the malevolent weather could, in theory, have covered some cart-ways with mud, broken down some bridges, clogged some waterways with debris, flooded some river banks, and formed occasional storms in the sea. These environmental hazards could have made, in some instances, the transportation of grain and other foodstuffs more challenging and, therefore, more expensive. But we cannot ignore some purely anthropogenic aspects, which made the transportation of foodstuffs challenging. Here, we have to account, first and foremost, for the ongoing pirate activities in the sea, in conjunction not only with the famine but also with the Anglo-Scottish war. Also, we have to bear in mind the lack of communal concern for public health and cleanliness in local environments, which undoubtedly made the blockage of waterways even worse. Likewise, managerial decisions of some lords, related to river-course diversion via the installation of weirs and flash-locks, prevented merchandise ships from transporting foodstuffs to local points of distribution and consumption and thus aggravating the supply crisis. This, of course, does not mean that these were ubiquitous phenomena. Pirate vessels were certainly not raiding all over, night and day; they were, as we have seen, spotted occasionally in some coastal regions (mostly along the North Sea coast), and, therefore, their contribution to the intensification of the famine was confined only to those few

[57] Kato, 'Uganda: Granary — The Solution to Drought And Famine'.

[58] Brady, 'The Sacred Barn', Appendix A.

coastal communities that were attacked by them. Likewise, it would not be possible to imagine all roads clogged with dirt and all river banks flooded.

The same thing can be said about storage methods and storage costs: it may have been a contributing factor to the famine, but certainly not the dominant one. Still, it is not to be neglected; and it is again the combination of natural forces and anthropogenic aspects that enhanced their ability to intensify the crisis. There is no doubt that the bad weather did significant damage to the stored crops, as the storminess, dampness, and humidity caused sprouting and mould and encouraged vermin activity. But human decision making, related to the duration of grain storage, could have been equally damaging. As we have seen in Chapter 4, withholding grain stocks during the famine years was a highly profitable enterprise, on the account of the high prices, which rose up with a remarkable speed. Apart from disrupting the grain supply to local markets, it also meant a physical destruction of a certain proportion of the same supply. During the crisis, better-off crop producers owing storage facilities had two options: (a) surrender their crops to markets and cut down the absolute storage costs or (b) withhold the stocks and profit from a fast price increase. As we have seen in the previous chapter, they chose the latter option, which was achievable only through some spoilage, in the context of the deplorable shape of grain storage facilities caused by the adverse weather. In other words, we can talk about deliberate destruction of *some proportion of* crops, through long-term hoarding, which implied, at least in the context of bad weather, *some* spoilage. This fact sheds a lot of light on the decision of grain producers and corn-mongers to hold on to their grain supplies and not release them until the potential profits reached their maximum. Long-term storage and the driving up the prices, however, entailed high-rate spoilage, and, as a result, a small proportion of the stored grain was doomed to destruction. In other words — it was impossible to profit without spoilage. This could have been yet another mechanism diminishing the meagre amounts of crops further and intensifying the crisis.

Chapter 6

Declining Food Security: Warfare and Taxation

Environmental Warfare and Famine

In his *Wealth of Nations* (1776), Adam Smith stated that '[a] dearth has never arisen ...from any other cause but a real scarcity, occasioned sometimes perhaps, and in some particular places, by the waste of war'.[1] Despite the ubiquitous influence of Smithian philosophy, it was not until the recent decade or so that we witness a growing scholarly awareness about the impact of warfare on food security.[2] Although the impact of warfare on the environment cannot be understated in any historical era or society, it is certain that environmental destruction tended to have a much greater impact on pre-Industrial societies living in the 'organic age' and depending on the available land resources much more than modern Industrial societies.[3] In the 'organic age', environmental destruction carried major long-term implications on landscapes and local human populations, and it would not be an exaggeration to state that pre-Industrial wars carried a pronouncedly environmental character, as Nature and its resources played the most crucial role in logistics, tactics, and outcome.

[1] Smith, *Wealth of Nations*, Ch. 4.5.

[2] For instance, Tucker and Russell, eds, *Natural Enemy, Natural Ally* and Charles E. Closmann, ed., *War and the Environment*; Bennett, *Natives and Exotics*; Zierler, *The Invention of Ecocide*; Brady, *War upon the Land*; Pearson, *Mobilizing Nature*, pp. 1–15; Slavin, 'Warfare and Ecological Destruction'.

[3] On this point, see Tucker, 'The Impact of Warfare on the Natural World', pp. 15–41.

There are several theoretical points to be borne in mind when talking about 'environmental warfare'. First and foremost, it is in this context that geographic entities (countries, regions, etc.) get divided into the 'war-free zone', largely unaffected by the ongoing warfare, and the 'war zone', where fighting takes place. War zones possess a series of peculiar socio-economic, geographic, and ecological traits, distinct from those characteristics of war-free zones, and those traits tend to undergo some dynamic processes in the course of fighting. One such process may be described as 'militarization of landscape' (to use Dorothee Brantz's and Chris Pearson's terminology).[4] According to this view, a landscape of (relative) peace became an environment of war. Thus, woodlands became a haven for local refugees of war, as well as sites of training, preparation for battle, ambush, and camouflage for soldiers and raiders rather than sources of lumber, fuel, shade, and peace of mind. Grains and hay are now fed to the agents of destruction (soldiers and their warhorses) rather than peaceful husbandmen, their families, and livestock. The militarization of landscape implies that Nature becomes both an aggressor and a victim, both an ally and an enemy, just as humans. Natural resources become the main objective of any such war — that is, assuring the victory for one's side and the neutralization or capitulation of the enemy's side. Because of the strong dependence of pre-Industrial societies on the organic economy of their natural landscape, the impact of warfare on the environment could indeed be devastating. On the one hand, the deliberate destruction of arable, pasture, and woodland inevitably creates a catastrophic dearth of food and mass impoverishment of local communities, who are cut off from their natural resources. On the other hand, plundering grain barns and carrying away livestock endows their enemy's side with a visible advantage, since they manage to secure their own access to the most important element for their well-being: food.

In other words, just as malfunctioning or failing grain markets during famine years are capable of depriving the worse-off socio-economic strata of their entitlement to food, so is warfare. Unlike market failure, however, warfare does not *a priori* imply the shift from FAD (food availability decline) to FED (food entitlement decline); here, the FED condition can be created without a harvest failure. But it is hard to envision any worse scenario than environmental warfare occurring in the context of harvest shortfalls, widespread food shortage, and animal murrain. Unfortunately for the inhabitants of the British Isles,

[4] Brantz, 'Environments of Death'. The process of militarization of late nineteenth-century French countryside has been studied in Pearson, *Mobilizing Nature*, pp. 1–15.

the Rider of the Red Horse joined hands with the other two horsemen of the Apocalypse, significantly aggravating the crisis and contributing to a further FED, especially in the 'war-zone' frontier, where communities were totally devastated by this adverse combination of exogenous and endogenous factors.

Britain c. 1310–1322: War Zones and War-Free Zones

The early fourteenth century was a period of intense warfare in the British Isles. Here, it is imperative to divide the British Isles into the 'war-free zone', largely unaffected by the ongoing warfare, and the 'war zone', where the fighting took place (Figure 6.1). The 'war zone' consisted of several theatres of war. First, one has to account for the Anglo-Scottish theatre of war. From 1296 onwards, Edward I attempted to impose his overlordship on Scotland. The First Scottish War of Independence (1296–1328) was costly for both sides. In the context of the Anglo-Scottish warfare, the 'war zone' in England encompassed the northern counties of Yorkshire, Durham, Northumberland, Lancashire, Westmorland, and Cumberland, and they were especially affected by frequent Scottish raids of 1312–1322 and 1327.[5] Within Scotland, the Anglo-Scottish warfare covered Carrick, Galloway, Strathclyde, Lothian, Dunbar, and Berwickshire in the south and all the way from Fife to as far north as Buchan, Moray, and south-eastern Ross along the east coast. Within Scotland proper, the newly inaugurated king, Robert I the Bruce, fought to impose and cement his rule among opposing chieftains between 1307 and 1309. The civil war between Robert Bruce and Scottish chieftains encompassed Lorn and Islay in the west and Garmoran and Ross in the north.[6] The Scottish theatre of war was closely linked to the Irish and Manx theatres of war (1315–1318), which started with the English reconquest of the Isle of Man and the Scottish invasion of Northern Ireland by Edward Bruce, Robert's brother.[7] Here, the 'war zone' stretched diagonally from Ulster in the north-east to Magh Luirg (Moylurg) in the south-west and included the hotly contested Isle of Man, situated in the Irish Sea half-way between west Cumbria in England and the

[5] The single best survey of these raids is in McNamee, *Wars of the Bruces*, pp. 72–122.

[6] On English attacks of eastern Scotland, see Watson, *Under the Hammer*; Robert the Bruce's reconquest campaign of 1307–1309, its background and outcome are described and analysed in a great detail in Barrow, *Robert Bruce*, pp. 188–241, and in Brown, *Bannockburn*, pp. 24–36.

[7] See Duffy, 'The Bruce Invasion of Ireland'.

Ards Peninsula in Ireland. The third epicentre of military hostilities was Wales, where Edward II's authority was challenged, first by Llywelyn Bren, a Welsh leader in the south (28 January–18 March 1316), and then by the marcher lords in the east in 1321–1322. Here, the 'war zone' included the southern counties of Glamorganshire, Monmouthshire, and parts of Carmarthenshire and Breconshire, and, to a certain extent, west Gloucestershire (in England proper). Finally, and in conjunction with the war between Edward and his contrariant barons, led by Thomas of Lancaster, the 'war zone' extended into Yorkshire. One could, perhaps, also add the rebellion of minor landlords, led by Gilbert de Middleton in the counties of Durham and Northumberland in 1317.[8] The 'war-free zone', on the other hand, encompassed all English counties south of the Humber (with an exception of west Gloucestershire, pillaged by the marcher lords in spring 1321), northern Scotland, the Scottish Isles, and most of the Gaelic parts of Ireland, as well as the Cambrian Mountains and northern parts of Wales. Here, the impact of war was indirect and socio-economic rather than environmental in nature. Thus, the English communities south of the Humber were undoubtedly pestered by royal officials, in the conjunction with war-related taxation and purveyance (forced sales of grain to provision armies), but local landscapes were neither ravaged, nor did they undergo the process of militarization.[9]

Landscape Devastation: Floral Resources

Although early fourteenth-century warfare had originated as a conflict about the control and authority over the Kingdom of Scotland, it soon developed into environmental warfare, resulting in destruction of landscape within the 'war zone', whereby all belligerent sides employed a *chevauchée* strategy, consisting of the deliberate devastation of landscape resources to weaken the enemy's position as much as possible. Our sources provide numerous references to pillages into the countryside, conducted by both sides, which resulted in the decline of the food supply of local communities. Plundering was not always committed by enemy forces: thus, in some instances, northern English villages were prone to attack by both English and Scottish soldiers. Similarly, Scottish natural resources could be ravaged by both English and Scots, especially in the

[8] Prestwich, 'Gilbert de Middleton'.

[9] On war-related taxation, see Jenks, 'The Lay Subsidies'. On purvayence, consult Maddicott, *The English Peasantry*.

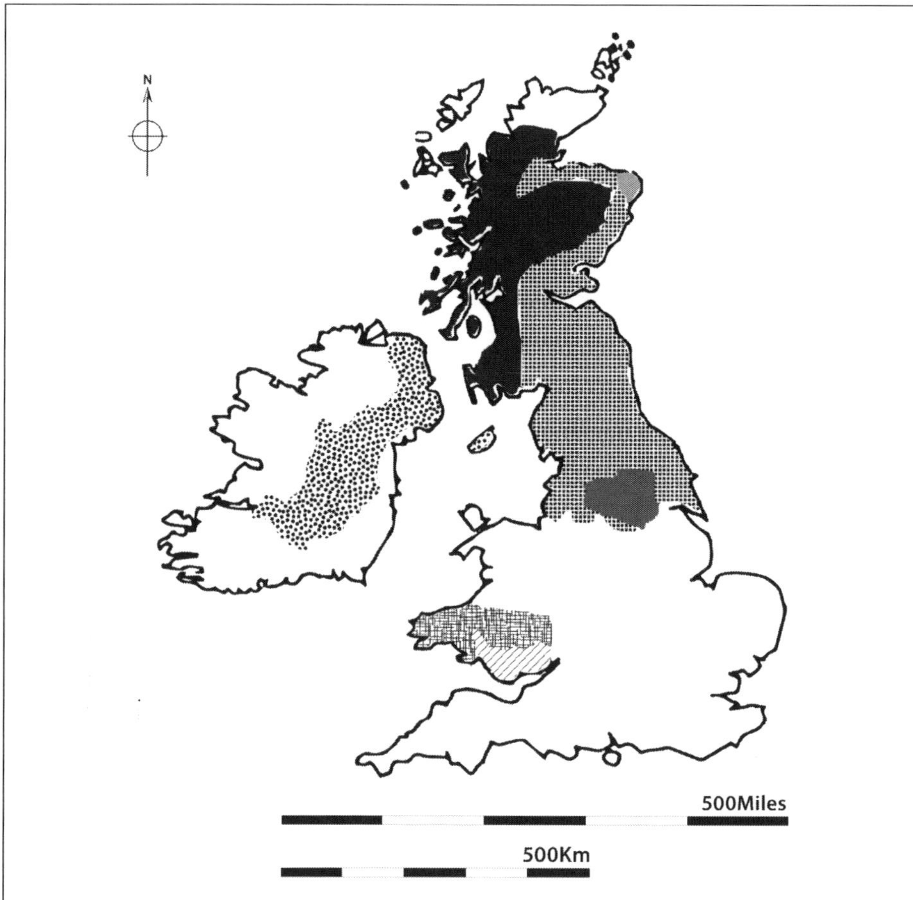

Figure 6.1. 'War zones' in the early fourteenth-century British Isles, *c.* 1296–1327.

Notes: Square dots = regions directly affected by the Anglo-Scottish warfare; black = territories affected by Robert I the Bruce's campaign of 1307–1309; light grey = earldom of Buchan, affected by both the Anglo-Scottish warfare and Bruce's campaign; round dots = Irish lands affected by Edward Bruce's campaign of 1315–1318; diagonal lines = south Welsh areas affected by Llywelyn Bren in 1316 and by the civil war of 1321–1322; crossed lines = south Welsh areas affected by the civil war of 1321–1322; dark grey = West Yorkshire areas affected by the civil war of 1321–1322; and white = 'war-free' zones.

context of the Scottish civil war of 1307–1309. Although we may consider the Edward I's invasion of Scotland, following the victory at Dunbar (27 April 1296) and the march to the north all the way to Inverness-shire, as a start point of the environmental warfare, the following discussion shall focus on 1310s and the early 1320s, when the intensity of ecological destruction seems to have reached its peak.[10]

Food plundering consisted of several techniques. The single most widespread and devastating technique was arable field destruction, either through ignition or trampling with cattle and horses. Setting fields on fire was certainly the most effective method of the two, but it was also challenging because of precipitation and dampness during the crisis years. Because of that, marauders had to resort to the second method, which often went hand in hand with frequent livestock raids. From 1311 on, and especially between 1311 and 1322, parts of the northern counties of England were badly devastated by marauding attacks of the Scots.[11] Thus, on 1 August 1315, after a nine-day siege, the Scottish army of Robert the Bruce invaded Carlisle and plundered the town. In the course of the assault, the Scots destroyed all the crops growing there.[12] Monastic houses, such as Durham Cathedral Priory, Bolton Priory (in the Yorkshire Dales), Hexham Abbey (Northumberland), St Bees Priory, and Holme Cultram Abbey (both in Cumberland), as well as their manorial estates, because of their wealth and vulnerability, were targeted repeatedly by the Scots.[13] Because late summer and early autumn were the campaigning season, most Scottish raids often coincided with the harvest time. Thus, at Stamford (Northumberland), the harvest of 1314 was interrupted by the Scots, who seem to have damaged crops growing there.[14] Crop fields were devastated outside the harvest times too. Thus, during Llywelyn Bren's revolt in south Wales (28 January–18 March 1316), the destruction of germinating grains and legumes by Welsh insurgents is reported at marcher lords' manors of Dowlais, Neath, and Rumney, all in Glamorganshire.[15]

In Scotland, local peasants, too, suffered from similar raids, conducted by both English soldiers and Scottish hosts. Thus, in August 1319, before the failed siege of Berwick-upon-Tweed, English garrisons were laying waste to the Berwickshire countryside.[16] In May 1308, Robert the Bruce burnt a large

[10] I have studied this in Slavin, 'Warfare and Ecological Destruction'.

[11] McNamee, *Wars of the Bruces*, pp. 72–122; Slavin, 'Warfare and Ecological Destruction', pp. 531–34.

[12] *CDRS*, no. 621.

[13] *The Bolton Priory Compotus*, ed. by Kershaw and Smith, pp. 456–509.

[14] TNA, DL 29/1/3, m.2.

[15] TNA, SC 6/1202/8.

[16] *Vita Edwardi Secundi*, ed. by Childs, pp. 164–65.

amount of the land of John Comyn, Earl of Buchan. According to Bruce's biographer, John Barbour, he devastated Buchan 'in such fashion, that for fifty years its men lamented about the *herschip* [hardship, destruction] of Buchan'.[17] Ireland, because of its political and socio-cultural fragmentation, seems to have suffered particularly from plundering by the Scottish invaders, English garrisons, and local Irish chieftains between 1315 and 1318.[18] In the course of his campaign, Edward Bruce's forces ravaged the Ulster lands in June 1315, and a year later the depredation spread to the south. Around August 1315, the armies of Richard Óg de Burgh, 2nd Earl of Ulster and 3rd Baron of Connacht, and Edward the Bruce, unable to fight each other because of the flooding, ravaged both banks of the River Bann in Ulster, sparing 'neither wood, nor plain, nor field, nor crop corn, nor residence, nor barn, nor church, without burning and wholly destroying'.[19] About the same time, the hosts of Feidhlim O'Conchobhair, king of Connacht (1310–1316), plundered all over his kingdom, burning local fields. The same policy was continued by his enemy and successor, Ruaidri na bhFeadh Ó Conchobair (1316–1317).

Grassland resources, consisting of permanent pasture and mowable meadow, were another type of natural resource targeted by the belligerents. Thus, between 1315 and 1317, the hinterland of Hexham (Northumberland), owned by the Archbishop of York, was severely devastated by the Scots.[20] In summer 1317, Scottish raiders overran the cattle pasture of Armathwaite Nunnery, an impoverished Benedictine house in Cumberland, to such an extent that Edward II granted the nuns pasture in his own forest of Inglewood, in the same county.[21] Some manorial accounts report the abandonment of destroyed tenements by local tenants, in conjunction with destruction of pasture.[22] An additional method of ecological destruction was devastation of woodland resources. In 1317, a large number of trees were burnt by the Scots at Lilswood

[17] *The Bruce by John Barbour*, ed. by Mackenzie, Bk 9: 294–300; *Scotichronicon*, ed. by Watt, VI, Bk 12:17, pp. 342–43. On the background to the '*herschip* of Buchan', consult Young, 'The Earls and Earldom of Buchan'.

[18] The Scottish intervention of Ireland and its impact is masterfully surveyed in Duffy, ed., *Robert the Bruce's Irish Wars*, pp. 9–43.

[19] *The Annals of Loch Cé*, ed. and trans. by Hennessy, p. 567

[20] TNA, SC 6/1144/3.

[21] *CDRS*, no. 578.

[22] TNA, SC 6/1144/3; TNA E 358/16.

Park, near Hexham (Northumberland).[23] It is unclear how the Scots managed to set Lilswood Park on fire, given the damp nature of northern English woods, and it is possible that the extent of the destruction was exaggerated by a local manorial reeve. Equally mystifying is the method by which the Scots cut down trees belonging to Durham Cathedral Priory at Bearpark (Durham) in the early 1320s, given the speedy pace of their raids in the northern counties.[24] Other arboreal resources, prone to environmental destruction, were gardens and orchards. Between 1317 and 1324, the Scots led a series of devastating attacks to Wark-on-Tweed (Northumberland), systematically destroying its orchard. Because of that, no fruits were produced in the course of these years.[25] At Hexham, in 1317, local orchards did not render any profit because fruit trees were cut and burnt by the Scots.[26] A similar destruction of woodland and arboreal resources was reported in the Lordship of Ireland. In late 1315, the troops of Edward II devastated the gardens and orchards of the manors of St Sepulchre and Finglas, as well as the woodland in Clondalkin, all in County Dublin and all belonging to the Dublin Archbishopric.[27] Although there is no doubt that the belligerents did inflict environmental damage, it is important to take contemporary descriptions with a grain of salt. As we have seen time and again, both chroniclers and manorial clerks may have somewhat exaggerated the scale of environmental destruction to emphasize (in the former case) the apocalyptic nature of the crisis or (in the latter case) to seek an excuse for poor management.

Landscape Devastation: Faunal Resources

The pillaging of faunal resources was another method of environmental warfare. The two main tactics here were the driving away large numbers of cattle and horses and the destruction of smaller livestock (sheep, swine, and poultry). The majority of references relate to cattle raids. This is hardly surprising, for several reasons. First, beef and milk were especially prized during the crisis years, when they could at least partially compensate for the dearth of cereal-based foods.

[23] TNA, SC 6/1144/3, m. 1a.

[24] *Correspondence, Inventories*, ed. by Raine, pp. 22–23.

[25] TNA, SC 952/12.

[26] TNA, SC 6/1144/3, m. 1a.

[27] National Archives of Ireland (NAI), RC 8/10, pp. 416–22. Printed in *Historic and Municipal Documents of Ireland*, ed. by Gilbert, pp. 366–71.

Second, cattle raids were undoubtedly a serious blow to agricultural producers, especially in the context of the bovine pestilence of 1319–1321 and the subsequent shortage of bovids.[28] Cattle raids were, therefore, one of the most efficient strategies of reducing the food resources of north English communities even further. Finally, cattle were considered to be a measure of wealth, socio-economic status, and prestige in medieval Gaelic societies.[29] Therefore, the cattle raids carried environmental, economic, social, and political aspects, all at once.

The volume of cattle raids reached its zenith between 1315 and 1319. Thus, in June 1315, the Scots attacked Bearpark (Durham) and, having sacked a local manor house, carried away 180 cows.[30] A month later or so, James Douglas, Lord of Douglas and a leading Scottish commander, drove away a large number of oxen from St Bees Priory.[31] During the Siege of Carlisle in late July 1315, a great multitude of cattle was seized away from the town's hinterland in Cumberland and Westmorland to feed Scottish soldiers besieging the town.[32] In the course of 1318 and 1319, the Scots conducted frequent and extensive cattle raids into Berwickshire, Northumberland, Cumberland, Westmorland, and Yorkshire, carrying away numerous animals back to Scotland.[33] Within Scotland proper, we hear about similar raids, especially in those lands loyal to the English crown. For instance, sometime in 1309, when Robert Bruce took Dunstaffnage Castle (in Argyll) from John MacDougall (or, Eóin MacDubhgaill), he took away 'such a multitude of cattle, that it was a marvel to behold'.[34]

The change came around September 1319, when a great pestilence broke out in bovine stocks, driven by Edward II's army to the failed siege of Berwick-upon-Tweed.[35] The Scots, unaware of the disaster, carried on their cattle raids

[28] Newfield, 'A Cattle Panzootic'; Slavin, 'The Great Bovine Pestilence'.

[29] For Scotland, see Britnell, *Britain and Ireland*, p. 10. For Ireland, see McCormick, 'The Decline of the Cow'.

[30] *CDRS*, no. 674.

[31] *The Chronicle of St Mary's Abbey, York*, ed. by Craster and Thornton, p. 68.

[32] *The Chronicle of Lanercost*, trans. by Maxwell, p. 213.

[33] *The Chronicle of Lanercost*, trans. by Maxwell, pp. 221, 226, and 227.

[34] 'Quhar men mycht se so gret aboundans cum off fe that it war woundir till behald'. *Bruce by John Barbour*, ed. by Mackenzie, Bk 10:108.

[35] *The Chronicle of Lanercost*, trans. by Maxwell, p. 228; Galbraith, 'Extracts from the Historia Aurea', p. 210.

into the northern counties and thus imported the virus into their lands.[36] As several chronicles narrate, the cattle murrain ravaged Scotland in 1320–1321.[37] The widespread transhumance of bovids across the 'war zone' undoubtedly assisted the pathogen in spreading, creating the optimal conditions for the dissemination of the cattle pestilence.[38] It is hardly surprising, therefore, that the volume of cattle raids shrank considerably after September 1319, as there were few animals to plunder.

Although most livestock raids targeted cattle, there are quite a few references to the plundering of other domestic animals. Because of their limited speed and range, sheep, swine, and poultry were usually slain and consumed on the spot, while horses were driven away. At Bearpark (Durham), in the early 1320s, the Scottish hosts drove away horses and sheep and 'forty of the better sheep were lost in the depredation'.[39] In the course of the civil war of 1321–1322 between Edward II and the contrariant barons, Andrew de Harclay, one of Edward II's commanders, conducted a series of livestock raids on the contrariants' estates, in the course of which he took a swine herd from Thomas of Lancaster's manor of Methley (West Yorkshire) and geese from Adam de Everingham's demesne of Birkin (North Yorkshire).[40] The contrariants paid to the king's supporters with the same token. Thus, the Welsh manor of Newport (Monmouthshire) belonging to Hugh Despenser, one of Edward II's closest allies, was plundered by the rebellious barons, who carried away large numbers of sheep, cattle, and horses.[41]

Domestic animals were not the only type of fauna that suffered from environmental warfare. There is also evidence of wildlife pillaging. Thus, according to a 1327–1328 account from Bamburgh Castle (Northumberland), a local rabbit warren had been destroyed by the Scots for a long time, and, therefore, it rendered no profit.[42] In the course of the 1322 campaign into northern England, the Scots destroyed dovecotes at Healaugh (North Yorkshire) and Dunford (West Yorkshire).[43] In addition, there is evidence of pillaging of aquatic resources. In the course of their frequent raids into Wark-on-Tweed

[36] *Bruce by John Barbour*, ed. by Mackenzie, Bk 17:225.

[37] Newfield, 'A Cattle Panzootic', pp. 168–69.

[38] Cooter, 'Of War and Epidemics'.

[39] *Correspondence, Inventories*, ed. by Raine, pp. 22–23.

[40] TNA E 358/16, m. 68.

[41] *Vita Edwardi Secundi*, ed. by Childs, p. 189.

[42] TNA, SC 6/950/3, m.3.

[43] TNA, E 358/16, m.68.

(Northumberland), the Scottish raiders destroyed a local salmon fishgarth (a river weir for raising fish). As a result, no salmon could be caught for several years, and the fish-garth could not be leased out.[44] It should be noted that although the destruction of salmon weirs could have some implications in a long run (as it is evidence in the Wark-on-Tweed accounts), it also would temporarily have allowed more fish into the spawning grounds and, thus, benefitted the aquatic ecosystem in a short run.[45] Although these wild natural resources may be considered minor in terms of their contribution to food and sustainability, they were still instrumental in providing complementary food and cash, especially in the context of widespread crop shortage and poverty. Thus, rabbits were prized for their meat and fur, as well as their fast reproductive cycle and large issues. Rabbit rearing was a particularly suitable type of enterprise in marginal lands that rendered lower crop yields.[46] Similarly, salmon are important sources of omega-3 fatty acids, protein, iron, calcium, and vitamin D. Doves, on the other hand, were consumed primarily by upper social strata in the summer and autumn.[47]

Although the majority of references talk about relatively small-scale depredations, there seemingly were instances when the proportions of physical destruction of animals were disturbing. One particularly illustrative instance comes from the Irish kingdom of Connacht, torn by a tribal conflict between several claimants to the local throne. In the course of warfare, local chieftains spent a great deal of energy raiding each other's livestock. In late 1315, Tadhg mac Domnall Ó Cellaigh, king of Uí Maine (1315–1316) and Ruaidri na bhFeadh Ó Conchobair, a Connacht prince and a claimant for throne, pursued their enemies Feidhlim O'Conchobhair, king of Connacht (1310–1316) and his foster-father Maelruanaidh Mac Diarmada (Mac Dermott), king of Magh Luirg (1294–1331). According to some Irish sources, en route, Tadhg and Ruaidri destroyed thousands of cattle, sheep, and horses belonging to Feidhlim and Mac Diarmada, in addition to killing many people, in several locations in Sligo.[48] Mac Diarmada reacted by invading Tadhg's domains along the Shannon and killed all his cattle and sheep. Then, he joined forces with his foster-son and

[44] TNA, SC 952/12.

[45] I am grateful to Prof. Richard Hoffmann, York University (Toronto) for this suggestion.

[46] Bailey, 'The Rabbit and the Medieval East Anglian'.

[47] Stone, 'The Consumption and Supply of Birds', pp. 148, 151, 159; Dyer, 'Seasonal Patterns in Food Consumption', p. 206.

[48] These included Letir-Luighne, Sliabh-Gamh and Glenn-Fathroimh.

raided all cattle and horses that were found in his kingdom, presumably to compensate for the losses of his own stocks. As some chroniclers remarked, 'never within the memory of man were so many cattle fruitlessly destroyed in one place'.[49] Again, while the language of the chronicles appears too apocalyptic, there can be little doubt that it still reflects the environmental damage inflicted by the fighting sides involved in a local tribal conflict. Manorial accounts, too, report (albeit without much dramatics) on anthropogenic destruction of livestock. Thus, at some point before 1324, the Scots burnt five large vaccaries of Barnard Castle (Durham).[50] Although the Barnard Castle account does not specify the number of cattle perished in the depredation, it could be several hundreds, if not more.

Landscape Devastation: Grain Storages and Mills

The destruction of food resources was not only about setting arable fields on fire or slaughtering flocks. Another method was related to the devastation of space reserved for the processing and preservation of crops, namely barns, granaries, and mills. Both mills and storage facilities became increasingly vulnerable targets during the ongoing warfare. Like the destruction of floral and faunal resources, the devastation of mills and storage facilities could have far-reaching implications for local populations and their economies. In some instances, belligerents would seize stored crops rather than destroy them, not the least because of obvious economic profits. Thus, sometime before 13 May 1318, during their raid into North Yorkshire, Scottish marauders broke into a royal barn at Boroughbridge and stole 6.75 quarters of grain, which were ready to be milled.[51] Two years later, the Scots drove away with an astonishing 270 quarters of crops from another royal granary at Holme Cultram Abbey (Cumberland).[52] In other instances, however, the raiders would deliberately set granaries on fire. The former tactic reflects the hasty nature of many of the *chevauchée* campaigns, undertaken by the fighting sides. While on the run, the marauders had neither the time to collect nor the possibility to carry the stored crops. Destroying mills and storage facilities with fire was the most efficient

[49] *The Annals of Loch Cé*, ed. and trans. by Hennessy, pp. 577–78; *Annála Connacht*, ed. by Freeman. pp. 238–41 (an online edition is available courtesy of the *CELT* Project).

[50] TNA, SC 6/835/1.

[51] *CIM*, no. 385.

[52] *CIM*, no. 408.

option in such circumstances, as it would not only deprive local communities of their food resources but also of the possibility to store and process food. As we have seen in the previous chapter, storage facilities were an essential backbone of late medieval economy, but it was especially during the periods of bad weather and food shortages that their role was especially vital. At some point before 26 July 1318, Scottish marauders destroyed vast granges of Fountains Abbey (Yorkshire), arguably the single largest and wealthiest Cistercian house in England.[53] Between 1315 and 1322, they conducted several raids on the Durham manors of Durham Cathedral Priory, driving away livestock and destroying its grain storages.[54] Shortly before the Battle of Byland (14 October 1322), the Scots burnt the granaries of Wath-Harrogate (North Yorkshire), destroying all the crops stored there.[55] In Ireland, in the course of the Bruce Campaign of 1315–1318, both belligerents and bands of robbers were involved in plundering local grain storage facilities. For instance, between 1315 and 1317 the granges of St Mary's Abbey, Dublin, were repeatedly robbed by both English and Irish bands. As a result, the house was greatly impoverished towards the end of the war.[56] Likewise, the granaries of St Mary de Hogges Nunnery and those of Abbey of St Thomas the Martyr (both Augustinian houses in Dublin), are reported to be plundered during the war.[57]

The destruction of barns and granaries could have some devastating consequences. As northern communities were already suffering from the disastrous effects of the torrential rain, ruining their meagre food supplies stored at local storages and destroying the same facilities undoubtedly aggravated the crisis even further. Constructing a new grain storage could be a long and costly enterprise, taking at least a few months, if not more. Given the omnipresent poverty of the northern communities, as indicated in both the lay subsidy rolls of 1315 and 1316 (which will be discussed at length below) and the account book of Bolton Priory (in the Yorkshire Dales), hiring storage facilities or securing grains at local markets for immediate consumption could be challenging or altogether unaffordable. The situation was especially dire in those instances when storages were destroyed alongside with the stored grains: in those cases, the victims would be deprived of both storage facilities and foodstuffs and

[53] TNA, E 179/67/9.

[54] *CDRS*, no. 133.

[55] TNA E 358/16.

[56] NAI, KB 2/8, p. 5; KB 2/12, p. 4.

[57] NAI, KB 2/7, pp. 47–48; NAI, KB 2/8, p. 23.

would incur painful losses, implying an apparent decline in their food entitlement, to use Sen's terminology.

References to mill destruction are also numerous and a few examples will suffice. At Wark-on-Tweed (Northumberland), a local mill, destroyed by Scots around 1316–1317, was not yet repaired by 1323–1324.[58] At Bamburgh Castle, in the same county, two windmills and a rabbit warren were described in 1327 as 'long destroyed'.[59] At Bransty and Wilton (both in Cumberland), watermills were burnt by Scots in or shortly before 1322.[60] Similarly, the destruction of aquatic mills at Temple Sowerby (Westmorland) is mentioned in 1315 and 1323, but it is unclear whether the sources talk about two recurrent acts of destruction or about the same event.[61] At Fishlake and Kempsell, two West Yorkshire manors belonging to Robert de Holland, one of the closest associates of Thomas of Lancaster, local mills were destroyed by the royal armies sometime in 1321–1322, in the course of the civil war.[62] There is also evidence of mill destruction in south Wales during Llywelyn Bren's revolt (28 January-18 March 1316). Thus, the destruction of mills and, consequently, the decline in mill revenues are reported at Pentyrch, Briton Ferry, Senghenydd, Miskin, Glynrotheny, and Rumney, all in Glamorganshire.[63] In Ireland, in the beginning of the Anglo-Scottish conflict, Edward II's armies destroyed manorial mills belonging to the Archbishopric of Dublin while marching in County Dublin en route to Ulster, as an extent of the archiepiscopal properties, compiled in December 1315, indicates.[64]

The omnipresent destruction of mills could be especially devastating in the context of local manorial customs. In the regions north of the Humber, manorial lords prescribed their tenants to keep the 'suit of mill' (*secta molendini*) and grind their crops at seignorial mills. Suit of mill was a manorial custom expecting tenants to mill their harvest produce at their lord's mill, for a grinding fee known as *multure*, usually representing between one-twelfth and one-

[58] TNA, SC 952/12.

[59] TNA, SC 6/950/3.

[60] TNA, E 372/172, m. 48.

[61] TNA, E 372/166, m. 32; E 372/172, m. 4.

[62] TNA E 358/16.

[63] TNA, SC 6/1202/8.

[64] National Archives of Ireland (NAI), RC 8/10, pp. 416–22. Printed in *Historic and Municipal Documents of Ireland*, ed. by Gilbert, pp. 366–71. The English translation can be accessed on the *CIRCLE* website.

twentieth of the ground amounts. Conversely, private hand mills were either non-existent or very few. This was in a stark contrast with the situation in the south, where tenant hand mills were not only plentiful, but also they competed with demesne mills, to the point that local lords did all they could to curtail the independent milling activities of their tenants.[65] In other words, in the north, because of manorial restrictions, local tenants strongly depended on their lords' mills, which were, in effect, the only existing type of milling facility. The destruction of demesne mills would have, in theory, deprived tenants of their ability to grind crops and, consequently, bake bread. Constructing a new mill could take at least a few months.[66]

It is unclear how northern tenants would have coped with this challenging situation — during the period between the destruction of an old mill and the construction of a new one. One possibility would be to bring their produce to neighbouring manors whose mills were spared by warfare. This, however, was often at odds with local manorial regulations. For instance, according to the 1326–1327 customal of Cockerham (just south of Lancaster), tenants were forbidden to mill their produce at other mills and were collectively responsible to maintain the dike of a local watermill.[67] This may echo the frequent Scottish incursions into Lancaster and its surrounding countryside between 1318 and 1322, during which many buildings were destroyed.[68] These raids may have forced the Cockerham tenants to seek milling opportunities outside of their manor, and the mill-related clause reflects the seignorial indignation towards their breaching of the custom. Another possible scenario could be that local tenants had to collectively lease their lord's mill outside of their manor. After all, mill farms (*firme molendini*) were a widespread practice both in the south and the north.[69] There is indeed some evidence that some lords attempted to lease out their demesne mills to local tenants in war-affected regions. For instance, shortly after the confiscation of former contrariant estates in West Yorkshire by Edward II in 1322, numerous mills were leased out.[70] In some cases, however, the lords had difficult times securing lessees, on account of depopulation and insecurity. Thus, at Learmonth (Berwickshire), mills rendered only £4

[65] Langdon, 'Lordship and Peasant Consumerism'.

[66] Langdon, *Mills in the Medieval Economy*, p. 253.

[67] Bailey, *The English Manor*, pp. 62–64.

[68] McNamee, *Wars of the Bruces*, pp. 84 and 90.

[69] Langdon, *Mills in the Medieval Economy*, pp. 56–63.

[70] TNA, SC 6/1145/21.

13s 4d, and no more, because nobody was ready to pay more due to fear of the Scots.[71] The length of lease-contracts in the North was, almost as a rule of thumb, shorter than in the South, averaging in the fourteenth century around five years.[72] The third possibility is that tenants had to temporarily resort to alternative crop-based food substitutes. The most obvious alternative would have been so-called *frumenty*, made of a mixture of broken-up wheat kernels boiled in hot milk. Similarly, broken-up grains could be added to legume pottage. Grain kernels could be easily broken up with the help of mortars; indeed, there is plenty of archaeological evidence of stone mortars found in late medieval peasant contexts.[73]

The Impact of Landscape Destruction on 'War-Zone' Communities

The devastation of landscape, through the destruction of biological organisms and buildings, had the harshest impact on some communities living within the 'war zone'. As noted in Chapter 3, crop-based products, mainly bread, ale, and pottage, may have contributed about 70 per cent to the total calorific intake of an average English commoner, around the same time.[74] Naturally, this is an average estimate and the degree of reliance on crop-based diets varied from region to region. Within the 'war zone', West Yorkshire, parts of North Yorkshire, Durham, and Northumberland had a strong reliance on arable, while Lancashire, Westmorland, and Cumberland focused primarily on pastoral husbandry. Nevertheless, even some upland manors and farms located on the infertile soil of the Pennines cultivated oats. The contribution of crops to the Scottish diet seems to have been lower (perhaps about 40 per cent), given the country's stronger reliance on livestock farming and, to a lesser degree, fishing.[75] All the same, arable farming was an important form of agricultural enterprise in the southern lowland parts, and this implies that the destruction of crop fields had some harsh implications on food security in lowland Scotland. Ireland, too, was a much more pastoral country than England, but even here there seems to have been somewhat of a contrast between the Lordship of

[71] TNA, SC 6/952/13.

[72] Langdon, *Mills in the Medieval Economy*, p. 189.

[73] Dyer, 'Did Peasants Need Markets and Towns?', p. 32.

[74] Dyer, 'Changes in Diet', and *Standards of Living*, pp. 134–35.

[75] Barrow, *Kingship and Unity*, p. 12; Driscoll, *Alba*, p. 53; Oram and Adderley, 'Lordship and Environmental Change', pp. 74–84.

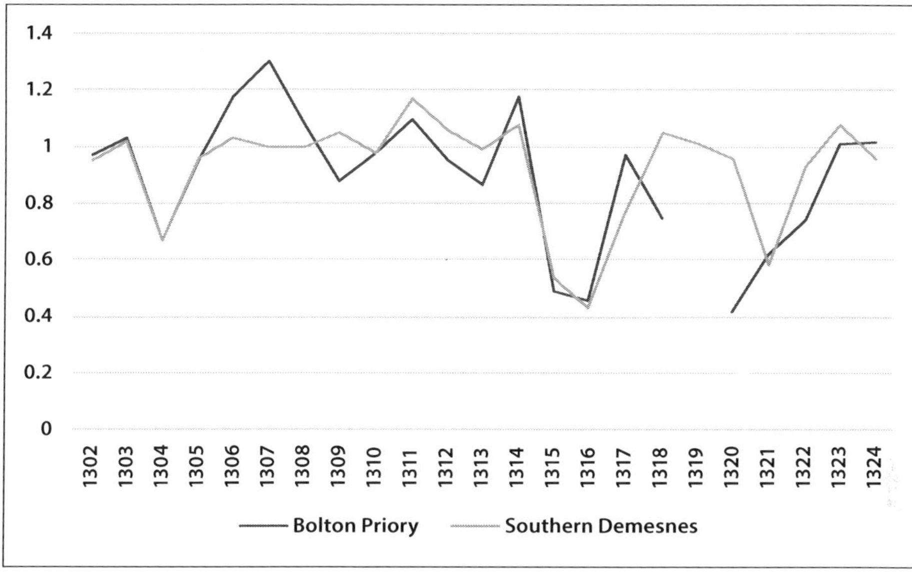

Figure 6.2. Crop yields on Bolton Priory demesnes (West Yorkshire) and southern demesnes, 1302–1324 (indexed on 1302–1314). *Source*: Slavin, Manorial accounts database (see Appendix 8); Campbell, *Three Centuries of English Crop Yields*; *The Bolton Priory Compotus*, ed. by Kershaw and Smith.

Note: The respective figures indexed on 1302–1314 (1302–1314 = 1.0), appearing on the vertical axis, are as follows: (1) Bolton Priory composite grain yields (wheat, rye, oats, and barley), net of seed and gross of tithe: 1.0 = 3.9:1.0; and (2) southern demesne crop yields (wheat, rye, oats, and barley), net of seed and gross of tithe: 1.0 = 4.1:1.0.

Ireland and the Gaelic regions.[76] In the Lordship of Ireland, long before the introduction of the potato in the eighteenth century, grains and legumes have undoubtedly played an important role, as is indicated in some late thirteenth-century accounts from the Irish estates of Roger Bigod, 5th Earl of Norfolk (*c.* 1245–1306) and the justiciar of Ireland's rolls.[77] Conversely, the Gaelic parts of Ireland relied strongly on pastoral husbandry, around the same time.

The patchy, descriptive, and, at times, dramatic nature of the sources does not allow a quantitative estimation of the sheer extent of the damage. All the same, the available evidence suggests, beyond any doubt, that the proportions were disastrous, and the damage done to the arable farming in northern England, southern and eastern Scotland, the Lordship of Ireland, and south Wales (affected by the revolt of Llywelyn Bren in early 1316) was indeed exces-

[76] Britnell, *Britain and Ireland*, pp. 10–11, 43–44.

[77] TNA, SC 6/1237–39; NAI, KB 2/7–12.

sive. For instance, at Wark-on-Tweed (Northumberland), almost 1200 acres of both the demesne lands and tenancy were burnt by the Scots in 1318.[78] Some sources indicate that entire manors with their fields were destroyed by belligerents. These included, for instance, the vills of Westwick, Middleton-in-Teesdale, Newbiggin, Langton, and Newsham, all in the hinterland of Barnard Castle (Durham), burnt entirely by the Scots in summer 1322.[79] The destruction of fields with fire depressed the crop yields beyond the years of the Great Famine, thus exacerbating the crisis and prolonging the dearth and food insecurity within the 'war zone'. This is exemplified in Figure 6.2, showing crop yields on the demesnes of Bolton Priory, situated in the Yorkshire Dales and attacked at least three times by the Scots between 1319 and 1322. Here, the crop yields were abysmally low in 1318–1322, despite the return of good harvests in the 'war-free zone' in the south (with the exception of the bad harvest of 1321). In 1320 they stood at 58 per cent below average, a figure lower than in 1316. The yields were 38 and 26 per cent lower than average in 1321 and 1322, respectively, and it was only in 1323 that they returned to their pre-1315 levels. Unlike in 1315–1317, the three back-to-back harvest failures of 1320–1322 were caused more by endogenous, than exogenous, factors. Surely, the Scottish raids did their job of destroying a part of the harvests.

There are omnipresent references to the destruction of other types of land-use — first and foremost, grassland and woodland. Invariably, pastureland was destroyed by trampling cattle, usually carried away by marauding Scots. The impact of warfare on grassland and woodland resources is not straightforward for interpretation. On the one hand, there is no doubt that the dramatic descriptions of contemporary chroniclers meant to exaggerate the apocalyptic aspect of the crisis, while local manorial reeves and clerks sought an excuse for poor management of their lords' estates. For instance, it is difficult to accept at face value that in early 1326, 360 acres of woodland at Harbottle (Northumberland) did not render any profit because of their destruction several years earlier.[80] After all, how many weeks would have been required by the speedily moving Scots to fell or burn 360 acres of woodland there? Similarly, one cannot treat without scepticism that all of Forest-in-Teesdale (Durham) was still in a ruinous state in 1325, owing to its devastation in late 1321 or early 1322.[81] On the other hand,

[78] That the attack occurred in 1318 is indicated in TNA, SC 952/12.

[79] TNA, E 372/172, m. 50.

[80] TNA, E 372/173, m.54.

[81] TNA, E 372/172, m. 52.

these descriptions, dramatic and, at times, unbelievable as they may be, should not be dismissed. It appears that they still reflect *some sort* of the environmental destruction of grassland and woodland *on some scale.* The destruction may have not been omnipresent; but it was undoubtedly devastating for those local communities where it occurred.

This was especially true in those regions where, because of ecological peculiarities and topographic attributes, grassland was the single most important component of land-use and livestock farming was the most paramount form of economy. Such a situation prevailed in the Pennines (a mountain range covering, within the 'war zone', north-eastern Lancashire, the Yorkshire Dales, the Cumbrian Fells, and the Cheviot Hills), the Vale of Pickering in North Yorkshire, and the lowland valleys of Yorkshire rivers Swale, Ouse, and Hull in north-eastern Yorkshire.[82] As we have seen, Scotland and Ireland were generally much more pastoral than England. In Scotland, regions such as Galloway, the southern Uplands (a belt running from South Ayrshire to East Lothian), Lorn, Garmoran, Argyll, Atholl, Moray, and Ross were particularly endowed with grassland and, as such, focused primarily on dairy husbandry.[83] In Ireland, pastoral husbandry was an omnipresent feature all over the country, while in Gaelic-speaking regions in the west it represented the core and backbone of local economies, with the vast majority of calories deriving from domestic animal products. Apart from the vital importance of grass as animal fodder, however, the destruction of pasture and sod (turf) had a serious impact on the availability of raw materials available for building and repairing houses and other structures. This was undoubtedly critical during the raids and pillaging, when many structures were destroyed by marauders.

As in the case of arable and pastoral destruction, deforestation, too, could have some series implications for local communities. Most species growing in northern forests, and particularly in those places mentioned in the sources, have long recovery cycles. Thus, oak (*Quercus robur*), dominating Bearpark and Lislwood Park, has at least a 19-year recovery cycle. Similarly, ash (*Flaxinus excelsior*) and hazel (*Corylus Avellana*), growing at Bearpark, and Scots pine (*Pinus sylvestris*) and Lodgepole (*Pinus contorta*), thriving at Harbottle, all have at least a ten-year coppicing cycle. It is uncertain if all these species were growing in those forests in the early fourteenth century, but if that was the case, then

[82] Campbell and Bartley, *England on the Eve of the Black Death*, p. 138.

[83] Barrow, *Kingship and Unity*, p. 12; Dyer, *Making a Living*, pp. 15–17; Britnell, *Britain and Ireland*, pp. 10–12.

we may speculate about the destruction and long regeneration of woodland resources and their impact on local communities and their economies, especially given their pronounced importance in the context of the early fourteenth-century crisis. First, the military conflict drove up a demand for timber to build weapons, vehicles, and ships. Second, the exceedingly cold winters of the famine years implied a pronounced need for timber as fuel material. Third, woodland could be exploited and harvested for alternative types of food to temporarily replace the standard menu during the famine years. Thus, hazel growing at Bearpark and the juniper found at Forest-in-Teesdale could be harvested for nuts and berries, respectively. Finally, woodland provided pasture for cattle and mast for swine. Here, dairy and meat products could at least partially compensate for the shortage of grain. Therefore, the negative implications of the depletion of woodland resources on both humans and animals cannot be overstated.

Livestock raids were devastating on several levels. Violent treatment, physical stress and abuse of animals tended to undermine animal health and reproductive cycles, as is indicated in the documents. For instance, in early 1322, at the height of the civil war, Edward II's supporters drove away thirty cows from Robert de Holland's vaccary in Marsden (West Yorkshire). As a result, most driven cows miscarried, and only four produced calves.[84] Also, the physical depletion of stocks implied depletion of food resources and muscle power, while the restocking process could be very slow and costly. This is exemplified in Figure 6.3, showing the slow pace of replenishment of animal stocks on the home farm of Bolton Priory and the manors of Durham Cathedral Priory. Here, stocks were drastically reduced in the course of the 1310s and the early 1320s, and it was not until the later 1330s that they started approaching the pre-warfare levels. The demographics of Bolton and Durham Priory livestock show similar trends, but their chronology, and sometimes causes, varied. Thus, the numbers of Durham animals started declining in 1312, namely since the beginning of Scottish raids on the Priory estates. The Bolton stocks did not start declining until 1314–1315, when local flocks were hit by an outbreak of a devastating disease, which may have been caused by a combination of scab and liver fluke.[85] To make things even worse, however, the bovine stocks of both houses were decimated in the closing years of the 1310s through both the outbreak of rinderpest and cattle raids.[86] To compensate for the omnipresent

[84] TNA 358/16, m. 63.

[85] Slavin, 'Landscapes of Panzootics', pp. 158–61, and 'Flogging a Dead Cow'.

[86] On rinderpest, see Newfield, 'A Cattle Panzootic'; Slavin, 'The Great Bovine Pestilence'.

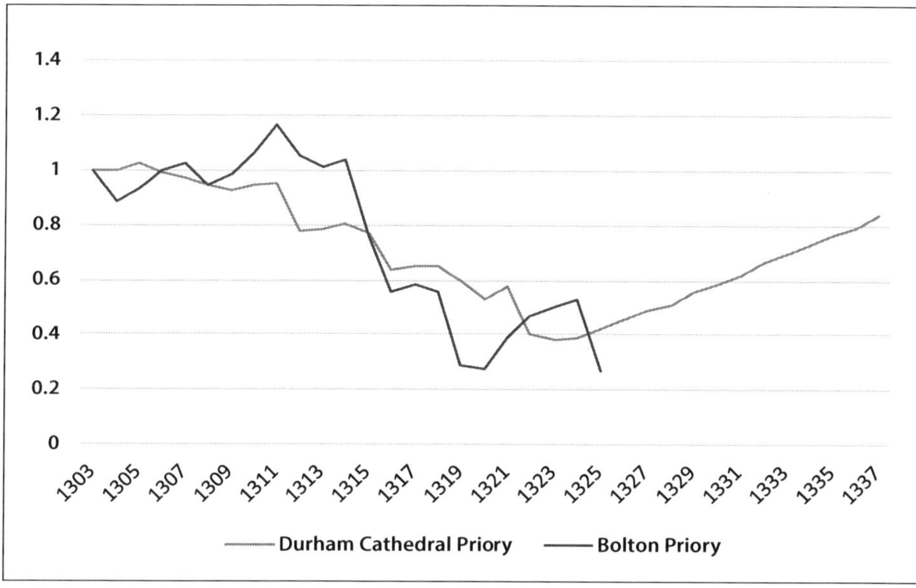

Figure 6.3. Livestock population (in units), on the estates of Durham Cathedral Priory and home farms of Bolton Priory, 1303–1337, indexed on 1303 (1303 = 1.0). *Source*: Slavin, Manorial accounts database (see Appendix 8); *The Bolton Priory Compotus*, ed. by Kershaw and Smith.

Note: The respective figures indexed on 1303 (1303 = 1.0), appearing on the vertical axis, are as follows: (1) Durham Cathedral Priory (fourteen demesnes) livestock units: 1.0 = 896 units; and (2) Bolton Priory livestock units: 1.0 = 759 units.

dearth of grain-based products, Bolton canons chose to resort temporarily to the consumption of pork. This resulted in virtually an entire swine stock getting butchered and eaten between 1315 and 1320. By 1320, the Bolton stocks amounted to some 27 per cent their 1303 levels, while in 1323, the Durham animal cohorts stood at 62 per cent below their 1303 levels. The impoverishment of Bolton Priory was undoubtedly one of the factors that prompted the canons to leave the house and disperse in 1325. On the Durham Priory demesnes, the pre-1312 figures were not achieved until the Black Death; in 1337, they still stood at about 84 per cent of their 1303 levels.

Apart from the loss of calories, livestock pillaging also meant a pronounced decline in income from pasturage sales and agistment (pasture rights) rents. A great number of manorial accounts from northern England, including those from Temple Newsam (West Yorkshire) and Errington (Northumberland), state that no profit derived from local pasturage because there were no animals to graze upon.[87] In those places, where pasturage was not destroyed by the Scots, depleted stocks stood in sharp contrast with abundant grassland. As we have seen in Chapter 2, the torrential rain of 1314–1316 produced a significant surplus of grass, especially in relation to the available animals and thus drastically decreased the demand for pasturage and, consequently, its value.

But, beyond the physical destruction of landscape and the decimation of animal stocks, war-zone ecological destruction had much wider and far-reaching implications. Here, we have to consider a long-term contraction (and stagnation) of local 'war-zone' economies, which show various signs of depression for at least a few decades. The long recovery of the northern economy is reflected in various sources. Thus, one chronicle stated that after the Great Famine, the lands of Northumberland remained waste for fifteen years.[88] Between 1313 and 1327, Cumberland, Northumberland, and Westmorland were all exempt from lay subsidies. In 1332, Northumberland was still exempt from taxation.[89] Although, as Chris Briggs has shown, these exemptions do not always reflect the damage from the Scottish raids, they certainly reflect the pronounced impoverishment of the northern counties, through war, weather, failed harvests, livestock diseases, and emigration.[90] As late as 1338, some Northumberland and Yorkshire demesnes of the Knights Hospitallers, formerly belonging to the Knights Templar, are reported to have been in a deplorable state because of Scottish incursions.[91]

No other document indicates the devastating impact of warfare on the northern counties as well as the *Nova Taxatio Eccelsiastica*, a detailed valuation for ecclesiastical taxation of the North, compiled in conjunction with the inability of local parish priests, vicars, and rectors to pay their taxes to the pope, in accordance with the earlier valuation of 1291–1292 (otherwise known as the *Taxatio Ecclesiastica* of Pope Nicholas IV).[92] The *Nova Taxatio* was compiled, at the royal request, at two main phases, in 1317 (covering parts of North Riding of Yorkshire, Durham, and Northumberland) and 1327–1328 (covering the

[87] TNA 358/16, m.68; SC 6/1144/3, m. 1a.

[88] *Chronica Monasterii de Melsa*, ed. by Bond, p. 333.

[89] Jenks, 'Lay Subsidies', pp. 35–39.

[90] Briggs, 'Taxation', pp. 668–70.

[91] Larking, ed., *The Knights Hospitaller in England*, pp. 133–43.

[92] The taxation has been printed as *Taxatio Ecclesiastica Angliae et Walliae Auctoritate*, ed. by Astle, Ayscough, and Caley. This outdated edition is now replaced with the fully searchable *Taxatio Database*.

remaining parts of Yorkshire, Lancashire, Cumberland, and Westmorland). As Table 6.1 shows, the values of ecclesiastical revenues in northern parishes have declined substantially in the wake of frequent attacks and harassment by Scottish marauders and local authorities, but also the crop failures and animal murrains, as well as population fall. Taken together, the sheer value decreased by some three-fourths — an overwhelming figure, indeed. Those figures, naturally, should be approached with much caution. For instance, according to the Northumberland portion of the *Nova Taxatio*, no parish priest or vicar could be taxed at all because of the utter destruction of the county. While there is no doubt that the raids caused significant devastation, and there is evidence that some parishes were destroyed and abandoned altogether, the view that the entire county would not pay anything is undoubtedly an exaggeration. But even Northumberland aside, the decline of revenue by 64 per cent may be an inflated estimate. All the same, however, there is no doubt that the *Nova Taxatio* reflects a colossal fall in agricultural output of the northern counties, as well as a visible decline in population levels and widespread poverty within these regions. Thus, of fifty-three parishes in the Carlisle deanery, as many as thirty-seven (that is, almost 70 per cent) could not be taxed at all, either because the revenue of local priests and chaplains did not (or hardly) suffice for their subsistence, or because they were 'entirely destroyed' (*totaliter destructa*) (thirteen parishes were described as such). Unsurprisingly, the worst decline in revenues is recorded in Northumberland, Cumberland, and Westmorland, which lay within the easiest reach of the attackers. Interestingly, the figures for Durham stand out as considerably lower (54 per cent). This may something to do with the ability of Bishop of Durham and Prior of Durham Cathedral Priory to occasionally pay off the Scots and thus avoid damage.[93]

Within Scotland, too, the ongoing warfare made its mark on local economies, both through the English invasions and Scottish tribal conflicts. Thus, according to John Barbour, Robert Bruce's campaign of 1307–1308 devastated Buchan 'in such fashion, that for fifty years its men lamented about the *herschip* [hardship, destruction] of Buchan'.[94] The *herschip* of Buchan was followed the by *herschip* of Galloway in June 1308.[95] The ecological and economic conse-

[93] McNamee, *Wars of the Bruces*, p. 135.

[94] *Bruce by John Barbour*, ed. by Mackenzie, Bk 9: 294 Roxburghshire –300; *Scotichronicon*, ed. by Watt, vol. VI, Bk 12:17 (pp. 342–43). On the background to the '*herschip* of Buchan', consult Young, 'Earls and Earldom of Buchan'.

[95] Oram, 'Bruce, Balliol and the Lordship of Galloway', pp. 38–40.

Table 6.1. Valuation of rural ecclesiastical property within the 'war zone', according to the *Nova Taxatio* (1317–1328) and Pope Nicholas' *Taxatio Ecclesiastica* of 1291–1292.

Deanery	**County**	**Year**	**No. of parishes**	**Value (£), 1317–1328**	**Value (£), 1291–1292**	**Difference**
Ainsty	Yorkshire, West Riding	1328	8	136.0	273.3	50.3%
Buckrose	Yorkshire, East Riding	1327	15	165.6	291.4	43.2%
Dickering	Yorkshire, East Riding	1327	13	185.0	326.2	43.3%
Boroughbridge	Yorkshire, North Riding	1318	34	364.3	727.7	49.9%
Catterick	Yorkshire, North Riding	1318	44	482.4	1,115.7	56.8%
Bulmer	Yorkshire, North Riding	1327	21	163.3	450.5	63.7%
Lydale	Yorkshire, North Riding	1327	17	167.0	428.8	61.1%
Cleveland	Yorkshire, North Riding	1327	19	238.6	436.0	45.3%
Durham	Durham	1318	32	489.1	1,027.0	52.4%
Aukland portions	Durham	1318	13	113.5	269.7	57.9%
Lanchester	Durham	1318	8	27.1	90.7	70.1%
Chester-le-Street portions	Durham	1318	8	64.7	146.7	55.9%
Darlington	Durham	1318	21	284.2	657.0	56.7%
Norton portions	Durham	1318	10	49.0	68.0	27.9%
Darlington portions	Durham	1318	5	45.0	94.7	52.5%
Newcastle	Northumberland	1318	36	0.0	1,032.5	100.0%
Corbridge	Northumberland	1318	33	0.0	955.3	100.0%
Bamburgh	Northumberland	1318	23	0.0	1,882.4	100.0%
Alnwick	Northumberland	1318	24	0.0	933.1	100.0%
Aymundern	Lancashire	1327	13	186.4	557.9	66.6%
Lonsdale	Lancashire	1327	16	157.6	781.3	79.8%
Westmorland	Westmorland	1327	28	108.5	702.2	84.5%

Deanery	County	Year	No. of parishes	Value (£), 1317–1328	Value (£), 1291–1292	Difference
Copeland	Cumberland	1327	8	34.0	160.0	78.8%
Carlisle	Cumberland	1327	41	36.3	694.4	94.8%
Allerdale (Alnedale)	Cumberland	1327	21	113.0	444.1	74.6%
Cumberland	Cumberland	1327	20	95.8	533.4	82.0%
Total (all deaneries)			531	3,706.2	15,079.9	75.4%
Total (without Northumberland deaneries)			415	3,706.2	10,276.6	63.9%

Source: *Taxatio Database*; TNA, E 179/67/9 (*Nova Taxatio*).

Note: The table includes only those parishes that appear both in the 1291–1292 and 1317–1327/8 valuations. The total number of parishes valuated in the *Nova Taxatio* was 593 rather than 531, as in the table.

quences of Bruce's punitive devastation of Buchan and Galloway were indeed far-reaching, stretching much beyond the troublesome second decade of the fourteenth century. The economic depression of Scotland was further exacerbated with the onset of the second phase of the Scottish Wars of Independence in 1332, with Edward III's armies conducting the systematic destruction of the Scottish countryside between the Forth estuary and Aberdeen from 1335 on. Unfortunately, the second phase of the Scottish War of Independence could not be examined in the present study. It may, nevertheless, suffice to note that when David II (1329–1371) conducted his national survey, known as *Verus Valor* (commissioned in 1357 and carried out before 1366), the value of estates in south-eastern and eastern Scotland were about 49 per cent lower than in Alexander III's reign (1249–1286) (Table 6.2).[96] According to the inquest, some borderland regions suffered in particular. For instance, the value of the sheriffdom of Dumfriesshire declined from £2666 13s 4d to £880 15s 4d (67 per cent). In the Selkirk sheriffdom, the decline was even more drastic: from £99 9s 10d to £18 6s (82 per cent). In, Berwickshire, and Peeblesshire, the respective values declined by 54, 37, and 32 per cent. Naturally, the figures of the *Verus Valor* account not only for the destruction during the First War of Independence but also for the subsequent Anglo-Scottish conflict of the

[96] The manuscript version of the *Verus Valor* is preserved in the National Archives of Scotland, Liber Niger, PA5/4, fols 44^{v}–47^{r}. The valuation can be consulted through the University of St Andrews database *The Records of the Parliaments of Scotland to 1707* (*RPS*), under the 1366/7/18 entry.

Table 6.2. Valuation of ecclesiastical and lay property in Scotland, according to the *Verus Valor* (1366) and the 'Ancient Taxation' in Alexander III's Reign (1249–1286).

		Verus Valor (1366)	Ancient Taxation	Decline
A. BISHOPRICS	Galloway	143.1	368.8	61.2%
	Argyll	133.3	281.3	52.6%
	Aberdeen	1,368.9	1,492.2	8.3%
	Ross	246.6	320.4	23.0%
	Dunblane	376.7	607.7	38.0%
	Brechin	321.8	441.2	27.0%
	Moray	559.4	1,418.6	60.6%
	Caithness	86.3	286.7	69.9%
	Dunkeld	602.7	1,206.3	50.0%
	Glasgow	2,028.5	4,080.6	50.3%
	St Andrews	3,507.0	5,414.0	35.2%
	Total	**9,374.4**	**15,917.7**	**41.1%**
B. SHERIFFDOMS	Clackmannan	243.7	331.0	26.4%
	Kinross	38.7	65.0	40.4%
	Fife	2,555.0	3,465.7	26.3%
	Perth	3,087.1	6,192.1	50.1%
	Forfar	2,240.3	3,370.3	33.5%
	Kincardine	722.0	1,088.5	33.7%
	Aberdeen	2,288.3	4,448.3	48.6%
	Banff	128.8	1,110.3	88.4%
	Inverness	1,080.5	3,113.6	65.3%
	Argyll	133.3	1,920.0	93.1%
	Wigtown	195.0	1,235.2	84.2%
	Dumfries	882.8	2,666.7	66.9%
	Dumbarton	96.5	1,442.5	93.3%
	Peebles	863.7	1,274.9	32.3%
	Selkirk	80.9	99.5	18.7%
	Roxburgh	523.9	1,133.8	53.8%
	Berwick-upon-Tweed	392.9	622.1	36.9%
	Stirling	687.7	1,750.0	60.7%
	Ayr	1,396.8	3,359.0	58.4%
	Lanark	1,756.0	4,057.5	56.7%
	Edinburgh	3,030.6	4,029.8	24.8%
	Total	**22,361.9**	**46,775.7**	**52.2%**

Source: *Records of the Parliaments of Scotland to 1707, RPS*, entry 1366/7/18.

Table 6.3. Valuation of rural ecclesiastical property in six Irish dioceses, according to the *Nova Taxatio* (1318) and an Earlier Valuation (*c.* 1303–1306) (in £ sterling)

Dioceses	***Nova Taxatio* (1318)**	**Earlier Valuation (*c.* 1303–1306)**	**Decline**
Meath	105.6	185.7	43.2%
Cloyne and Ross	62.8	112.9	44.4%
Limerick	39.2	69.5	43.6%
Cork	28.4	37.8	24.9%
Ossory	42.8	116.3	63.2%
Total	278.7	522.2	46.6%

Source: TNA, E 101/233/21; RCBCIA, Red Book of Ossory.

1330s, 1340s, and 1350s. Therefore, it reflects the extent of damage done by both wars, and it is impossible to estimate how much damage was caused by the early fourteenth-century conflict alone.

The extent of the environmental devastation in Ireland is harder to gauge because of comparatively thin documentation. Nevertheless, the records of ecclesiastical taxation, levied *c.* 1303–1306 and 1318 survive for six provinces: Meath, Cloyne, Ross, Limerick, Cork, and Ossory.[97] As the Ossory portion of the taxation indicates, the Irish *Nova Taxatio* inquest of 1318 was carried out at the request of Edward II 'after the Scottish war', apparently to assess the damage done by the war.[98] The value of parish churches, reflecting agricultural producers of local parishioners, declined by some 63 per cent at the diocese of Ossory; the comparable figures for Meath, Cloyne, Ross, and Limerick fell by 44 per cent, while the situation at Cork was somewhat better, with the values declining by about 25 per cent (Table 6.3).

Unlike in Scotland, no large-scale warfare has returned to Ireland in the pre-Black Death era (with the exception of the Burgh Conflict of 1333–1338). Yet, the combination of the harvest failure, warfare, and the subsequent outbreak of cattle plague in 1321 had undoubtedly depressed the Irish economy and decreased its population. The *Annals of Loch Cé* narrate that the death of Edward Bruce was the best thing 'performed since the beginning of the world', for he was 'the destroyer of all Erinn in general, both foreigners and Gaeidhel', and because of him 'theft, and famine and destruction of men occurred

[97] TNA, E 101/233/21.

[98] Representative Church Body of the Church of Ireland Archives (RCBCIA), Red Book of Ossory (the *Liber Ruber*), fols 17^r–27^r. Printed in Lawlor, 'Calendar of the Liber Ruber'.

throughout Erinn during that time, for the space of three years and a half' — to the extent that 'people used to eat one another, without doubt, throughout Erinn'.[99] This decline in the Irish economy is reflected in the accounts of Irish government revenues. As the accounts indicate, average annual figures fell from £4219 in 1299–1314 to £2526 in 1314–1321 and £2456 in 1321–1326, plummeting further to £1906 in 1326–1331, temporarily recovering to £2503 in 1331–1337, and falling down again to £1268 in 1337–1341.[100]

Military Purveyance: English Counties

The food supply of local communities was reduced not only through illegal transfers, be it crop raids or pirate attacks, but also through what may be considered 'legal' or 'institutionalized' transfers, connected with the provisioning of royal garrisons. In order to provide its armies with necessary food supplies, the Crown has been relying on the institution of *purveyance*. Roughly speaking, purveyance was a royal prerogative to require subjects to submit or sell food supplies (primarily crops, but also livestock and fish), at the king's request and for his use. In legal theory, at least, the king's officials were not allowed to force local producers to sell their produce against their will and for arbitrary prices; each transaction had to be properly conducted and witnessed. This view was indeed stipulated in both *The First Statute of Westminster* (1275) and *Articuli super cartas* (1300).[101] In reality, however, royal officials tended, in many instances, to abuse their position and compel local communities and individuals to surrender or sell their produce at disadvantageous prices and against their will, as much as against royal injunctions. This fact earned purveyance a highly negative connotation in the historiography of late medieval England. Thus, Jones, Maddicott, and Krug argued that purveyance was a very cumbersome institution, especially from the last decade of the thirteenth century onwards, in the context of warfare and royal taxation.[102] On the other hand, Masschaele offered a more balanced view, arguing that the intensity of purveyance varied across places and socio-economic statuses.[103] A close analysis of all the available evidence regarding purveyance during the famine years may contribute to this debate.

[99] *The Annals of Loch Cé*, ed. and trans. by Hennessy, pp. 596–97.

[100] Calculated from Richardson and Sayles, 'Irish Revenue, 1278–1384', pp. 99–100.

[101] *The Statutes of the Realm*, ed. by Raithby, I, 76–77.

[102] Jones, 'Purveyance for War'; Maddicott, *English Peasantry*; Krug, 'Wartime Corruption'.

[103] Masschaele, *Peasants, Merchants, and Markets*, pp. 219–24.

Overall, despite occasional complaints against the sheriffs' arbitrary actions, a close analysis of the surviving sheriffs' accounts and royal writs from the reigns of Edward I and Edward II reveals that purveyance had a very limited impact on English society as a whole. First, the sheer scale of all crops purveyed never seemed to have exceeded 0.5 per cent of the total agricultural capacity of those counties that were requested to surrender or sale their produce. Even in the most intense years of fighting, such as 1296–1297, when England was fighting in Gascony, Flanders, and Scotland, and 1324, during the War of St Sardos, the total amounts of crops purveyed were around 31,500 quarters in both cases. Second, as a general rule, the western and the northern counties were purveyed less frequently than the eastern ones. In every case, purveyance touched only a part of each provisioning county, and even within each vill, for not every household was expected to sell its produce. In many instances, sheriffs managed to acquire large portions of grain from great landlords rather than from local smallholders. For instance, between March and August 1319, the sheriff of Kent procured the required total of 3382 quarters of crops from five substantial landlords (the Archbishop of Canterbury, Canterbury Cathedral Priory, St Augustine's Abbey Canterbury, Aymer de Valence, and Bartholomew de Badlesmere).[104] Third, only in very few instances were grain producers required to surrender their produce for free, and, in the vast majority of cases, they sold their produce on a voluntary basis. Finally, despite occasional complains about forced sales for excessively low prices or non-payments,[105] the 'purveyance' prices were, at least in theory, not very different from the market ones: on average, the ratio between the two was 1.04 for wheat, 0.90 for both legumes and oats.[106] In other words, the institution of purveyance does not seem to have been as burdensome as some might think, at least not in the days of the first two Edwards. The system seems to have indeed become more cumbersome during the early years of Edward III, whose political ambitions and military campaigns commanded astronomically high expenditure.[107]

But does this view hold true for the famine years? Unfortunately, the only surviving sheriff's accounts from the famine years are the 1315–1316 one for Devon and the 1315–1316 and 1316–1317 ones for Norfolk.[108] In addition,

[104] TNA, E 101/566/9.

[105] Maddicott, *English Peasantry*, pp. 26–27.

[106] Calculated from the sheriffs' accounts. For the archival shelf-marks, consult Appendix 8.

[107] Krug, 'Wartime Corruption'; Krug, 'Purveyance and Peasants'.

[108] TNA, E 101/555/3–4 and E 101/574/22–24.

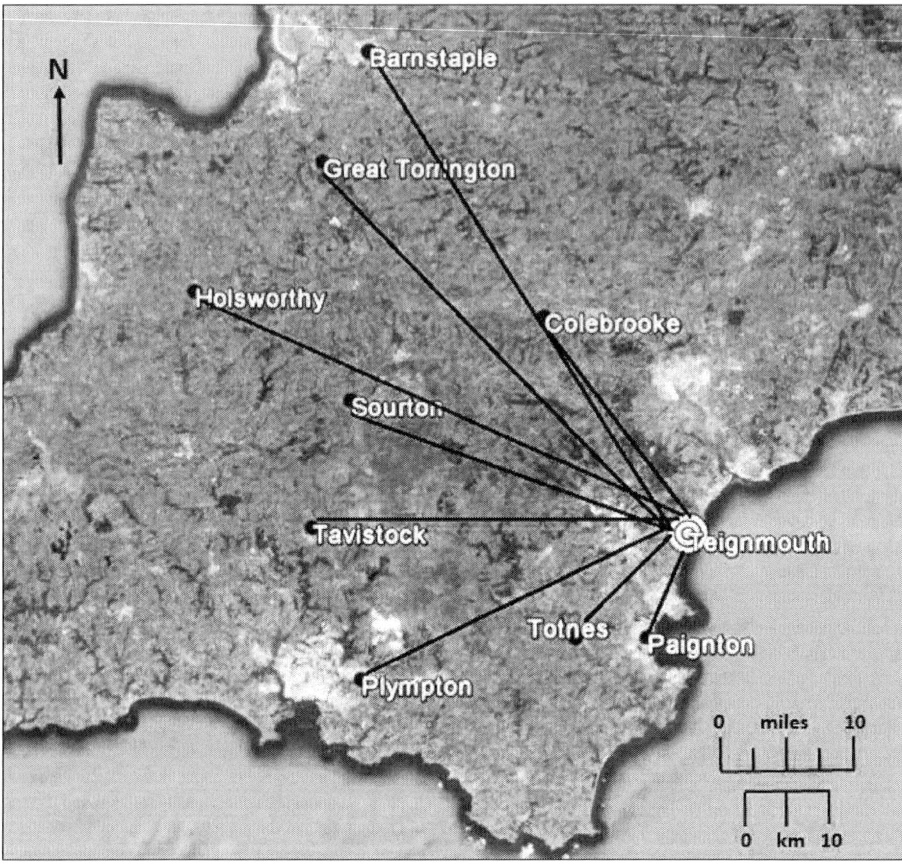

Figure 6.4. The Purveyance Zone in Devon, 1316. *Source*: TNA, E 101/555/3–4.

the 1315–1316 and 1316–1317 Wardrobe accounts record the shipment of foodstuffs to the north, on an earldom basis. To this, we might add a royal letter of appointment of thirty-six purveyors to all three ridings of Yorkshire on 10 September 1316, giving the total amounts of crops to be purveyed.[109] As the Devon account indicates, at some point in 1315–1316 (perhaps, May–June 1316, judging by the high price levels of wheat), the sheriff acquired 280 quarters of wheat and 200 quarters of oats from nine loci in that county. The 'provisioning zone' of Devon had an impressive coverage, stretching all the way from Barnstaple in the north-west to Paignton in the south-east (Figure 6.4).

[109] *CPR, 1313–17*, pp. 543–44.

Table 6.4. Grain prices paid by purveyors in 1316 (presumably, spring/summer).

Crop	No. of transactions	Quarters bought	Weighted average price (s per quarter)	Minimum price (s per quarter)	Maximum price (s per quarter)
Wheat	32	1,648	16.8	10.0	24.0
Barley	13	1,153	7.3	6	9
Oats	10	820	4.3	4.0	8.0

Source: TNA, E 101/555/3, E 101/574/23, E 101/376/6; SAL, MS 120, fols 13–20.

The collecting point was the port of Teignmouth, where the grain was loaded and shipped to Skinburness in Cumberland (and eventually carted to Carlisle). Unfortunately, unlike the Devon account, neither the Norfolk-Suffolk accounts nor the wardrobe books specify the places of the transactions.

In many instances, there is a detailed break-down of transactions, indicating the prices paid by the sheriffs and their assisting purveyors for the purveyed grain (Table 6.4). The weighted average figures for wheat, barley, and oats were 16s 10d, 7s 4d, and 4s 4d a quarter, respectively. It appears as if the purveyors had to pay wheat prices at their market level, while the spring crops were purchased at considerably lower prices. Indeed, in some instances, the purveyors paid remarkably high prices for wheat. Thus, the sheriff of Devon purchased 280 quarters of wheat at 19s 1d per quarter, while his Nottinghamshire-Derbyshire and Cambridgeshire-Huntingdonshire counterparts each acquired 100 quarters of wheat at an overwhelming 22s and 24s per quarter, respectively. The lowest recorded price was 10s a quarter, paid by the sheriff of Norfolk and Suffolk. Conversely, the highest barley price paid was 9s a quarter. While it is possible that the considerable difficulties of finding any wheat for sale — because of its chronic scarcity — made any bargaining or coercion impossible, the acquisition of the more readily available barley at lower prices was more doable, thanks to their higher yields. A closer consideration of the chronological context, however, may result in a different interpretation. Sadly, the accounts recorded neither the seasonality of transactions nor (with one exception) the duration of storage of the crops in barns. Some idea can, nevertheless, be gleaned from indirect evidence. The *terminus ante quam* of transactions can be established from the dates of victual shipments from local ports, recorded in the same accounts. In the case of 1316, most ships left the ports in July, namely, shortly before harvest and when the effects of the famine were felt most acutely. Second, the sheriff of Devon's account also indicates that the crops were being stored at a Teignmouth granary for ten weeks. This is, largely, a comparable figure with non-famine years, when the average duration of storage was in the area of eight to twelve weeks.[110] Given the dire circumstances of the northern garrisons and communities in the context of famine and warfare, it is likely that the sheriffs, who, unlike producers and corn-mongers, could not profit from the rapidly rising prices, would avoid long-term storage and do their best to ship the victuals at their earliest convenience. If we, therefore, accept the estimate of two to three months as a guideline, then it appears that most transactions may have occurred between April/May and July 1316. By April 1316, the average market price for wheat was 16s; by May it rose to almost 18s, while in June it stood at 20s, rising to almost 22s by the time of the grain shipment. The average purchase price of wheat was 16s 10d, matching the April levels. Unless most transactions took place before May, it appears that a good proportion of wheat may have been purchased at lower-than-market prices. The gap between purveyance and market prices was much more apparent within the barley and oat sector. If our chronological interpretation is correct, then it turns out that barley and oat prices paid by the purveyors may have been, respectively, 40 and 50 per cent lower than their market levels. The purveyors' ability to procure the spring crops for such low prices undoubtedly reflects the fact that these crops were more readily available than wheat, owing to their higher yields. Whatever extractive mechanisms were used by the purveyors, they undoubtedly contributed to the disruption of what Amartya Sen called 'exchange entitlement' between local producers and royal officials.[111] By underpaying, say, 40 per cent, the purveyors cut the food supply of local crop producers by at least the same number of calories. After all, peasants, unlike sheriffs and their assistants, had no means to secure crops for the same purveyance prices. To compensate for their calorific loss, they would have to buy crops for market prices, which very few could afford, especially during the 'hungry gap' period. This, once again, stresses the impact of institutional factors on transforming shortage into famine.

Fortunately, however, the sheer amounts of grains purveyed by local purveyors were marginal, in relative terms, to the total arable production capacity of respective counties in 1315 and 1316. Let us take Devon, for instance. Even with the abysmally low per-acre yields, the total arable land of Devon (consisting of some 418,000 acres) was still capable of rendering approximately 284,000 quarters of crops in the harvest of 1315.[112] The purveyed amount of

[110] Claridge and Langdon, 'Storage in Medieval England', pp. 1249–51.

[111] Sen, *Poverty and Famines*, pp. 3–4 and 167–73.

[112] The estimated arable acreage derives from Broadberry and others, *British Economic Growth*, pp. 66–67.

Table 6.5. Total amounts of purveyed crops in 1315 and 1316.

	In quarters		*As approximate share of total output*	
Counties	**1315**	**1316**	**1315**	**1316**
Cambridgeshire + Huntingdonshire		624.9		0.24%
Devon	480		0.17%	
Derbyshire + Nottinghamshire		310		0.05%
Gloucestershire		110		0.05%
Hampshire + Wiltshire		527		0.11%
Lincolnshire	604	1,490.5	0.09%	0.28%
Norfolk + Suffolk	855	880	0.09%	0.12%
Somerset + Dorset		525		0.13%
Sussex + Surrey		210		0.03%
Yorkshire		2,940		0.04%
Total		7,617.4		0.08%

Source: SAL, MS 120, fols 13–20; *CPR, 1313–17*, p. 600.

Note A royal writ requested that Yorkshire purveyors collect 2940 quarters of crops, which were never been delivered to the local sheriff, so it is uncertain what percentage of this figure was actually collected.

480 quarters of crops represented about 0.17 per cent of the total arable production of that county. Overall, it appears that the purveyed crops accounted for, on average, no more than some 0.1 per cent of the total arable output in the purveyed counties (Table 6.5). In absolute terms, these are indeed marginal figures, hardly capable of intensifying the dearth on a national level. At the same time, however, it is vital to distinguish between the impact on *entire counties* and *local communities*. As we have seen, in the case of Devon, only nine townships supplied the local sheriff with grains. But even within each provisioning vill, the purveyance levels were anything but cumbersome: each vill sold between 20 and 100 quarters of grain. For instance, Barnstaple sold 50 quarters of wheat, while its total population size was in the area of 1500–2100 inhabitants. Plympton sold the same amount of wheat; it is unlikely that its population exceeded 800–1000 people. Tavistock, with a comparable population size to that of Plympton, purveyed only 20 quarters of wheat. Totnes, whose total population unlikely exceeded approximately one thousand people (and proba-

bly, it was slightly lower than that), sold 90 quarters of crops.[113] In other words, these were insignificant amounts of grain in relation to the demographic size of local communities and their consumption levels. In the instance of Barnstaple, 50 quarters of wheat accounted for no more than three per cent of the total grain intake of the town (allowing about 1 quarter of wheat per capita, contributing about 70 per cent of kilocalories to the total dietary intake of the community of about 1800 people). Even in the case of Totnes, 90 quarters of grains represented about 10 per cent of the town's total annual supply of grains. Still, it may well have deprived some poorer townspeople of their access to food, but, overall, it is unlikely that purveyance (or, at least its 'official' facet) had a strong impact on the intensification of famine on the people of Devon, as a whole.

Unfortunately, no detailed sheriffs' accounts survive for other counties, and so it is unclear whether Devon reflects the situation elsewhere in the country. In fact, there is some evidence pointing quite to the contrary. Unlike in Devon, where purveyance seems to have conducted with no excessive abuse or complaints, it appears that the Yorkshire purveyors took a clear advantage of their position and the chaotic situation in their county, whose northern parts were under a constant threat of Scottish raiders. As early as 12 December 1316, namely three months after their appointment, the Yorkshire purveyors were accused of diverting the king's money to themselves, seizing crops and livestock from local people, chronically underpaying for their produce, and supplying the Scottish Marches with smaller amounts than instructed.[114] The Yorkshire purveyors seem to have never acquired the targeted 2940 quarters, and whatever sum they have seized for themselves was never delivered to the local sheriff, let alone to the northern garrisons. In other words, they made an easy and fast buck at the expense of the hardships of the local population.

The corrupt ways of the Yorkshire purveyors in autumn 1316 seem to have been only the tip of the iceberg, reflecting the corrupt part of the institution of purveyance. Contrary to Devon (and some other counties), where grains were purveyed on a limited scale and, as far as the sources can tell, in a legal manner, the situation became increasingly problematic elsewhere in the country, at least in regard to those communities and households that were unlucky to have been picked by the purveyors. This is reflected in the number of com-

[113] The population estimates derive from the 1327 lay subsidy rolls, on the assumption that about 60 per cent were exempt from taxation and that an average household consisted of 4.25 members. TNA, E 179/95/6. The upper limit for Barnstable has been suggested in Campbell, 'Benchmarking Medieval Economic Development', p. 909.

[114] *CPR, 1313–17*, p. 600.

plaints against local purveyors, which has increased a great deal during the crisis years to the point that on 14 March 1317, Edward II commissioned a special inquiry into frequent instances of forced surrender of grain and livestock in Derbyshire, Nottinghamshire, and Yorkshire.[115] In Northumberland and elsewhere, the taking of foodstuffs was conducted by imposters posing as royal officials collecting victuals for the royal campaign to Scotland.[116] On 16 August 1316, the king commanded an inquiry into the names of those false officers and ordered that they be arrested.[117] On 22 November 1317, he declared that those officials suspected of forcing local producers to sell their crops or livestock against their will would be prosecuted as thieves.[118] Over the course of 1324 and 1325, the king requested several general *Oyer and Terminer* inquiries to investigate various misconducts of royal officials during his reign. Although the surviving rolls, covering Gloucestershire, Herefordshire, Lancashire, Norfolk, Suffolk, and Yorkshire, deal primarily with the events of 1322, in conjunction with the double levy of the tenth and the sixth and the contrariants' revolt of 1321–1322, there is still a good deal of evidence about the despotic malfeasance of royal officials going back to the famine years. Thus, on 13 June 1315, the king's purveyors of livestock carried a large number of young cattle worth £15 from the village of Wheldrake (North Yorkshire).[119] Similarly, at some point in 1315–1316, William de Tiderynton, the itinerant bailiff of Richard de la Riviere, the sheriff of Gloucestershire, seized eight oxen and 10 quarters of beans from Robert Romeyn and his wife at Tewksbury. In addition, the couple was forced to thresh the said beans for no wages.[120] Similarly, in 1314, several communities in the Blackburnshire, Salford, and the West Derby hundreds (Lancashire) were badly extorted by the officials of a local sheriff.[121] These were indeed large-scale robberies, whose proportions and, consequently, impact are impossible to quantify, not only because of the fragmented nature of sources, but also because not all instances of malfeasance may have been reported and not all complains may have been taken seriously. Finally, as some later *Oyer and Terminer* inquiries indicate, in some instances local purveyors bought crops on

[115] *CCR, 1313–18*, p. 460.

[116] *CPR, 1313–17*, p. 237.

[117] *CPR, 1313–17*, p. 534.

[118] *CCR, 1313–18*, p. 584.

[119] TNA, JUST 1/1117, membr. 5.

[120] TNA, JUST 1/296, membr. 4.

[121] TNA, JUST 1/425, membr. 26.

credit rather than with cash. In those cases, the payments were either delayed (in some instances for a decade) or never delivered. This seems to have been an ongoing issue during Edward II's reign, as it is indicated in some later *Oyer and Terminer* inquiries. Although those inquiries do not cover the famine years, there can be little doubt that the failure to make payment to vendors was yet another manifestation of omnipresent corruption related to purveyance during the famine. In other words, *recorded* prices, as they appear in the sheriffs' accounts, do not necessary mean *paid* prices, and there is no way to tell when and if local vendors received cash in exchange for their merchandise.

Rural producers were not the only victims of the corrupt purveyors. In his writ from 11 June 1315, Edward II complained that religious communities were suffering from the illicit seizure of grain and livestock by the king's purveyors and encouraged any such victim to visit the royal chancery in order to seek remedy.[122] The vulnerability of the landed estates of conventual institutions prompted some of them to seek royal protection and apply for a special exemption from purveyance. Between early 1314 and late 1317, no less than 103 letters of protection, mostly for one year, were granted to various religious houses, as well as to individuals of high standing, primarily the king's clerks and prominent merchants.[123] In those letters, Edward II forbade his officers to take any foodstuffs from the manors of those applicants.

Military Purveyance: The Lordship of Ireland

Although most of the purveyed grain came from England, parts of the Lordship of Ireland contributed their share to the provisioning of the royal garrisons. Because of its geographic proximity to Scotland, Ireland played an important role in supplying the king's army with food during the Scottish campaigns. Here, Dublin and Drogheda were two main ports, from which food-laden ships would sail to Skinburness in Cumbria, the Isle of Man, Galloway, and Carrick. The Irish contribution to the food supply became especially vital following Edward the Bruce's invasion of Ireland in late May 1315. Our main source of information regarding detailed food purchases derives from a purveyance account of the royal army, recording the purchase of grains and wine in the Dublin area in the period between May and October 1315, and the

122 *CCR, 1313–18*, p. 235.

123 Calculated from *CCR, 1313–18* and *CPR, 1313–17*.

Justiciar of Ireland's rolls.[124] Throughout that period, royal officers purchased 667.3 quarters of wheat and 399.6 quarters of oats to provision the English garrisons involved in the Irish campaign. Unlike the purveyance of English counties, where crops were purchased often directly from local producers, the Irish system seems to have been more centralized, whereby foodstuffs were bought from corn-mongers. As the purveyance account indicates, wheat was purchased in fifty transactions, while the number of oat transactions was twenty-five. An average transaction sold 13 quarters of wheat and 11 quarters of oats; the median figure for the respective grains was 3 and 8 quarters. It appears that grain trade was concentrated in the hands of a number of prominent corn-mongers, known to and trusted by royal administration. The names of some traders appear repeatedly in the purveyance account. Thus, William Brailes and his associates sold 14.4 quarters of grains in five transactions. Similarly, Thomas Burgeys and his fellows sold 16.5 quarters of crops in three transactions. It is unclear why the local administration preferred to rely on the corn-mongers rather than buy grains directly from local producers. One possibility is that the imposition of compulsory grain sales would have increased the tensions between the native population and the colonial authorities, and this fact may have had some far-reaching consequences in the context of the Anglo-Scottish warfare. After all, the relationships with the royal government have never been easy and straightforward, and the pro-Scottish sentiment among both the Gaelic and Anglo-Norman natives of Ireland was quite a widespread phenomenon following Bruce's invasion. Compelling local tenants to sale their produce in the context of the ensuing dearth and omnipresent violence would certainly not endear the colonial government to the native population.

In some instances, local purveyors took an even more extractive approach. Thus, on the Easter week of 1317 (3–9 April), Richard de Clare, sheriff of Cork and a major Anglo-Norman landlord, commanded the tenants of his manor of Agni (Limerick) to surrender eighty cows, which would be delivered to the royal army stationed near Loddyn Hill, next to Limerick. In addition, he seized three heads of cattle from Robert Boudyn.[125] Around the same time, a certain Maurice son of David unjustly seized a cow from Roger Fartebury at Wallstown (Cork) and delivered it to Ulster mercenaries fighting for Edward II against the Scots.[126] Such extractive methods undoubtedly contributed to the

[124] TNA, E 101/14/40; NAI, KB 2/7–12.

[125] NAI, KB 2/9, p. 71.

[126] NAI, KB 2/9, p. 124.

indignation of local communities towards the royal officials and garrisons. To make things even worse, just as in England, some payments were either made in part, or delayed, or may have not been carried out at all. Unfortunately, there is no direct evidence about that from the famine years, but several references from non-famine years may indicate that partial and delayed payments, as well as non-payment, were, just as in England, one manifestation of local administrators' corruption. Thus, in 1308, Nicholas le Saltere, a Drogheda merchant, complained that the king's officials paid only for three out of twenty-two casks of wine sold by him for the Scottish campaign.[127] During the so-called Riot of Louth (early 1312), two Anglo-Norman rebels, Thomas de Stanley and Nicholas de Verdun, would systematically underpay local tenants for their victuals and livestock.[128] In 1325, Robert de Napton, a minor Anglo-Norman landlord, stated that he was yet to receive payment for wheat and oats taken from his demesne at Dysart (near Drogheda) at an unspecified point.[129]

At times, the indignant individuals would retaliate. Thus, we hear about thefts of victuals from English soldiers. For instance, on 27 February 1316, one Michael de Trym of Drogheda was accused of buying and selling cattle stolen from the king's army in different parts of Meath and Louth.[130] In some instances, local tenants would openly support the enemy's side and supply them with provisions. Thus, on 27 February 1316, Robert Prior of Louth Priory was accused of receiving Edward Bruce and his men in his house and supplying them with food, bought from men loyal to the king, and sending it to the Scots hiding in the woods.[131] Similarly, on 20 January 1317, William Payn, cellarer of St Mary's Abbey, a Cistercian house in Dublin, was accused of receiving Irish rebels and felons hiding on Mount Leinster (the highest mountain in the Blackstairs Mountains area, between the counties of Carlow and Wexford), treating them, and supplying them with food and drinks.[132] It is possible that Payn's motives were practical rather than ideological: according to one entry from 26 April 1316, St Mary's Abbey was described as devastated and greatly impoverished by frequent raids conducted by both English and Irish sides, in the course of which its granges were robbed and livestock driven

[127] Smith, *Colonisation and Conquest*, p. 152.

[128] NAI, KB 2/7, p. 21.

[129] Smith, *Colonisation and Conquest*, p. 152.

[130] NAI, KB 2/7, p. 58.

[131] NAI, KB 2/7, pp. 58–59.

[132] NAI, KB 2/8, pp. 49–50.

away.[133] In supporting the king's enemies, the abbey may have sought to secure a physical protection in the face of frequent plundering of its possessions.

Food Extortions of Encastellated Communities

The malfeasance of the corrupt royal officials was not the only aspect of the impact of war on food supply of the commoners. A similarly extortive behaviour was exhibited by local castle constables in regard to nearby rural communities. Because of the ongoing warfare and constant threat of raid or invasion, the 'encastellated' communities (that is, parishes or religious houses situated within the vicinity of a castle or fortified structure) within the 'war zones' found themselves in a fierce struggle for dwindling food resources. The triple threat of military conflict, environmental shocks, and economic crisis, which decreased the available food supply a great deal, pushed those regions into a quasi-Darwinian competition for food, and hence subsistence and survival, between vulnerable, starving, and weakened rural communities, on the one hand, and, for most part, protected, well-fed, and fit men-at-arms and their households. According to the logic of the Darwinian dictum, the outcome of this competition would be obvious.

Indeed, there are numerous instances of local castle officials taking advantage of the vulnerability of nearby communities and extorting or pillaging their food resources. A few examples would suffice. In 1315, the men of Bamburgh Castle plundered its surrounding Northumbrian villages.[134] In May 1315, Edward II learned that the constable of Builth Castle (Breconshire, Wales) and his men were abusing local tenants by forcing them to overwork, extorting their victuals and livestock, killing their swine, compelling them to sell their produce for excessively low prices, and making them grind their grains for a heavy fee at bailiff's mills rather than their own hand mills. Such treatment caused significant misery and impoverishment among the local community.[135] By September 1316, the Cumbrian villages of Papcastle, Broughton, and Crosby were devastated by both the extortions of the keeper of Cockermouth Castle and the Scottish raids to such a degree that many local inhabitants had to abandon their tenements and move away.[136] There are numerous references to the robbery and

[133] NAI, KB 2/8, p. 5; KB 2/12, p. 4.

[134] TNA, E 159/93, membr. 14^{v}.

[135] *CPR, 1313–17*, pp. 322–23.

[136] TNA, C145/78, membr. 11; *CIM*, no. 297.

destruction of tithes within encastellated communities, paid by parishioners to local parish priests and vicars. In the course of 1314–1316, the sheriff of Northumberland repeatedly seized the crop tithes of William de Airemynn, proctor of Whiteberne and Wearmouth (both in Northumberland), which, in any event, were undoubtedly meagre on the account of the harvest failures and the Scottish incursions.[137] Between 1313 and 1318, Nostell Priory (Yorkshire) was being repeatedly robbed of its tithes in Yorkshire and Northumberland parishes by constables of Bamburgh, Alnwick, and Norham.[138]

The Final Blow: The Taxation of 1315 and 1316

While the institution of purveyance and occasional harassment of tenant communities by the officials of nearby castles might be on a local and limited scale, there was one real burden that seems to have affected all of England: royal taxation. Between 1188, the year of the 'Saladin's Tithe', and 1334, the English population was taxed no less than twenty-four times at rates varying from one sixth to one fortieth applied to the value of a taxpayer's 'movables' (mostly livestock and crops, but also other goods).[139] Nineteen taxations (also known as 'lay subsidies') out of the total twenty-four were imposed in the period of 1275–1334, all in conjunction with the Edwardian wars. Following the disastrous defeat at Bannockburn, Edward II was determined to strike back and lead a major campaign against Scotland. In this context, he imposed one levy in 1315 (one-twentieth for the countryside and one-fifteenth for towns and the Ancient Demesne, namely royal manors) and one in 1316 (one-sixteenth for the countryside and one-fifteenth for towns and the Ancient Demesne). The war-devastated northern counties of Cumberland, Lancashire, Northumberland, and Westmorland were exempt from the tax because of their economic depression, while both Durham and Cheshire were exempt because of their palatine status. In each case, the famine-year taxations had a threshold at 10s of movables (mostly livestock, but also grain and other goods). In other words, it is inevitable that a large proportion of population — some 50–60 per cent — was exempt from those levies, particularly on account of poverty during the famine years. For those not exempt, however, the 1315 and 1316 levies could not have

[137] *CCR, 1313–18*, pp. 147, 251–52, 360, 365.

[138] McNamee, *Wars of the Bruces*, p. 144.

[139] The period between 1275 and 1334 is studied in Jenks, 'Lay Subsidies'. A general introduction to royal taxation is Jurkowski, Smith, and Crook, *Lay Taxes*.

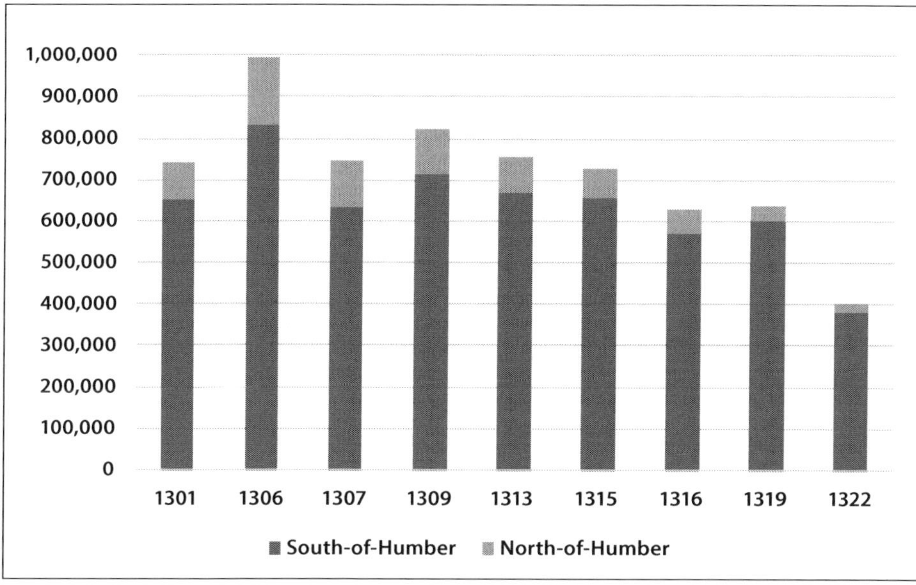

Figure 6.5. Estimated taxed wealth of English population, 1301–1322 (in £ Sterling).
Source: Jenks, 'Lay Subsidies', pp. 31–39.

Note: The figures appearing on the vertical axis stand for the total taxed wealth, not the amounts paid in taxation.

possibly come in a worse time. The twentieth of 1315 was to be collected, in two instalments, by 1 June and 15 September, namely before the prices began to soar.[140] Nevertheless, it assured that the taxed households would be deprived of some cash, which they could have used towards purchasing some food during the disastrous agricultural year of 1315–1316. It was, however, the sixteenth of 1316 that proved to be really cumbersome for the tax payers. On 5 August 1316, amidst the disastrous harvest, the Parliament agreed to grant the king a new subsidy, which was to be collected, in two parts, by 3 November 1316 and 25 April 1317. In the context of high prices and a ubiquitous dearth of crops, seizing cash from starving tax payers made them considerably more vulnerable to poverty and, consequently, famine.

Figure 6.5 establishes the approximate total wealth of taxed individuals in England between 1301 and 1322. This is achieved by multiplying the total taxed sums by the fraction factor of each levy (for instance, in 1316, one-sixteenth and

[140] Jurkowski, Smith, and Crook, *Lay Taxes*, pp. 32–33.

one-fifteenth were collected, respectively, from rural settlements and towns; consequently, the factors here are 16 for rural wealth and 15 for towns and the Ancient Demesne). As the table indicates, the total estimated taxed wealth in 1315 was only slightly below those rendered in 1313 (about £729,810 and £756,680, respectively), while the 1316 figures were about 17 per cent below the 1313 ones (about £628,142). At first glance, the 1316 figures may appear surprisingly high, given the omnipresent dearth and poverty. But it should be borne in mind that crops and livestock were two most main components of taxed wealth, and it was the price level of those commodities that determined the tax levels, in each case. Basically, the higher were the prices, the heavier was the burden of taxation because the value of taxed possessions would be driven up accordingly. According to this logic, 1316 was the single worst year to be taxed, with the value of crops rising considerably because of the high prices.

In theory, the omnipresent poverty would have reduced the total number of potential tax payers, because more people would have found themselves below the threshold of 10s, which entails exemption. Indeed, as the Bedfordshire, Hertfordshire and Oxfordshire rolls (the only surviving rolls for individual counties) reveal, the number of taxed households in 1316 fell by 10–20 per cent, in relation to the 1306–1315 period (Table 6.6). At the same time, however, taxed households had to pay more *per capita*, because of the increased value of their goods. Thus, an average taxed Bedfordshire household had to pay 3s 2d in 1316, compared to 2s 4d between 1306 and 1315. In Hertfordshire, an average sum levied on an average tax-paying household rose from 2s 6d in *c*. 1306–1315 to 2s 10d in 1316. Similarly, in Oxfordshire, the respective figures rose from 1s 9d to 2s 10d. Obviously, the situation varied across vills and households. For instance, at Henlow (Bedfordshire), the goods of Matilda Sloot were assessed at 93s 9d in 1309, at 130s 9.5d in 1315, and 145s 11d in 1316. Judging by the value of her assessed possessions, Matilda Sloot was among the better-off tenants in her community and may have been somewhat more resilient to taxation and food shortage than some of her neighbours. John de Caldwell, whose goods were assessed at 16s in 1309, was among the poorest taxed individuals of the same vill. Yet, he was still taxed in both 1315 and 1316, with his possessions valued, respectively, at 26s 7 1/4d and 20s. In other words, John de Caldwell seems to have possessed fewer crops during the famine years compared to 1309, yet these crops were assessed at a higher value — high enough for him not to be exempt from taxation. Some households seem to have had too little during the famine years to be taxed. Of eighty-one taxed individuals of Henlow, recorded in the 1309, 1315, and 1316 lists, only twenty-one appear in all three cases. For all of Bedfordshire, only some 22 per cent of all taxed individuals appear

Table 6.6. Absolute and relative taxation levels in Bedfordshire, Oxfordshire, and Hertfordshire, 1306–1316.

Years	**1306–1315**	**1316**
Taxed households, Bedfordshire	6,081	5,425
Per household (s)	2.32	3.17
Taxed households, Oxfordshire	6,858	5,523
Per household (s)	1.72	2.81
Taxed households, Hertfordshire	5,188	4,200
Per household (s)	2.53	2.83

Source: TNA, E 179/71/10–12; TNA, E 179/20/8–10; and TNA, E 179/8 and 10.

in all three lists. In the case of Hertfordshire and Oxfordshire, whose taxation records survive, respectively, for 1313 and 1316, and 1306 and 1316, about 25 per cent of households were taxed in both instances.

The taxation of 1316, coupled with the acute dearth of food and excessively high prices, meant that the taxed households and individuals suffered various degrees of decline in their food entitlement. Handing over sums of money to local tax-collectors was, in effect, handing over calories. This, of course, was true in every instance of taxation. For instance, during the collection of the thirtieth and twentieth of 1306, which was paid in three instalments between 3 February and 3 November 1307, an average taxed household in Oxfordshire paid about 1s 9d, which could buy about 0.44 quarters of wheat in that year. Because of the sharp decline in the purchasing power of money during the famine years, 2s 10d collected from an average Oxfordshire household in 1316–1317 could buy only some 0.18 quarters of wheat at that year's prices. Allegedly, the dramatic fall in the nominal value of money meant that the relative burden of the 1316 taxation was lighter compared to the pre-famine subsidies. This reasoning, however, is incorrect. It should be borne in mind that in 1307 the crop prices were low, as the result of the abundant harvest of 1306 (in fact, all harvests between 1305 and 1314 were bountiful, standing, on average, at about 17 per cent above average). Under these circumstances, fewer landholding households had to purchase additional crops from local markets to supplement their dietary needs. Conversely, because of the disastrous harvest of 1316, very little could be extracted from tenant fields, and there was a dire need to make up for the loss of calories claimed by the vagaries of Nature. Even though 2s 10d could buy less food in 1316–1317 than 1s 9d in 1307, it was much more likely that an average commoner would need to resort to the market in the former than in the latter case. Besides, 0.18 quarters represented, roughly speaking, over two months of grain diet for an average commoner or about a half-month for his entire household of 4.25 people in the early fourteenth century (assuming that about 1400 kcal out of the total 2000 kcal derived from crop-based diets). In the dire circumstances of the famine, these amounts could potentially be consumed over a longer period of time.

In addition, the surviving taxation records do not always reflect reality accurately and, as such, should be approached with care. First, it is impossible to estimate what percentage was exempt from taxation because of their *poverty and inability* to pay and what percentage managed to evade on the account of their *poverty and unwillingness* to pay. As we have seen, only about one quarter of all tax payers appear in both the pre-famine and in the 1316 taxation rolls. Second, and more importantly, the taxation rolls may not always reflect the true sums of money taken from local tax payers. As with purveyance, some tax-collectors may have abused their authority to extort local communities and seize money beyond the taxation levels. One may also think of some imposters wandering around and posing as tax-collectors, before the arrival of real taxers.

Tax-collectors' corruption is reflected in later *Oyer and Terminer* rolls, as well as in several political poems composed in the early fourteenth century. Obviously, we have to keep in mind that these literary pieces were highly stylized and biased to paint a highly dramatic and perhaps exaggerated picture of the widespread misery of the commoners. As such, these should be approached with caution and regarded as the reflection of popular indignation and protest, as well as collective memory rather than omnipresent malfeasance of royal officials and sufferings of peasants. All the same, they can be used as a tool to understand how the officials' corruption, in the context of the famine crisis, was perceived and remembered by the survivors, or at least by the composers of the poems. Indeed, the poems paint a gloomy picture. For instance, in the *The Song of the Husbandman* (composed around 1340, but reflecting the reality of the Great Famine), the Husbandman complains that now that good years and grain harvests are all gone, 'fourth penny has to go to the king for good' (*for ever the furthe peni mot to the kynge*). There seems to be no end to the burden, even though the Husbandman has already paid his taxes ten more times ('Mo then ten sithen told I my tax'). To pay the king's subsidy, the Husbandman had to sell his seed corn ('to seche selver to the kyng I mi seed solde').[141] This undoubt-

[141] *Song of the Husbandman*, in *Medieval English Political Writings*, ed. by Dean, lines 43–46.

edly reflects the omnipresent scarcity of cash among the worse-off elements of the society during the years of dearth. Another early fourteenth-century poem, tentatively titled *Against the King's Taxes* and written in a macaronic combination of French and Latin around 1340, in conjunction with wool taxation, is even more reflective. Just as in the *Husbandman*, the narrator bemoans that many people are forced to sell their goods to pay their taxes. As the result, they end up with very little property and no ready cash. To make things even worse, however, only half of the collected sums end up reaching the king, while the other half is being seized by corrupt officials, who also seize the property of the poor against their will. Similarly, wool-collectors constantly use false weights and measures to rob local communities. The main taxation burden, therefore, befalls the poor and needy, living in pain and sweat, rather than rich and powerful, who are enjoying their lavish lifestyles. The poor are reduced into such misery that the only reason that they have not yet revolted is because they have no leader of their own.[142] Finally, the so-called *Simonie or the Poem on the Evil Times of Edward II* (*c.* 1325) contains an instructive verse about the impact of taxation. The king's officials, driven by the madness of their greed, tend to overtax the poor masses well above what is prescribed. About half is stolen before it is collected and counted. Furthermore, they often tax people who ought to be exempt, including smallholders and landless labourers. Conversely, some rich people are taxed disproportionately low, and they pay as little as some poorest among paupers.[143]

Conclusions

The impact of warfare on the food supply reflects a complex interaction between exogenous and endogenous factors, which act together and deprive people of their natural entitlement to food and thus transform shortage into famine. Unlike market- and storage-related issues, war is capable of contributing to famine *without* a 'natural' harvest shortfall. For instance, a prolonged siege of a town or a castle can cut a besieged community off from its regular food supplier, be it local markets or agricultural hinterland. Likewise, *chevauchée* campaigns may destroy grain fields and, thus, cause a disastrous harvest failure. No pre-Industrial warfare, however, can be more devastating than if it coincides with a period of short-term climatic shocks and bad harvests. Such

142 Tyson, 'Against the King's Taxes'.

143 *The Simonie*, in *Medieval English Political Writings*, ed. by Dean, lines 43–46.

an unfortunate combination of Nature's vagaries and human violence had most devastating impact on local communities living within war zones. The example of the early fourteenth-century war in the British Isles, thoroughly analysed in the present chapter, illustrates this point very well.

The conflict initially commenced in 1296 as a struggle for political power in Scotland, but it quickly acquired an additional aspect — that of ecological warfare, when both sides used natural resources in order to quicken their triumph and weaken their enemy. True, *chevauchée* campaigns began as early as 1296, when Edward I, following his victory at Dunbar (27 April 1296), marched into Scotland, all the way into Inverness-shire in the north. But it was not until 1311 that the environmental warfare became truly intense. The incessant Scottish raids reached their peak between 1314 and 1319; namely, they overlapped with the torrential rain, harvest failures, livestock mortality, and poverty. As such, they could not have come in a worse time. Theoretically, the destruction of food resources between *c.* 1311 and *c.* 1314 could have been partially compensated by grain trade. These were years of good harvests in 'war-free zone', implying that there were grain surpluses for sale on local markets. In other words, the 'war-free zones' could have used their comparative advantage of being unaffected by the conflict and come to rescue. From 1315 onwards (and especially in 1315–1317), such a scenario was no longer possible: all of England was devastated by the harvest failures, and each community had to take care of itself, in a dire struggle for survival. Although there were some signs of crippling interregional grain trade, it was done on a very small scale. Moreover, most of the foreign grain secured by Edward II either through commercial privileges or contracts with his chief grain supplier, Antonio Pessagno, were meant to feed soldiers and besieged burgesses of frontier towns rather than starving peasants. Because of all that, warfare was yet another factor that exacerbated the crisis and assured that the harvest shortfalls and dearth would be transformed into a full-fledged famine.

The war-related subsistence crisis was created by two complementing mechanisms: (1) environmental destruction, entailing the devastation of natural resources, and (2) institutional extraction, namely, compulsory purveyance, food extortion, and taxation. The former mechanism was restricted to the 'war zone', with the 'war-free' communities (primarily, those living south of the Humber) being largely unaffected by the environmental destruction in the north. Although, as we have seen, both chroniclers and manorial officials tended to exaggerate the scale of environmental destruction (each group for its own reason), there can be little doubt that it still contributed a great deal to the crisis of local communities in the north. The latter mechanism, on the other hand, was applied in both the 'war zone' and the 'war-free zone'. It is true that some northern counties were exempt from taxation on the account of their devastation and poverty, but they were not always spared of corrupt governmental officials and greedy castle keepers. As we have seen, Yorkshire purveyors were accused of diverting the king's money to themselves, seizing crops and livestock from local people, and chronically underpaying for their produce. In Northumberland, some imposters, posing as royal officials, seized the last crumbs from impoverished communities. In other words, the situation was much more stressful and taxing in the 'war zone', where local inhabitants had to put up not only with failed harvests and high prices but also with food extraction by royal officials (or imposters) and food destruction by marauders.

In that sense, the implications of the crisis within the 'war zone' were by far harsher and enduring than within the 'war-free zone'. Within the 'war-free zone', the crisis was more or less over by late 1317 and had certainly finished with the abundant harvest of 1318. An altogether different situation prevailed within the 'war zone', where the situation was anything but normal in 1318. With the exception of the Lordship of Ireland, where the fighting came to its end by October 1318, the food crisis continued in northern England and, to an extent, in south-eastern Scotland at least into the early 1320s, with raids, skirmishes, and battles inevitably entailing the destruction of natural resources. If the fighting and raiding were long, recovery was even longer. As we have seen, both the arable and livestock sectors of local economies contracted for several decades, at least until the Black Death of 1348–1351. The demographic and economic decline and stagnation continued into the post-Black Death period because of recurrent outbreaks of borderland conflicts, going into the sixteenth century. When Henry VIII commissioned the famous *Valor Ecclesiasticus* in 1535, numerous northern communities in Cumbria, Northumberland, and Durham were reported to have been rendering low or no rents 'in the times of war...because of Scottish invasions'.[144] This echoes the *Nova Taxatio* of 1318, which similarly recorded numerous communities of the north as unable to pay papal taxes because of Scottish raids. To be sure, the environmental and institutional peculiarities of the North hindered the speedy recovery of its economic life. The comparatively poor soils, the adverse climate, and the under-marketized nature of the economy undoubtedly contributed to the long-term environmental and economic downturn, which implied that the 'war zone', and chiefly the northern English counties, experienced a prolonged food crisis, stretching far

[144] *Valor Ecclesiasticus Temp.*, v, 278, 300–04.

beyond the traditional chronological framework of 1315–1317 (or, according to some, 1315–1322) and, namely, well beyond the harvest failures and agrarian crisis of those years. In effect, the agrarian crisis of 1315–1317 was just *one of several episodes* within a long and complex chain of events making the food security crisis of the early fourteenth century, which affected the northern parts of the British Isles all the way until at least the Black Death of 1348–1351.

Although the experiences of famine and food insecurity of the northern communities of the 'war zone' were undoubtedly harsher and longer than elsewhere, one should not downplay the impact of warfare on the 'war-free zone'. The Scots did not penetrate any deeper than West Yorkshire, and therefore the effects of warfare south of the Humber were indirect and restricted to purveyance and taxation. Here, the repercussions of purveyance varied from region to region and from place to place, depending on local context. Although the purveyed sums of grains accounted for a meagre fraction of the sheer agricultural output (even in the context of the failed harvests), and some regions and vills were spared altogether, purveyed communities had to pay a high price. Apart from the fact that the compulsory nature of grain sales was undoubtedly at odds with the will of the vast majority of small-scale agricultural producers, the purveyance prices of spring grains, in contrast with wheat, were well below their market equivalent. But even if tenants were *offered* a 'just price' for their produce, there is no way to assert that they were actually *paid*. There are numerous complains indicating that royal officials received foodstuffs on credit, often delaying and sometimes never delivering the payments. As I have suggested above, *recorded* and *entered* sums do not necessarily mean *paid* sums. Furthermore, some communities became victims of harassment by royal officials or those posing as royal officials. The increasing number of complaints against their corrupt ways prompted Edward II to commission a special inquiry into frequent instances of forced surrender of grain and livestock in Derbyshire, Nottinghamshire, and Yorkshire, on 14 March 1317.[145] In the context of famine, when everything was dear and few, seizing 1 quarter of grain from a starving household could have had extremely negative consequences. If purveyance alone was not bad enough, there also was taxation. As the surviving rolls for Bedfordshire, Hertfordshire, and Oxfordshire reveal, although the sheer number of taxed households decreased because of widespread poverty, the average sum paid by the averaged taxed household rose because of high grain prices, driving up the value of taxed goods. Just as the purveyance accounts, the taxa-

[145] *CCR, 1313–18*, p. 460.

tion rolls may not always record the true sums of money taken from local tax payers. There is no doubt that some tax-collectors, just as purveyors, tended to abuse their authority and extort money and goods far beyond the prescribed levels of taxation.

The early fourteenth century was by no means the only instance when eco-biological shocks went hand in hand with warfare. One may think, for instance, about the First Crusade (1096–1099), originating in the context of disease, drought, and the food shortage of 1094–1095. Similarly, the Second Crusade (1147–1149) is to be viewed against a short-term weather anomaly of the 1140s in western Germany, France, and the Low Countries, marked by the freezing winter of 1142–1143, coastal inundations in spring 1143, and gale and flooding in winter/spring 1144, all of which resulted in several back-to-back harvest failures.[146] The Nine Years War (1594–1603), over the control of Ulster, was fought in circumstances of short-term climatic anomaly, disease, and famine of the 1590s and early 1600s.[147] The Thirty Years War (1618–1648) and the Wars of the Three Kingdoms in Britain (1639–1651) were fought in the context of the deteriorating climatic and environmental conditions of the early seventeenth century (and, particularly, the onset of the Maunder Minimum), while the wars of Louis XIV (1661–1715) occurred during the Maunder Minimum.[148] Intriguingly, the reign of the 'Sun King' coincided with a period of very little solar irradiance. The connection between natural shocks and warfare is a fascinating topic, which is yet to receive its fully deserved scholarly attention.

[146] Slavin, 'Crusaders in Crisis', and 'Ecology, Famine and Religious Violence'.

[147] Morgan, *Tyrone's Rebellion*, p. 180.

[148] The impact of the environmental and climatic crises on political crises of the seventeenth century (and in particular, warfare), has been studied in Parker, *Global Crisis*.

Chapter 7

Poverty, Malnourishment, and Famine Demography

Overpopulation and Poverty

No famine, or any food crisis, can be fully appreciated without considering the link between population levels, poverty proportions, and the available food resources. As we have seen, one obvious consequence of harvest failures is a rise in food prices; the degree of the rise in prices depends on market efficiency/failure and the willingness/reluctance of better-off producers and traders to release their hoarded stocks on local markets. The excessively high prices inevitably mean the decline of collective purchasing power of commoners because of the lack of financial resources or any other means of exchange. Fewer and fewer people can now afford to buy food, while the gap between the commodity prices and the available funds to buy them becomes wider and wider. The degree of compromise of the collective purchasing power is largely determined by the real wages and by the size of a deprived population, *in relation to the available resources*. It is crucial, therefore, to establish not only the sheer size of population but also a relative proportion of poor people within that population group. The connection between population size, poverty, and vulnerability to famine according to Malthus' 1798 *Essay on the Principle of Population* has been already discussed in Chapter 1.[1] According to Malthus, economic growth is determined by the growth of two variables: human population and natural resources. Because of the passion between the sexes, human population grows geometrically, while natural resources grow arithmetically. It is inevitable,

[1] Malthus, *An Essay on the Principle of Population*.

therefore, that human resources outstrip natural resources. Overpopulation, demographic congestion, poverty, low living standards, socio-economic tensions, and — most importantly — vulnerability to shortages and famines are all natural outcomes of human demographic growth. According to Malthus, the higher the proportion of poor people is, the more likely that shortage will become an acute famine. Famines themselves, alongside with wars and diseases, are regarded by Malthus as a 'providential' or 'positive' check, preventing the population from growing endlessly and assuring that the problem of overpopulation and poverty is temporarily solved. Although not adhering explicitly to the Malthusian theory, Amartya Sen has convincingly shown, along the same lines, that poverty is a decisive factor as far as famines are concerned: the bigger the proportion of poor people is, the more vulnerable the society becomes to various socio-economic stresses.[2]

Poverty and Calorific Decline

Perhaps nothing stresses the connection between poverty and vulnerability to subsistence crises as clearly as the decline in calorific supply and, thus, calorific intake. While better-off tenants might have hardly felt any change, especially during the first year of the famine, the vast majority of population, living on small holdings, and at times on infertile land, most certainly experienced the catastrophic food deficit. As we have seen in Chapter 3, on the eve of the famine, about 60–65 per cent of England's population was living on less than a virgate (thirty acres or so), and perhaps as much as half of her population was holding ten acres of land or less.[3] The problem of overpopulation and land-hunger implied that England was too poor to have remarkable inequality: according to the most recent estimate, about 41 per cent of England's population lived below the poverty line *c.* 1300.[4] These figure, however, seems to be an under-estimate, given that a great proportion of poor and landless people were under-recorded. This issue cannot, unfortunately, be dealt with in the present study. In any event, in the most densely populated regions, rural congestion and overpopulation were especially apparent. Thus, in eastern Norfolk, most tenants had to get by with just a few acres of land. Thus, at Hevingham and Martham (both in eastern Norfolk), the majority of local tenants (about 73

[2] Sen, *Poverty and Famines*.

[3] Broadberry and others, *British Economic Growth*, pp. 118–27.

[4] Broadberry and others, *British Economic Growth*, pp. 317–18, 321.

per cent) held between one and three acres of land by the end of the thirteenth century.[5] Similarly, the absolute majority of tenants at Coltishall (Norfolk), held less than four acres of land in the early fourteenth century.[6] Similarly small holdings, standing on average between two and four acres per family *c.* 1300, were recorded in other parts of eastern Norfolk.[7] It should be borne in mind, however, that these smaller farm areas were intensively cultivated and, thus, may have been significantly more productive, in terms of yields per acre, than elsewhere in England.[8] Nevertheless, the demographic growth brought about a non-negligible proportion of landless rustics, getting by either as daily wage-earners or as vagrants and beggars. According to a recent estimate by Bruce Campbell (which assumes the total population of about 4.7 million *c.* 1300, which is at odds with our present estimate of 5.25 million), about 37 per cent of the English population was made up of cottagers, landless peasants, and non-agricultural rustics. To these we should also add town dwellers, representing some 15–20 per cent of the total population. It would be incorrect to assert that these people never had any direct access to food supply: some burgesses held gardens and, in some cases, very small agricultural allotments within city walls.[9] Also, we have to account for the fact that in many instances there were agricultural fields attached to the city walls, as it was in Colchester, Nottingham, Norwich, Cambridge, and other towns. But there is no doubt that these would have not been sufficient to sustain an average urban household, even in the most abundant years of plentiful harvests and low grain prices. One thing in common between landless rustics and town dwellers was the fact that both groups were not self-sufficient, as far as food supply is concerned, and had to rely on outside channels of food provisions (markets, charity institutions, employers, etc.). Thus, on the eve of the famine, a disturbingly large proportion of the English population had no means to sustain itself, and it was that share that was especially vulnerable to any tipping point or disruption in the food supply. Therefore, it was the landless masses that would have borne the heaviest burden of the famine, and there can be little doubt that famine-related diseases and mortality would ravage, first and foremost, that socio-demographic group.

[5] Campbell, *The Great Transition*, p. 185.

[6] Campbell, 'Population Pressure, Inheritance', esp. pp. 104–06.

[7] Campbell, 'Population Pressure, Inheritance', p. 106.

[8] Campbell, 'Agricultural Progress'.

[9] Dyer, 'Gardens and Garden Produce', p. 37.

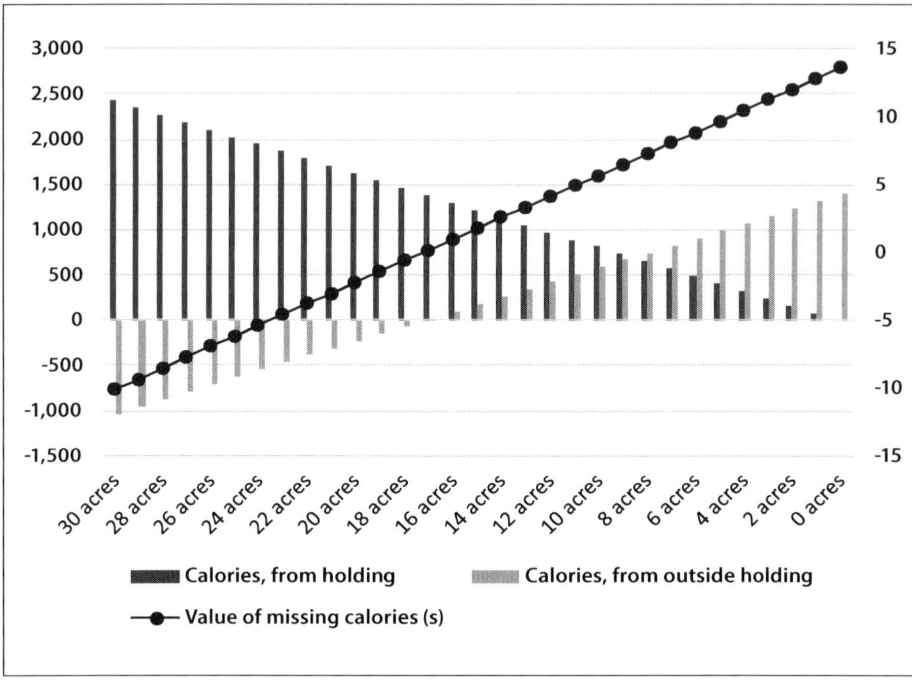

Figure 7.1. The 'missing calories per-capita' curve,
based on various landholding-size scenarios in 1316–1317.

Note: The estimated figures, appearing on the left vertical axis, are 'net processed' calories, after all losses to seeding, tithe payment, spoilage, and extraction rates. I have estimated the extraction rates to have been somewhat higher in the famine years, based on the fact that more barley, dredge, and oats were converted into bread, and less ale was being brewed. The estimated extraction rate here is 28 per cent (namely, 28 per cent of the calories were lost in baking and brewing), which is higher than the estimated 38 per cent for non-famine years. The estimated figures, appearing on the right vertical axis, are the value of missing calories (in shillings).

What about landholding tenants? What would have been the bare minimal number of arable acres necessary in order to avoid the crisis? Figure 7.1 makes a bold attempt to establish a 'missing calories per-capita' curve, which estimates the capacity of different landholding sizes to supply tenant families with the necessary crop-based calories in 1316–1317. The model assumes that (1) a daily requirement was 2000 kcal per capita; (2) about 70 per cent of calories were derived from crop-based products; (3) about one third of arable land would have been fallow every year; and that (4) one unit of grain would have rendered as many calories as the same unit would have yielded in non-famine years. As the figure shows, in 1316–1317, a typical yardlander (a holder of about thirty acres) would have had sufficient amounts of food for each household member and would, in addition, have produced a surplus of over 1000 kcal per capita. This was a luxurious scenario, quite detached from the cruel reality of 1316–1317: in that year, yardlanders constituted a small minority of population — perhaps about 16 per cent, according to Bruce Campbell's recent estimate. An average holder of a fifteen-acre holding (a 'half-yardlander') would not have produced enough for each family member and would have had to secure over 180 kcal per capita from outside of his holding. But even such scenario was a bit too idyllic: the proportion of half-yardlanders shrunk considerably over the course of the thirteenth century, and it is unlikely that the share of tenants holding parcels within the range of ten to twenty acres exceeded 10 per cent of the total population on the eve of the famine. Far larger was a group of smallholders, whose proportion would increase in inverse proportion to landholding size as it progresses towards the bottom of landholding scale. Ten-acre holdings were by far more common than half-yardlands; yet five-acre holdings would have been more common than ten-acre tenancies. Ten-acre holders would have produced about 58 per cent of the required calories to sustain their families, while their five-acre-holding neighbours would have supplied only about 29 per cent of what was needed. Naturally, landless rustics and most of urban population would have been in the most dire situation.

In other words, the vast majority of tenants could not secure enough calories from their holdings. In theory, at least, the only immediately licit way to compensate for the missing calories was to resort to an alternative channel of food supply. The two most obvious choices would have been either the grain market or livestock resources. Neither possibility, however, would have been easily achievable. As we have seen, the grain market was failing because of the abysmally high prices and the disruption of the food supply, caused by hoarding and transaction costs. As far as reliance on livestock is concerned, it would, as discussed in Chapter 3, require a considerable augmentation of animal numbers. With the exception of landlords or substantial tenants, such expansion of livestock would not have been possible on such a short notice. It would have been much easier for a half-yardlander to secure the remaining 180 or so calories from the local market or his livestock than for a five-acre holder to procure the deficient 1000 or so calories from the same sources. All this implies the inevitable: that a large proportion of England's population was doomed to starvation, with all its harsh derivatives.

The Story of Bad Bread

In estimating the loss of crop-based calories, it is tempting to equate the harvest shortfalls with calories. In other words, one may assume that the fall in crop yields by 50 per cent entailed calorific decline by 50 per cent, too. In reality, however, the situation may have been even worse. John Trokelowe, an eye-witness of the famine, provided some fascinating details about the quality of harvested grain and, consequently, baked bread. According to Trokelowe, grains were devoid of their usual nutrients because of the lack of the summer sun. Once baked, the poor-quality bread could not nourish people, who remained in the state of permanent hunger, even if consuming large quantities of that bread.[10] The *Chronicle of St Mary's, York*, also comments on the quality of bread, noting that in Yorkshire the rain corrupted the germinating seeds to the extent that bread came out 'malted and corrupted' (*braseatus et corruptus*). The same chronicle reports the high prices of grains on Yorkshire markets and adds that hardly any people could eat local bread, and some left the county in search of better-quality bread.[11]

Although any interpretation of these reports may be speculative, modern nutritional science has established the disastrous effects of excessive rain on maturing grains. According to several studies, flooding can damage the germination process and reduce the calorific value of matured grains in several ways. First, it can create mould, namely microfungi, which encourage biodegradation through mycotoxins.[12] Second, it can also reduce the quality and size of kernel contents (germ and endosperm).[13] Finally, and perhaps most importantly, it causes unwanted sprouting of pre-harvested seeds (a process known as 'preharvest sprouting'), whereby alpha-amylase enzymes break down the kernel starch and reduce the calorific value, nutrients, and quality of flour.[14] In particular,

[10] 'Nec habebat panis robur nutritivum, seu virtutem substantialem more solito in se, pro eo quod grana a calore solis aestivi nutrimentum non habebant. Unde comedentes ex eo, licet magnum exinde sumerent quantitatem, brevi elapso intervallo famelici remanebant'. *Johannis de Trokelowe Annales*, ed. *by Riley*, p. 92.

[11] *The Chronicle of St Mary's Abbey, York*, ed. by Craster and Thornton, p. 67.

[12] Didwania and Joshi, 'Mycotoxins'.

[13] Labuschagne, Elago, and Koen, 'The Influence of Temperature Extremes on Quality and Starch Characteristics', and 'The Influence of Extreme Temperatures during Grain Filling'; Beckles and Thitisaksakul, 'How Environmental Stress Affects Starch Composition and Functionality'.

[14] Trethowan, 'Preharvest Sprouting Tolerance'; Thomason and others, 'Understanding Pre-harvest Sprouting of Wheat'.

excessive alpha-amylase activities in starch render flour thinner and more liquid than it should be. As a result, the final product (that is, bread) tends to be stickier than usual.[15] Whatever the correct interpretation may be, the reports of bad bread, incapable of feeding people, seem to corroborate nicely our nutritional understanding of the impact of disastrous weather on the quality of grains and flour. It appears, therefore, that the disastrous rain of 1314–1316 not only ruined harvests but also damaged harvested grains. Although impossible to quantify, it is likely that relative calorific loss was greater than relative volume loss. In other words, if net crop yields stood at, say, 50% below average, the calorific loss may have reached an even higher figure, which, alas, cannot be estimated.

From Field to Table: The Gradual Loss of Harvested Calories

To appreciate the remarkable challenges of food security that most English commoners were facing during the famine years, it is necessary to follow the process of calorific loss from the moment the crops were harvested to the moment they were served (the 'field-to-table' process). In the course of the process, a certain amount of calories would be lost to seeding (about 24 per cent of harvest in good years, 38 per cent in 1315 and 1316, and 31 per cent in 1317); tithe payment (10 per cent of harvest in all instances); spoilage in barns (about a further 10 per cent when stored *c.* 1300, as well as in 1316–1317 and 1317–1318, and about 15 per cent during the torrential rain in 1315–1316); and extraction rates (about a further 38 per cent *c.* 1300, about 28 per cent in 1315–1316 and 1316–1317, and about 33 per cent in 1317–1318). Also, we may, in a speculative manner, deduct a further 10 per cent or so, resulting from the nutritional and calorific loss of bad kernels and under-nourishing flour in 1315–1316. If the final figure fell short of the minimal requirement of 1400 kcal per capita, or about 5950 kcal per family (70 per cent of the total daily calorific requirement), the remainder had to be supplied from outside the holding (primarily, the market). Tables 7.1–7.2 estimate, in a very crude and speculative manner, the reduction in calorific value in the course of the 'field-to-table' process and the number of missing calories to be supplied from outside the holding and their financial equivalent, on the basis of the annual crop requirements of a ten-acre-holding family. Obviously, I have not taken a ten-acre-holding family to be a 'typical' one: peasant society was strongly stratified, and its famine experiences differed markedly. Those peasants who held eleven to fifteen acres

[15] Chereji, Cretescu, and Cărpiță, 'The Influence of Alpha Amylase'.

Table 7.1. Speculative levels of calorific supply per household from a ten-acre holding and the levels and value of missing calories.

	c. 1300	1315–1316	1316–1317	1317–1318
Gross yield (per diem)	13,787	11,868	9,556	12,030
After tithing (per diem)	12,408	10,681	8,601	10,827
After seeding (per diem)	9,430	6,622	5,332	7,471
After spoilage (per diem)	8,487	5,629	4,799	6,724
After food/drink conversion	5,262	4,053	3,455	4,505
Missing calories (per diem)	688	1,897	2,495	1,445
Value of missing kcal (per annum, in shillings)	2.6	18.5	24.7	7.9

Source: Slavin, Manorial accounts database (see Appendix 8); Appendices 3 and 9.

were certainly faring better than those holding one to seven acres. Rather, ten-acre holding was taken as a purely hypothetical model, without any intention to generalize.

Tables 7.1–7.2 establish two scenarios: (a) accounting for tithing, seeding, spoilage, and food conversion, *but not* for further nutritional and calorific reduction caused by the quality of flour and starch (Table 7.1); and (b) accounting for tithing, seeding, spoilage, and food conversion, *and* for further nutritional and calorific reduction caused by kernel corruption (Table 7.2). Table 7.1 compares the crisis years to an average year (*c.* 1300).

As Table 7.1 and 7.2 indicate, the famine years saw an alarming crisis in food security, which would have been experienced by a hypothetical English smallholder of a ten-acre holding. The malignant combination of abysmally low crop yields, insufficient land, astronomic prices and omnipresent poverty deprived English smallholders of their ability to secure sufficient calories for themselves and their family members. In contrast with an average year *c.* 1300, when ten arable acres (that is, about 6.7 sown acres) could, at least in theory, provide about 88 per cent of the required calories for the entire family *per annum* (the remaining 12 per cent could have been easily afforded through purchases at local markets for about 2s 7d), the situation in 1315–1316 and 1316–1317 was more than desperate. The same amount of arable land would yield no more than some 68 and 58 per cent of the total required crop calories, not counting the further calorific reduction because of kernel corruption. Taking into account kernel corruption, the figures are reduced to 61 and 52 per cent of the required levels. In order to purchase the remaining calories at local markets, a

Table 7.2. Speculative levels of calorific supply per household from a ten-acre holding and the levels and value of missing calories, allowing losses of a further 10 per cent to kernel corruption.

	1315–1316	1316–1317
Gross yield (per diem)	11,868	9,556
After tithing (per diem)	10,681	8,601
After seeding (per diem)	6,622	5,332
After spoilage (per diem)	5,629	4,799
After further nutritional losses (per diem)	5,066	4,319
After food/drink conversion	3,648	3,110
Missing calories (per diem)	2,302	2,840
Value of missing kcal (per annum, in shillings)	22.4	28.2

Source: Slavin, Manorial accounts database (see Appendix 8); Appendices 3 and 9.

hypothetical ten-acre-holding family had to pay between 18s 5d and 22s 4d in 1315–1316 and between 24s 7d and 28s 2d in 1316–1317, depending on each scenario. These were truly astonishing sums, representing about a half the value of the taxed movables of an average taxed household in 1316, as indicated in the surviving lay subsidy rolls from Bedfordshire, Hertfordshire, and Oxfordshire, discussed in the previous chapter. The situation was undoubtedly gloomier in the 'war zone': because of environmental destruction, ten arable acres were able to provide even less calories, while the value of the missing calories would have conversely been even higher than the estimated figures above. In other words, it became remarkably difficult for most English commoners to secure sufficient calories both from their holdings and from outside their tenancies. In the case of those families holding less than five acres, such as the tenants in eastern Norfolk, such a task became, more or less, mission impossible, and the only two options for them were, in effect, either to starve or to resort to illicit channels of food security.

This gloomy situation bears much resemblance to other historical famines, as far as the levels of calorific supply are concerned. For instance, during the worst phase of the Siege of Leningrad (November 1941–January 1942), the official amounts of daily *per capita* rations stood at about 768 kcal for workers and just 400 kcal for clerks.[16] At the height of the Dutch famine, between December 1944 and April 1945, the *per capita* ration figures varied from 400

[16] Kislitsyn, *Leningrad ne sdayotsya*, p. 124 (in Russian).

to 800 kcal per diem.[17] During the North Korean Famine of 1994–1998, and especially during its peak (1997–1998), the *per capita* levels of daily rations allocated through the public distribution system stood at about 634 kcal in the Pyongyang area and 563 kcal in other areas.[18] More examples can be added. In all cases, these 'official' rations fell well short of the minimal 'bare-bone' calorific requirement and starving populations had, just as in the early fourteenth century, to either resort to 'alternative' famine foods or starve.

To appreciate the situation better, it is instructive to consider it through the prism of the concept of 'calorific currency'. The term 'calorific currency' denotes here the idea that in most pre-Industrial agrarian societies based on subsistence economies, calories were not only the most demanded types and units of energy, but also the single most demanded and sought-after commodity. Although the degree of dependence and demand would vary across places and social strata, the vast majority of people living and working within the framework of subsistence economy would undoubtedly exchange or convert most of their available resources (land, goods, labours, services, etc.) into calories to ensure their physical existence. Calories should be regarded, in effect, as some sort of a universal currency, into which the majority of available resources were being converted. The widespread dearth of grain implied that the financial value of calories rose. Taking into account each crop's relative contribution to the arable land and selling price, it is possible to estimate the financial value of crop calories (Appendix 5). The financial value of one million crop calories rose from about 10s 4d in an average non-famine-year to 26s 2d in 1315–1316 and 26s 11d in 1316–1317.

Alternative Food Consumption

The most obvious reaction to the alarming deficit of the required grain-based calories was a shift to an alternative schedule of food consumption. Initially, this meant that a certain proportion of the missing calories could have, at least in theory, been extracted from livestock, through large-scale slaughter. This tendency is reflected in some household accounts. Thus, the canons of Bolton Priory, ravaged by the adverse combination of harvest failures, animal murrain, and the Scottish threat, had to slaughter nearly all of their swine stocks (some 95 per cent) in late 1316.[19] Of all domestic animals, swine were the single most

[17] Burger, Sandstead, and Drummond, eds, *Malnutrition and Starvation*.

[18] Kim, Lee, and Sumner, 'Assessing the Food Situation in North Korea'.

[19] Kershaw, *Bolton Priory*, p. 107.

efficient producers of meat, and thus of calories, per relative weight. On average, around 1300, one sow produced two litters of piglets annually, rendering together ten piglets. An average fattened porker yielded about 50 lb of meat, rendering approximately 77,270 kcals.[20] It is hardly surprising, therefore, that the Bolton canons chose to resort to pork. Because of its being a target for Scottish attacks, Bolton Priory was far more unfortunate than most monastic communities of its day, and its desperate strategy to take up the slack does not by any means reflect a general trend within aristocratic households. On the other hand, some sources clearly indicate that this strategy was widely undertaken by poorer commoners, at least by those who had *some* livestock to be slaughtered and consumed. According to John Trokelowe (and later affirmed by Thomas Walsingham), pigs were eaten in large numbers because of grain deficiency. Trokelowe also adds that 'common meat licit for consumption was hardly to be found'.[21] This is corroborated by some tithe accounts associated with animal tithes. At East Meon Church (Hampshire), each year between 1310 and 1314, eight piglets and twenty-two geese were tithed, on average. In 1314–1315 the number of piglets declined to five. In 1315–1316 no piglets or geese were paid. In the course of the following year, the respective numbers were two and four, while in 1317–1318 they rose to just three and seven.[22] In other instances, the numbers of tithed animals in both non-famine and famine years were too small to identify any trend. The temporary shift to meat consumption is also reflected in the changing ratio between cash heriots and animal heriots. Thus, on five manors of the Winchester Bishopric (Taunton in Somerset, East Meon, Bishop's Waltham, Wargrave, and Fareham, all in Hampshire), cash heriots accounted for 63 per cent of all heriots — in contrast with the more-or-less equal ratio in the non-famine years between 1300 and 1349.[23]

The small size of holdings and omnipresent lack of capital, which characterized a large proportion of English commoners, meant that there were clear limits to their ability to secure animal-derived calories *in lieu* of the missing grain calories, either from their own stocks or from outside their holdings. Some tenants could potentially secure the missing calories through the employment of credit or through selling parts of their holdings. As we shall see in the following chapter, the famine years saw a remarkable spike in land transactions, while the

[20] Harvey, *Living and Dying in England*, pp. 228–29.

[21] *Johannis de Trokelowe Annales*, ed. by Riley, p. 92.

[22] HantsALS, 11M59/B1/66–73.

[23] Calculated from Postan and Titow, 'Heriots and Prices on Winchester Manors'.

volume of credit transactions was unusually high during the first year of the famine. Although credit and landed markets were instrumental in mitigating the famine (as will be discussed in the next chapter), they could by no means prevent it. True, some tenants undoubtedly saved their lives by securing food in credit and converting their land into cash, but not every tenant had access to either credit or the land market. Some poorer elements lacked the means, reputation, and trustworthiness to secure credit from their better-off neighbours. Similarly, as we have seen, about 37 per cent of the rural population was made of cottagers and landless rustics, who had no land to sell. But even those holding a few acres could offer correspondingly little to local land-markets. In the case of those poor and deprived masses, the only option to procure the missing calories in a legal manner would be to resort to what historians and anthropologists refer to as 'famine foods', often made up of otherwise repugnant comestibles.

According to one source, the people of Northumberland ate horses and dogs.[24] Horse-flesh consumption is forbidden in Christianity. Despite this ban, however, there were instances of illicit trade in and consumption of horse-meat in non-famine years in late medieval England and other parts of the British Isles.[25] This is also corroborated by some archaeological evidence: remains of a piece of butchered horse bone have been unearthed at West Cotton (Northamptonshire). Apart from being butchered, the bone piece was broken, presumably to extract marrow for human consumption.[26] Obviously, the date of butchery cannot be narrowed down to the famine years, and it is unclear whether it comes from the famine context at all. John Trokelowe noted that horse-flesh was too expensive for some people.[27] A chronicler from Bermondsey reports the consumption of mice, dogs, and pigeon dung (*stercus columbarum*).[28] The *Brut* chronicle adds that starving people were consuming cats.[29] Although dogs, cats, rats, and mice have never been a part of native European culinary tradition, they have been consumed in parts of East Asia (chiefly, in China, South Korea, Vietnam, and Thailand).[30] In the European context, dogs, cats, rats, and mice are reported as 'famine foods' on numerous occasions, includ-

[24] *Vita Edwardi Secundi*, ed. by Childs, pp. 120–21.

[25] Claridge, 'England's Sacred Beast?'.

[26] Dyer, 'The Material World of English Peasants', p. 3.

[27] *Johannis de Trokelowe Annales*, ed. *by Riley*, p. 95.

[28] *Annales Monasterii de Bermundeseia*, ed. by Luard, p. 470.

[29] *Castleford Chronicle*, ed. by Eckhardt, p. 210.

[30] Podberscek, 'Good to Pet and Eat'.

ing the Finnish Famine of 1696,[31] Great Irish Famine of 1845–1852,[32] and the Ukrainian *Holodomor* of 1932–1933.[33] Neither dog meat nor mouse-flesh nor cat meat can boast high calorific value, rendering, respectively, about 262, 216, and 240 kcal per 100 gram. 'Pigeon dung', on the other hand, may be a biblical cliché, related to the famine in Samaria during the siege by Ben-Hadad, king of Aram, when one fourth of the *cab* of 'pigeon dung' (Biblical Hebrew: *hireyonim* or *divioynim*) was sold for 5 *shekels* (2 Kings 6:25).[34]

The deficiency of ale and cider, caused by both grain and fruit failures and the cuts in brewing in favour of baking, would likely have prompted people to resort to fresh water consumption. This, in turn, could have encouraged the infestation of parasitic worms in human bodies. Although long before bacteriological knowledge, early fourteenth-century people certainly had a clear idea about dangers associated with fresh water consumption.[35] In normal years, fresh water was reserved to the very lowest social echelons, who could not afford ale. The consumption of fresh water became especially dangerous during the famine years because of the incessant rainfall. Rainwater is a reservoir of numerous bacteria, parasites, and chemicals, all harmful for human health. To make things even worse, however, excessive flooding tends to contaminate water resources and encourage the activity of viruses, protozoa, and parasitic worms.[36] One recent palaeoecological study of medieval York has established the omnipresence of roundworms, whipworms, and other parasites in local human populations. As the authors noted, water pollution could be one potential agent of their dissemination.[37] These types of worms cause numerous pathologies, including trichuriasis, whose symptoms are weight loss, dehydration, constant hunger, incessant diarrhoea, rectal prolapse, slow growth, and anaemia.[38] Surely, the further loss of nutrients and calories to intestinal parasites would have made the survival of the starving populations even more difficult.

[31] Myllyntaus, 'Summer Frost', p. 82.

[32] Cathal Poirteir, *Famine Echoes*, pp. 60 and 65.

[33] Holodomor 1932–1933: Famine Genocide in Ukraine, 'Holodomor Survivor Accounts and Memoirs'.

[34] Marvin, 'Cannibalism as an Aspect of Famine', pp. 80–81.

[35] Jørgensen, 'Local Government Responses to Urban River Pollution'.

[36] Crum-Cianflone, 'Bacterial, Fungal, Parasitic, and Viral Myositis'.

[37] King and Henderson, 'Living Cheek by Jowl'.

[38] Kaminsky, Valenzuela Castillo, and Abrego Flores, 'Growth Retardation and Severe Anemia'.

Cannibalism?

No other manifestation of famine food consumption, however, is as bizarre — and disturbing — as the consumption of human flesh. Instances of cannibalism are reported in several English and Irish textual sources. The *Annals of Bermondsey* reports that 'paupers used to eat their children' (*pauperes enim pueros suos manducabant*).[39] Robert de Greystanes, the chronicler from Durham Priory, adds that 'because of the magnitude of the famine, women used to eat babies' (*mulieres parvulos prae famis magnitudine comedebant*).[40] The *Brut* chronicle reports that hungry people stole children and ate them.[41] John de Trokelowe notes that: [42]

> As many assert, both men and women, in many places, used to secretly eat their own and others' babies. But what is especially horrible — and let it be passed along to the knowledge of the future generations — also incarcerated thieves would devour their new inmates, half-alive, in a matter of moments.

Thomas Walsingham, relying on Trokelowe, tells the same story of child-eating, adding that the thieves and 'those held in chains' would first mutilate their new inmates before eating them.[43] In Ireland, as the *Annals of Ulster* and the *Annals of Loch Cé* report, 'people undoubtedly used to eat each other throughout [the country]'.[44] According to another source, hungry and exhausted Ulster soldiers were seen digging corpses to eat them.[45] Outside the British Isles, the instances of child-eating and corpse digging are also reported in Poland, Bohemia, the German Empire, and the Baltic lands.[46] The authenticity of these reports (as well as similar reports from other historical famines) has long been debated among historians. Some accept cannibalism reports as the extreme outcome of harvest failures and food scarcity.[47] Some other scholars dismiss them outright

[39] *Annales monasterii de Bermundeseia*, ed. by Luard, p. 470.

[40] *The Chronicle of Robert Greystanes*, ed. by Raine, p. 96.

[41] *Castleford Chronicle*, ed. by Eckhardt, p. 210.

[42] *Johannis de Trokelowe Annales*, ed. *by Riley*, p. 95.

[43] *Thomae Walsingham Historia Anglicana*, ed. by Riley, p. 147.

[44] *The Annals of Loch Cé*, ed. and trans. by Hennessy, p. 596.

[45] Lydon, 'The Impact of the Bruce Invasion', p. 285.

[46] *Joannis Długosz senioris canonici Cracoviensis Opera omnia*, ed. by Przezdziecki, VII, 83; Jordan, *The Great Famine*, pp. 148–50.

[47] Newfield, 'The Contours of Disease', p. 235; Garnsey, *Famine and Food Supply*, pp. 28–29, 35; Stathakopoulos, *Famine and Pestilence*, pp. 81–85.

as unsubstantiated hearsays, *curiosa*, or literary/biblical *topoi*.[48] Some remain undecided.[49]

In my search for possible reports of cannibalism in judicial or legal sources, I have gone through every single surviving gaol delivery and coroner's roll for the period between September 1315 and September 1318, totalling 3989 gaol delivery roll entries for fourteen counties and over two hundred coroner roll entries surviving for five counties.[50] There is not a single entry to indicate one instance of cannibalism. However, there is one Bedfordshire case of a particular interest, which, while not indicating man-eating, reveals a rather bizarre incident from the height of the famine. On 18 August 1317, William Nonne de Harewold admitted to having been a criminal (*latro*), and who had committed a long series of crimes with his accomplices. One of his accomplices was John Cuddyng de Carleton, who, together with William Nonne, was accused of 'murdering a certain unknown stranger' (*interfecerunt quendam hominem extraneum*) in the house of John de Harewolde in Carlton (Bedfordshire) and 'skinning the same man' (*et dictum hominem excoriaverunt*). In addition, John Cuddyng was accused of 'tanning the skin' (*pellem eius tannavit*) of the victim.[51] Nonne, convicted of a large number of crimes, was subsequently hanged. We will never know if and why Nonne and Cuddyng flayed their victim and tanned his skin and if his flesh was actually consumed in the context of widespread famine. After all, the gaol delivery roll entry does not say anything about the consumption of flesh. Even if this bizarre episode was, in any way, connected to a possible case of cannibalism, then it only strengthens the notion that this phenomenon was extremely rare and confined to social margins. After all, no other gaol delivery roll or coroner roll entry reports anything remotely related to man-eating.

In other words, the Carlton incident cannot be used to corroborate the chroniclers' reports of cannibalism. This, however, does not mean that famine-time cannibalism is fiction. There are numerous eye-witness testimonies, crimi-

[48] Marvin, 'Cannibalism as an Aspect of Famine', pp. 73–86; Braid, 'Economic Behavior, Markets and Crises', p. 344.

[49] Kershaw, 'The Great Famine and Agrarian Crisis in England', pp. 9–10; Jordan, *The Great Famine*, pp. 149–50; Ó Gráda, *Eating People is Wrong*, pp. 11–37.

[50] The surviving gaol delivery rolls are for the counties of Bedfordshire, Buckinghamshire, Essex, Herefordshire, Hertfordshire, Kent, London and Middlesex, Norfolk, Northamptonshire, Somerset, Suffolk, Surrey, Suffolk, and Yorkshire. The coroners' rolls survive for Bedfordshire, Berkshire (Wallingford Castle), Kent, London and Middlesex, and Northamptonshire.

[51] NRO, JUST 3/1/2, membr. 5^{r}.

nal prosecution, and journalists' reports in later historical famines. In particular, instances of cannibalism (although never widespread) have been recorded during the Ukrainian *Holodomor* famine of 1932–1933, during the first winter of the Siege of Leningrad (1941–1944), and during the Moldovan famine of 1946–1947, as well as during the Great Leap Forward Famine (1958–1961) in China.[52] More examples can be cited here.

Famine Pathologies and Mortality

It is rare that famine victims die of starvation; in most cases, as far as historical evidence tells, they tend to succumb to various hunger- and malnourishment-related diseases. Thus, fever (usually, typhoid fever), dysentery, and diarrhoea are three most common famine-related infectious diseases in different geographic and climatic settings. These three diseases ravaged the starving populations of Ireland in the late 1840s, India during the 1897–1900 famine, and Russia during the 1918–1922 famine. Similarly, during the Ethiopian and Sudanese famines of the mid-1980s, the major causes of death were infectious diseases, especially dysentery, diarrhoea, malaria, and measles. Conversely, no more than 10 per cent of all deaths resulted from starvation, in all these cases.[53] In his study of famine in late sixteenth-century Cumberland and Westmorland, Andrew Appleby conceded that the main cause of elevated mortality was typhus.[54] The Great Famine was no exception to this rule. As John de Trokelowe tells, people would develop dysentery (*morbus dysentericus*), which would claim almost all the starving because of corrupted food intake (*ex corruptis cibis*). This disease would then be followed by either acute fever (*acuta febris*, most likely typhoid fever) or tumorous throat disease (*pestis gutturuosa*, most likely scarlet fever).[55] Later on, Trokelowe adds the following:[56]

> the wisdom of doctors could not find any suitable remedy against those diseases in its craft, as it used to in the ancient times. For, medical herbs, which they used to apply in times of that mortality, in mitigation [of sufferings] because of the mad-

[52] Boriak, 'Sources and Resources'; Reid, *Leningrad: Tragedy of a City*, pp. 286–92; Ó Gráda, *Eating People is Wrong*, p. 14; Várdy and Várdy, 'Cannibalism in Stalinist Russia'.

[53] Ó Gráda and Mokyr, 'What do People Die of During Famines'; Ó Gráda, *Famine: Short History*, p. 112.

[54] Appleby, *Famine in Tudor and Stuart England*, pp. 102–03.

[55] Carole Rawcliffe identified it with scarlet fever. See Rawcliffe, *Urban Bodies*, p. 361.

[56] *Johannis de Trokelowe Annales*, ed. *by Riley*, p. 94.

> ness of weather and the extraordinary collision of the elements, degenerated against their nature and rendered a virus, instead of a virtue (*virus pro virtute reddebant*).

Although these diseases hit any age group, it is infants, small children, and old people who bear the heaviest burden.[57]

Some evidence suggests that starvation affected the mental health of starving people. The coroner's roll for Bedfordshire recorded two instances of people suffering from 'frenzy disease' (*infirmitas vocata Frenesye*). One person, William Rungefer of Eaton, hanged himself on 9 March 1317, while another one, Emma le Bere, killed her entire family, including an unspecified male relative and three daughters, before killing herself on 15 June 1316.[58] The impact of famine on the mental health of starving individuals has long been noted in both historical sources and scholarly literature, while suicide is one sad manifestation of despair and madness, despite the negative cultural and religious attitudes to self-killing. Thus, at the time of the Great Famine of 1845–1852, Ireland, despite its strong Catholic identity, suffered from an increase of suicide deaths.[59] Suicide rates rose a great deal in India during the famines of 1876–1878 and the late 1890s, despite both Dharmic religions and Islam considering suicide a sin.[60] During the Holocaust, some Jewish prisoners, driven to despair by stress, humiliation, and starvation, found their way out by killing themselves, notwithstanding the negative views on suicide in Judaism.[61] As recently as 2011, famine-related suicide instances have been reported in Somalia, a predominantly Muslim country.[62]

Famine diseases led to famine mortality. Excess deaths are noted in a wide array of sources, both narrative and statistical. As expected, chronicles provide vague descriptions of the proportions of famine mortality, usually talking about 'a great pestilence', 'dire mortality', or 'most cruel pestilence'.[63] One chronicle recorded that 'such great mortality has never been heard of since

[57] Crump, Luby, and Mintz, 'The Global Burden of Typhoid Fever'; Kotloff and others, 'Burden and Aetiology of Diarrhoeal Disease'.

[58] TNA, JUST 2/6, membr. 4 and 2.

[59] Laragy, 'Narratives of Poverty in Irish Suicides'.

[60] Mayer, *Suicide and Society in India*, pp. 85–88.

[61] Ryn, 'Suicides in the Nazi Concentration Camps'.

[62] Radnedge, 'Somalia Drought Victims Attempting Suicide to Escape Worst Ever Famine'.

[63] *Annales de Monasterii Osenia*, ed. by Luard, p. 344; *Annales Londonienses*, ed by Stubbs, p. 236; *Vita et Mors Edwardi Regis Angliae*, ed. by Stubbs p. 301.

the beginning of ages'.[64] Another narrator noted that people were dying so fast that people could no longer care for the dying and bury the dead.[65] According to Robert de Greystanes, people were dying everywhere: in towns, fields, and roads. In London alone, 20,000 people died.[66] Some sources specified that it was predominantly poorer commoners that succumbed.[67] The vulnerability of poor people is reflected in coroners' rolls, surviving for the famine years for Bedfordshire, Kent, London, and Northamptonshire, as well as Wallingford Castle prison (Berkshire).[68] In particular, the London rolls contain several entries recording homeless paupers and beggars who had died 'of famine and cold' (*fame et frigore*).[69] One case is particularly sad. On 17 December 1316, a certain homeless beggar, Agnes de Goldelane, searching for food, collapsed by the wall of St Mary's chapel in the parish of St Mildred and died around the following dawn 'because of her body's weakness and lack of food' (*ex debilitate corporis sui et defectu victus*).[70]

The fact that famine mortality targeted the poorer and frailer individuals is supported by palaeopathological evidence. Although the 'archaeology of famine' is still a nascent field, there are some exciting developments in more recent years. The new data derives from several Black Death cemeteries. The bulk of the information derives from 636 Black Death skeletons at the East Smithfield Cemetery in London (excavated back in 1986–1988 and currently archived at the Museum of London).[71] In addition, 185 skeletons were unearthed at another Black Death burial at Hereford Cathedral (excavated in 1993 and currently archived at the Biological Anthropology Research Centre (BARC) at the University of Bradford).[72] In 2013–2014, during the London Crossrail construction works, twenty-five further skeletons were unearthed at another Black Death burial site at Charterhouse, Farringdon (in the care of the

[64] *Gesta Edwardi de Carnarvan*, ed. by Stubbs, p. 48.

[65] *Annales Monasterii de Bermundeseia*, ed. by Luard, p. 470.

[66] *The Chronicle of Robert Greystanes*, ed. by Raine, p. 96.

[67] *Flores Historiarum*, ed. by Luard, p. 174; *Le livere de Reis de Brittanie*, ed. by Glover, pp. 332–33; BoL, MS Devon Top. 5, fol. 5v (History of Abbots of Newman Abbey).

[68] TNA, JUST 2/6; BerksRO, W/JCm 1–3; TNA, JUST 2/47; TNA, JUST 2/94/A-B; TNA, JUST 2/108/A–B.

[69] TNA, JUST 2/94/A, membr. 2 and 3.

[70] TNA, JUST 2/94A, membr. 3.

[71] Kausmally, 'East Smithfield Black Death Cemetery Summary'.

[72] Boucher, Craddock-Bennett, and Daly, *Death in the Close*.

Museum of London Archaeology, or MoLA).[73] These data will be augmented in the foreseeable future by further evidence from West Smithfield Cemetery (London), Thornton Abbey (Lincolnshire) with forty-eight skeletons, and St John's Hospital, Cambridge.[74] In particular, much light has been shed by recent work of Sharon DeWitte and her colleagues.

The current *status quaestionis* on famine pathologies can be summarized as follows. It appears that the famine mortality was selective, targeting frailer individuals, who were more prone to diseases than their healthier peers. This can be gathered from the fact that the pre-famine cohorts tend to exhibit less pathological signs and higher stature than those born after 1317. In other words, the Great Famine weeded out weaker and frailer individuals, while the famine survivors may have been healthier than both the famine victims and the post-famine cohort.[75] Conversely, the post-famine cohort (and in particular those born shortly after the famine) exhibit various pathological signs, including shorter stature, cribra orbitalia (lesions on orbital roofs), and porotic hyperostosis (lesions on cranial vault bones). As the Hereford evidence suggests, the highest frequency of porotic hyperostosis (seven out of twelve, or 58 per cent) occurs in individuals aged 20–29 (namely, those born between *c.* 1320–1329). By contrast, within the pre-famine cohort, only two out of ten individuals (20 per cent) exhibit this pathology.[76] The increasing incidence of cribra orbitalia is also detected in the post-famine cohorts buried at East Smithfield and the recently excavated Charterhouse cemetery.[77] As DeWitte and I have argued elsewhere, these pathologies may be explained by the chronic deficiency of dairy products in the 1320s, in conjunction with the impact of the Great Bovine Pestilence (1319–1320), which killed about 62 per cent of English and Welsh cattle.[78] That famine mortality tends to target frailer individuals is also confirmed by the analysis of three mass burials from St Mary Spital, London

[73] Dick and others, 'Detection and Characterisation of Black Death Burials'. A monograph on the Charterhouse excavations has been recently published by Crossrail Archaeology project series; see Pfizenmaier, ed., *Charterhouse Square*.

[74] Cessford and others, 'The St John's Hospital Cemetery'.

[75] Slavin and DeWitte, 'Between Famine and Death', pp. 37–60.

[76] Kacki, 'Influence de l'état sanitaire', pp. 286–87.

[77] Slavin and DeWitte, 'Between Famine and Death', p. 45; Pfizenmaier, ed., *Charterhouse Square*, pp. 132–33.

[78] Slavin and DeWitte, 'Between Famine and Death'.

(including, most likely, the 1258–1259 famine).[79] As Yaussy, DeWitte, and Redfern have recently shown, these mass burials, including the mid-thirteenth-century famine victims, exhibit a higher frequency of early life stressors than those buried in other contexts at the same site.[80]

Although palaeopathological evidence on the selectivity of famine mortality comes from urban contexts, there can be little doubt that frailer rustics were coping with the crisis as badly as their urban counterparts. This is partially reflected in manorial court rolls and, in particular, in entries dealing with heriot payments — death dues owed by tenants to their lords upon entry on their deceased relatives' plots of land (and in some instances, dues paid on transfers of land).[81] Assuming a good degree of correlation between frailty, living standards, and poverty, one expects to see an inverse correlation between the size of peasant holdings and mortality frequency. That is, poorer tenants were more likely to succumb to famine-related diseases, on account of their poverty, low living standards, and, consequently, frailty. Indeed, as Table 7.3 demonstrates, about 57 per cent of all heriots paid on eight manors during the famine years came from very small holdings, amounting to less than two acres each. About 14 per cent of all heriots were paid for landless properties, consisting of cottages and tofts. This is somewhat in contrast with non-famine years, when about 47 per cent of all heriots were collected from holdings of less than two acres, while landless properties accounted for 7 per cent. It is true that this sample is based primarily on evidence from East Anglia, where the population density was the thickest in England and where such small holdings were an omnipresent reality and substantial tenants were few. All the same, the disproportionate predominance of heriots related to small holdings undoubt-

[79] Connell and others, *A Bioarchaeological Study of Medieval Burials*.

[80] Yaussy, DeWitte, and Redfern, 'Frailty and Famine'.

[81] Heriots could be paid either in cash, or in the 'best beast', usually oxen or cattle. As such, heriots represent proxy for mortality trends on a manorial level. This is not to say that heriot entries are not without their pitfalls. First, some court rolls under-record heriots, and, hence, deaths of tenants. This is largely because, in some instances, heriots were only collected from tenants upon entry on large holdings (See Schofield, 'Dearth, Debt and the Local Land Market'). Second, heriot entries do not indicate the dates of tenants' demise. Instead, they were recorded in the nearest subsequent court, which allows us to establish the *termini ante-* and *post-quos* of tenants' deaths. Third, some heriot entries do not specify the size of peasant holdings. Finally, they were only collected from unfree peasants. For the purpose of the present exercise, I have confined the analysis only to those court rolls that meticulously record heriot payments, specify the names of tenants and the size of their parcels, and have no chronological gaps for the famine years.

edly points out that it was the poorest who had to pay the heaviest price of starvation and disease. This is especially apparent in the case of Wakefield, a huge manor extending over many townships in West Yorkshire, which did not suffer from overpopulation, yet almost half of all inherited holdings contained less than five acres. The inverse relationship between the number of heriots and landholding sizes indeed agrees with the palaeopathological evidence discussed above. We can only speculate that landless rustics and urban paupers, who held no properties and, thus, left no trace in either manorial or urban records, were dying in even higher proportions.

Partially because of higher living standards and partially because of being much less numerous, better-off tenants appear much less frequently in heriot records than their poor neighbours. Still, about 15 per cent of all heriots were related to larger holdings, amounting to more than fifteen acres each. At less densely populated manors, such as Wakefield (West Yorkshire), the deaths of half-virgaters and virgaters accounted for about 28 per cent of all tenant deaths. Naturally, there is no way to know for certain if these better-off peasants died of some famine-related causes or of something else. At the same time, it should be borne in mind that famine-related sicknesses, such as dysentery, typhoid fever, and scarlet fever, which seem to have been ravaging Britain's population, are highly infectious diseases, respecting neither wealth nor status.

As expected, due to the infectious nature of famine diseases, mortality would, in some cases, ravage among close relatives and within the same households. Thus, at Scunthorpe (Lincolnshire), two brothers (most likely bachelors or widowers living together), Roger and William, sons of William, both died sometime between late November and late December 1316, and their parcel of land was inherited by their two sisters, Alice and Matilda.[82] Similarly, at Pilham (also in Lincolnshire), two brothers, Robert and Richard, sons of John, both perished sometime in April 1317, leaving their holding to be inherited by their younger brothers, William and Thomas.[83] At Hindolveston, two sisters (and, judging by a lack of filial heirs, most likely spinsters), Emma and Huelyna, daughters of Geoffrey Carpenter, holding in common a meagre parcel of just half an acre, died either in May or June 1317.[84] Their property was inherited by their relative of an unspecified degree, Margery Knyght (possibly a married sister or niece). More instances can be added. A similar trend may

[82] LA, KR 83, membr. 3.

[83] LA, KR 82, membr. 6.

[84] NRO, DCN 60/19/3.

Table 7.3. Distribution of size of holdings inherited by local tenants on eight English manors, 1311–1321, famine and non-famine years.

	Number of heriots		*As a percentage of the total*	
Size of holdings	**Famine years**	**Non-famine years**	**Famine years**	**Non-famine years**
no land (cottages, messuages, etc.)	39	20	14%	7%
<2 acres	118	109	43%	40%
2–5 acres	54	71	19%	26%
6–10 acres	29	41	10%	15%
11–15 acres	19	18	7%	7%
>15 acres	18	16	6%	6%
Total	**277**	**275**	**100%**	**100%**

Source: NRO, NRS 19505 (Horsham St Faith, Norfolk); NRO, Lest DA/6–8 (Heacham, Norfolk); NRO, DCN 60/19/3 (Hindolveston, Norfolk); NRO, DCN 60/32/9 (Sedgeford, Norfolk); RHM, Langham Court Rolls (Langham, Norfolk); Chicago University Library, Special Collections Research Center, Bacon Rolls 11–14 (Redgrave, Suffolk); and BoL, Gough Camb 1 (Chesterton, Cambridgeshire). Also, *Court Rolls of the Manor of Wakefield*, III and IV, ed. by Lister; Leeds University Library (Brotherton Library), Special Collections, MD 225/1/37–48 (Wakefield, West Riding of Yorkshire).

Note: Although the differences may not appear meaningful when eyeballing, a chi-square test rendered the p-value of 0.05, meaning that it may still be considered statistically significant.

also be reflected in similar mortality trends of parish priests in some communities, as reflected in episcopal registers recording clerical appointments. Thus, at Stawell (Somerset), no less than three priests died between about July 1316 and January 1317.[85] This trend, however, was by no means ubiquitous: of the total of 677 post-mortem clerical appointments made between July 1315 and August 1317 (covering 631 parishes), only 36 (5.7 per cent) were for the same parish. Nevertheless, the high frequency of clerical deaths may indicate the contagious nature of the famine diseases.

It is unfortunate that the available evidence cannot shed any light on the gendered aspect of famine mortality. Although some heriots were collected upon deaths of female tenants (usually, widows), these instances clearly fall into minority, given the male predominance in land inheritance customs. Taken as a whole, only about 16 per cent were paid to enter upon deceased women's properties. This certainly cannot serve even as a remotest proxy to the ratio of

[85] *The Register of John de Drokensford*, ed. by Hobhouse, pp. 115, 119, and 122.

famine mortality between genders. Nor can it be speculatively inferred from other historical famines because of contradictory and variable evidence. On the one hand, there is a great deal of evidence supporting 'female advantage mortality', whereby women succumb in smaller proportions than men because of their biological and physiological mechanisms. First, women tend to be less susceptible to famine diseases (chiefly typhoid). Second, female bodies are smaller and lighter, with a slower metabolism and a larger proportion of natural fat content.[86] The phenomenon of 'female advantage mortality' has been spotted in late nineteenth-century famines in India, during the Bengali famine of 1942–1943, in the course of the terrible War World II famines in Greece, the besieged Leningrad, the Warsaw Ghetto and the Netherlands, during the Great Leap Forward Famine of 1958–1961 in China, the Bangladeshi famine of 1974–1975, and the Sudanese famine of 1983–1985. On the other hand, during the 1590s crisis in England and the Great Irish Famine of 1845–1852, the differences in mortality ratios between the two sexes was barely noticeable.[87] It is unclear, therefore, if early fourteenth-century men were more susceptible to the famine than their wives, sisters, and daughters, or if mortality proportions were roughly equal across the genders.

Famine Mortality Trends: Excess Deaths

The evidence of heriots can also be used to estimate the excess death rates among unfree tenants. For the purpose of this exercise, I have used all the available court rolls recording heriots numbers for the years 1311–1321, with no or as few gaps as possible. The sample consists of court rolls for nineteen manors, most of which survive with a complete run of courts for that decade and heriot entries for a further twenty-eight manors in southern England belonging to the Bishop of Winchester and enrolled into the Winchester Pipe Rolls. This makes the sample a total of forty-seven manors.[88] Given its frustrating geo-

86 Healey, 'Famine and the Female Mortality Advantage', p. 155.

87 Ó Gráda, *Famine: Short History*, pp. 94–95; Ó Gráda, *Ireland Before and After the Famine*, p. 109; Healey, 'Famine and the Female Mortality Advantage'.

88 The manorial court rolls include the following manors: Heacham, Hindolveston, Horsham St Faith, Langham, Sedgeford (Norfolk); Hinderclay, Redgrave, Rickinghall, South Elmham (Suffolk); Chesterton (Cambridgeshire); Alsacoe, Corringham (Lincolnshire); Wakefield (Yorkshire); Halesowen (Worcestershire-Shropshire border); Norton, Bramfield, Hexton, Ridge and Sandridge (Hertfordshire); the Winchester Bishopric manors, for which I have extracted heriot payments, are Waltham St Lawrence, Woodhay (Berkshire); Ivinghoe,

graphic skew towards East Anglia and south England, this is far from being an ideal and representative sample. But then again, given the fact that other court rolls for the 1310s either survive with significant chronological gaps or do not record heriots at all, this is probably as far as one can get when building a database of famine heriots. To get a relative figure for excess mortality rates, it is necessary to calculate a ratio between the heriot numbers in famine and non-famine years. This can be done by averaging the annual non-famine figures (that is, for 1310–1315 and 1318–1321) and the corresponding figures for 1315–1316 and 1316–1317 for each manor.[89] Because of the difference in population size and, hence, heriot numbers across manors, it is necessary to calculate a weighted average for all forty-seven manors for both the non-famine and famine years. Each manor is assigned its proportionate weight, according to the relative share of its heriots in the total sample. This renders the respective figures of about 2.25 and 2.94 for 1315–1316 and 1316–1317, in relation to the average of the non-famine years (1310–1315 and 1318–1321), which is indexed on 1 (Figure 7.3, see below). The figures varied from place to place.

In his 1998 study of pre-Black Death peasant welfare, Mark Bailey conceded that 'a volatile marketing system and a landholding structure dominated by smallholders...exposed the mass of the peasantry to a considerable degree of risk without providing them with the means to absorb that risk'.[90] Although it is true that in many instances the landholding structure, dominated by thick population densities, may have influenced excess mortality levels, the connection and dependence between the two variables should by no means taken for granted. Indeed, as we have seen, Norfolk (and some other parts of East Anglia) was the single most densely populated and most arable-intensive area in England (and seemingly, the British Isles), where land-hunger was an ever-pressing issue. The crisis is clearly reflected in the heriot entries at Horsham St Faith, where the 1315–1316 and 1316–1317 figures were 3.9 and 3.4 above their non-famine levels. Conversely, at Coltishall, where an average household held less than four

West Wycombe (Buckinghamshire); Alvestock, Burghclere, Bishopstoke, Bitterne, Bishop's Waltham, Bentley, Droxford, East Meon, Fareham, Hambledon, Havant, Merdon, Overton, Twyford, Wargrave (Hampshire); Witney (Oxfordshire); Holway, Poundisford, Staplegrove, Trull and Bishop's Hull (Somerset); and Farnham (Surrey).

[89] Because seasonal mortality patterns were depended upon food availability, I have opted to calculate the figures based on agricultural, rather than calendar years. In other words, each year runs between two Septembers.

[90] Bailey, 'Peasant Welfare in England', p. 246.

acres of land, there is hardly any increase in the number of deaths.[91] It has been claimed that the resilience of the Coltishall community to the famine may have something to do with the dynamic land market, which allowed local tenants to liquidate their assets in order to buy food.[92] Indeed, as we shall see in the next chapter, the land market did play a role in mitigating the crisis, although only in those instances when peasants had the means to participate in it. At the same time, at Halesowen (on the Worcestershire-Shropshire border), where there is no evidence of overpopulation, the 1316–1317 figure amounted to 5.33. Likewise, at Corringham (Lincolnshire), where population density levels could not have possibly matched those of Norfolk, the 1316–1317 figure stood at an overwhelming 6.0. At West Wycombe (Buckinghamshire), where, likewise, the population density was far less remarkable than in Norfolk, the respective figures for 1315–1316 and 1316–1317 stood at 1.45 and the exceedingly high 12.4. In other words, overpopulation did not mean increased vulnerability — and the other way around. The lack of straightforward correlation between the two variables has already been noted by Stephen Devereux and most recently repeated by Phillipp Schofield in his study of the Great Famine in Wales.[93]

It should also be borne in mind that those figures relate to tenants aged about twenty and over, and they do not account for younger people (as well as the older cohorts within the same communities). Therefore, the calculated heriots may reflect excess deaths of the adult share of local communities rather than entire communities. To project the heriot figures downwards and upwards, in order to create counterpart mortality rates for younger and older people, I have established, in a speculative manner, a tentative 'age structure' of England *c.* 1315, based on the 'Group Mortality Level 6, West' model, included in the Princeton Life Tables. This model is based on a population with an expectation of life at birth (e0) of thirty, an annual growth rate of 0.5 per cent, a crude birth rate (CBR) of 35 per 1000, and a crude death rate (CDR) of 30 per 1000.[94] The choice of this particular group is justified by its life-expectancy proximity to the situation in pre-Black Death England. Each age group is assigned its hypothetical excess mortality levels based on age-distribution mortality data from early modern English food shortage crises and the Great Irish Famine and is then

[91] Campbell, 'Population Pressure', pp. 98–99.

[92] Campbell, 'Population Pressure', p. 120.

[93] Devereux, *Theories of Famine*, p. 54; Schofield, 'Wales and the Great Famine'.

[94] Coale and Paul, *Regional Model Life Tables and Stable Populations*, p. 110. I am thankful to Prof. Richard Smith for this methodological suggestion.

indexed on the group of ten–forty-nine years. During the Great Irish Famine, and early modern English food crises, excess mortality rates were scattered disproportionally across different age groups, with infants, children aged seven or less, and old people being particularly vulnerable to famine-related mortality. This was also the case during the Finnish famines of 1831–1832 and 1857, the Russian famines of 1868, 1882, 1905, the Soviet famine of 1946–1947 and the Bangladeshi famine of 1974–1975. On the other hand, during the Russian famines of 1872, 1891–1892, the Soviet famines of 1918–1822 and 1932–1933, and the Greek famine of 1941–1944, infants and children were faring better than adults.[95] For the purpose of the present exercise, I have assumed that the early fourteenth-century famine exhibited the first type of age-distribution of mortality and opted to rely upon the data from early modern English shortage-related mortality crises and the Great Irish Famine. This choice can be justified by the fact that heavy mortality among infants and young children was a recurrent trend in all early modern English food crises and by the fact that of all the famines and shortages listed above, the mid-nineteenth-century Irish famine was the only one whose proportions may have been as alarming as those of the early fourteenth-century catastrophe. Moreover, the early fourteenth-century catastrophe took place in the same natural environment and similar climatic settings as early modern outbreaks of local famines and general food shortages. Finally, both early fourteenth-century England and mid-nineteenth-century Ireland were Malthusian societies suffering from overpopulation.

If we apply this age-distribution mortality trends to the Great Famine, then we arrive at the estimate that the overall excess mortality levels of 1315–1316 and 1316–1317 may have stood at about 3.80 and 5.02 relative to the non-famine years; namely, they were in the region of 114 and 151 per 1000 (compared with about 30 per 1000 in non-famine years) (Table 7.4). These estimates, crude and speculative as they may be, are considerably higher than crisis mortality ratios from the early modern period, deriving from post-1500 parish registers. For instance, during the 1550s subsistence crisis, the respective figures for 1557, 1558, and 1559 stood at 1.64, 2.08, and 1.82 relative to their 1547–1556 levels. In 1597 and 1603, the mortality ratios were slightly above 1.3 their 1587–1596 levels. In 1623, 1624, and 1625 the figures stood at 1.24,

[95] Dyson, 'On the Demography of South Asian Famines Part I' and 'On the Demography of South Asian Famines Part II'; Ó Gráda, *Ireland Before and After the Famine*, p. 109; Adamets, 'Famine in Nineteenth and Twentieth Century Russia'; Pitkänen, 'Famine Mortality in Finland', pp. 89–90; Hionidou, '"Send us Food or Coffins"', pp. 195–97, and *Famine and Death in Occupied Greece*, pp. 171–72.

Table 7.4. Speculative excess mortality rates within different age groups, relative to the 10–49 group during the famine (10–49 = 1.00) and relative to all age groups during non-famine years (1310–1315, 1318–1321 = 1.00)

Age group	Break-down within the sample (as % of total)	Excess mortality, relative to 10–49 group, famine years (10–49 = 1.00)	Excess mortality, 1316, relative to non-famine years (1311–1315, 1318–1321 = 1.00)	Excess mortality, 1317, relative to non-famine years (1311–1315, 1318–1321 = 1.00)
0–4	12.9%	3.74	8.42	11.11
5–9	10.97%	2.65	5.96	7.87
10–19	20.04%	1.00	2.25	2.97
20–29	17.21%	1.00	2.25	2.97
30–39	14.10%	1.00	2.25	2.97
40–49	10.89%	1.00	2.25	2.97
50–59	7.63%	1.00	2.25	2.97
>60	6.26%	3.48	7.83	10.34
Weighted average			3.80	5.02
National, adjusted			2.94	4.26

Source: As in Table 7.3; Coale and Demeny, *Regional Model Life Tables and Stable Populations*, p. 110; Scott and Duncan, 'The Mortality Crisis of 1623', pp. 15–16; Scott, Duncan, and Duncan, 'Infant Mortality and Famine'; Dobson, *Contours of Death and Disease*, pp. 160–220; Ó Gráda, *Ireland Before and After the Famine*, p. 109.

Note: For the conversion of 'manorial' rates into 'national' rates, see Appendix 6.

1.13, and 1.75 of their 1613–1622 levels. In 1727, 1728, 1729, and 1730, the mortality ratios stood at 1.22, 1.37, 1.54, and 1.25 above their 1717–1726 levels.[96] The remarkable gap in mortality proportions between the early fourteenth-century famine and early modern subsistence crises has been discussed by Cormac Ó Gráda, where he conceded that early modern England stood well above the Malthusian threshold of subsistence levels, thanks to the rising levels of agricultural output and literacy.[97] Here we may add, in a simplistic manner, that those early modern subsistence crises simply did not evolve into full-fledged and ubiquitous famines.

[96] Calculated from Wrigley and Schofield, *The Population History of England 1541–1871*, pp. 531–32.

[97] Ó Gráda, 'Neither Feast, Nor Famine'.

Even with all these caveats and re-adjustments, these figures cannot be taken as reflective of all of England, let alone all the British Isles. First, they account mostly for unfree tenants. This means that free tenants, most landless rustics (apart from some occasional entries related to cottage-heriots), vagrants, and town dwellers are left out. Likewise, they do not account for the tiny proportion made up of landlords and clergy. Converting all these unrecorded groups into an approximate share within the total population of England, we arrive at a rather frustratingly high figure of about 77 per cent.[98] According to Marxist logic, the extractive policies of landlords would have placed unfree tenants in a disadvantageous position compared to their free neighbours and made them more vulnerable to famine. In reality, however, the situation was more likely inverse. As we have seen in Chapter 1, unfree peasants were not necessarily worse-off than their free counterparts. Almost invariably, free tenancies were smaller in size (however, less expensive, in terms of rent value per unit of land).[99] Likewise, there is no doubt that landless peasants and non-agricultural rustics, as well as town dwellers, were more prone to starvation and would, thus, exhibit higher levels of excess mortality.

Second, as we have noted, because of the frustratingly thin coverage of surviving court rolls with complete heriot records for the 1310s, the sample captures only a small number of manors, with a strong preponderance of East Anglian and southern (primarily, Hampshire) estates (in addition to two Lincolnshire manors, one manor situated on the Worcestershire/Shropshire border, and an additional manor in the West Riding of Yorkshire). The predominance of meagre holdings, chronic overpopulation, and the total dependence on arable agriculture in Norfolk and other parts of East Anglia may imply that this region was especially vulnerable to famine. But then again, as we have seen, the remarkably high levels of heriot payments at Halesowen (the Worcestershire/Shropshire border) and Corringham (Lincolnshire), on the one hand, and no significant change in mortality levels at Coltishall (Norfolk), on the other, indicate that the two variables may not have always been correlated and the situation may have been more complicated, varying from place to place, and depending on local environmental and structural conditions.

[98] This is on the assumption that about 50 per cent of all tenants were free and that landed tenants accounted for 45 per cent of all England's population *c.* 1300. These estimates derive from Campbell, 'The Agrarian Problem', p. 36; Broadberry and others, *British Economic Growth*, pp. 317–18.

[99] Campbell, *The Great Transition*, pp. 182–98.

Both East Anglia and south England were strongly arable regions, allocating a large proportion of their cultivable acreage to tillage and depending on grain as the most basic staple food. Thus, both Norfolk and Suffolk devoted about 60 per cent of their land to arable farming. Although arable farming was the predominant type of economy within the 'Central Province', elsewhere more pastoral types of agricultural prevailed. Here, local tenants derived a larger proportion of their calories from non-farinaceous products and, thus, depended less on grain harvests. It is highly unlikely that these 'pastoral' regions fared in the crisis as badly as their 'arable' counterparts, and, therefore, they unlikely suffered equally high excess mortality rates. For instance, there is nothing to suggest that the crisis hit Cornwall, where arable accounted for only about one quarter of all cultivable land.[100] The only exception to this rule would be the war-torn counties of the north-west, precisely where population densities were the thinnest and where pastoral economy was predominant. Nonetheless, this region suffered severe depopulation, not only because of incursions and destruction of landscape but also presumably because of an aggressive spread of diseases by invading belligerents and fleeing residents. There is no direct evidence to support the latter assumption, but several Irish annals mention famine and diseases in conjunction with the ongoing warfare. Thus, the *Annals of Loch Cé* speak, *sub anno* 1315, of '[n]umerous wonderful diseases throughout all Erinn this year, viz., a destruction of people in great number occurred in it, and famine, and various distempers. Slayings of people, and intolerable, destructive bad weather also happened in it'.[101] A similar entry is found in the *Annals of Connaught*.[102] The negative impact of armed conflict on collective health of starving communities has been well documented in numerous famines, historical and contemporary alike.[103] The examples of the Siege of Leningrad, the Bengali famine of 1942–1943, and the Dutch Hungry Winter of 1944–1945 — let alone the infamously tragic episodes of famine among Holocaust victims — are too well known to be expanded upon. During the 1992 Somali famine, the ongoing civil war had a tremendous impact on famine-related diseases.[104] Most recently, famine mortality has been exacerbated by an armed conflict in Somalia in 2010–2012 and by civic violence in

[100] Hatcher, *Rural Economy and Society*, p. 85.

[101] *The Annals of Loch Cé*, ed. and trans. by Hennessy, p. 579.

[102] *Annála Connacht*, ed. by Freeman, p. 241.

[103] Leaning and Guha-Sapir, 'Natural Disasters, Armed Conflict, and Public Health'.

[104] Green, 'The Course of the Four Horsemen', p. 38.

Haiti following the devastating earthquake and tsunami of January 2010 and Hurricane Matthew of October 2016.[105]

In other words, these regional differences pose numerous challenges. One potential way to take up the slack for the missing data is to tentatively convert the manorial excess death rates into their comital equivalents. The conversion is done on the basis of assigning each county a different relative weight based on the share of arable land in its total cultivable acreage (as estimated by Broadberry and his colleagues) and its population size, with Suffolk serving as a benchmark.[106] It then adjusts the calculations by augmenting the figures for the northern counties that suffered from war and depopulation. The particulars are outlined in Appendix 6. The conversion from manorial to comital rates eventually renders the 'national' estimates of excess death rates of 2.94 and 4.26 for 1315–1316 and 1316–1317 relative to non-famine years, respectively (namely, 88 and 128 per 1000 deaths, respectively). Although these estimates are crude (and perhaps speculative), they are certainly much more realistic than 3.80 and 5.02 (namely, 114 and 151 per 1000 deaths, respectively), when accounting for famine mortality on a national level. We shall return to this point later in this chapter.

Seasonality of Death

Another aspect of famine mortality to be considered is its seasonality. As a rule, famines involve the amplification of a normal year's seasonal distribution.[107] As numerous studies of famines have shown, each famine tends to exhibit peculiar trends in its seasonality patterns, dictated by various factors, primarily biological and environmental. The seasonality would vary from case to case. For instance, in the course of the crises of the late 1590s, mortality reached its peak in winter and early spring of 1596–1597 and 1597–1598, while during the shortages of the early 1620s, the figures peaked between the end of 1622 and late summer 1623, and between autumn 1623 and spring 1624.[108] To a large degree, it was an exaggerated replication of the mortality seasonality of 'normal' years in early modern England, when all age groups tended to experi-

[105] Maxwell and Fitzpatrick, 'The 2011 Somalia Famine'.

[106] Broadberry and others, *British Economic Growth*, pp. 23–24 and 51–52.

[107] Dyson and Ó Gráda, 'Introduction', p. 10.

[108] Healey, 'Famine and the Female Mortality Advantage', pp. 165–67, 182–83.

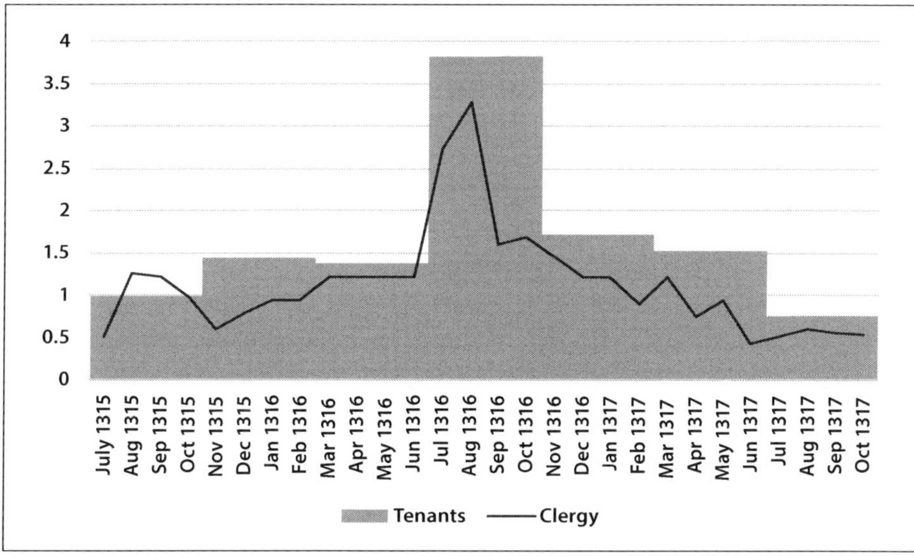

Figure 7.2. Seasonality of famine mortality, July 1315–October 1317 (indexed on July–October 1315). *Source*: Court rolls: as in Table 7.3; episcopal registers: *The Register of John de Drokensford*, ed. by Hobhouse; *Registrum Ricardi de Swinfield Episcopi Herefordensis,*, ed. by Capes; *The Register of Adam de Oreleton*, ed. by Bannister; *The Registers of John de Sandale and Rigaud de Asserio*, ed. by Baigent; *The Register of Thomas de Cobham*, ed. by Pearce; *The Register of the Diocese of Worcester*, ed. by Bund ; *The Register of John de Halton,* vol. II:, ed. by Thompson ; *The Register of Walter Langton*, vol. I:, ed. by Hughes; *The Registers of Roger Martival*, vol. I: *The Register of Presentations and Institutions*, ed. by Edwards; NRO, DN/Reg/1; Lambeth Palace Library, Register Reynolds; WAAS, b. 716–093-B.A.2648/1; and LA, REG/2.

Note: (a) In those instances when episcopal registers supplied no dates of clerical deaths, the latter are assumed to have occurred approximately two months before the appointment of new clerks. (b) Heriots were paid in the next court after the tenants' death, and court rolls did not record the dates of tenants' deaths. Therefore, a monthly reconstruction is impossible here. Instead, since most courts would sit at least once every few months, I have reconstructed a seasonal trend, on the basis of four-month periods. This has been achieved by calculating the total number of heriots paid between month 1 and month 4 (July 1315–October 1315, November 1315–March 1316, etc.). Every four-month period is then indexed on the period July–October 1315. Thus, the total number of tenant heriots recorded between November 1315 and March 1316 was 45 per cent higher than the corresponding number of heriots collected between July and October 1315; hence, the indexed figure for that four-month period is 1.45 (namely, 45 per cent above average).

ence peak mortality in late winter and early spring.[109] The same amplification of normal mortality seasonality is found in other famines. Thus, during the Great Irish Famine of 1845–1852, starving victims would contract a disease during

[109] Wrigley and Schofield, *Population History of England*, p. 293; Wrigley and others, *English Population History*, pp. 326–27.

the summer months (when both grain and potato prices would be at their highest), leading to excessive mortality in late autumn and winter.[110] During the Indian famines of 1896–1897 and 1899–1900, mortality levels (resulting primarily from fevers, dysentery, cholera and malaria) peaked, respectively, around June–August and August–September. This went hand in hand with the appearance of monsoons (producing green vegetables, often of bad quality, consumed by starving communities in their unripe form) and mosquitoes.[111] During the Bengali famine of 1942–1943, the level of excess death was at their highest in December of both years, coinciding with malaria activity around that time.[112]

What about the Great European Famine? Fortunately, the seasonality trends can be gathered from two main sources: heriot records in court rolls and episcopal registers (registering clerical appointments). One would, perhaps, add the *Inquisitions Post Mortem*, recording deaths of royal tenants-in-chief, but given the uniquely high socio-economic status and the obviously small size of the latter group, its mortality patterns can hardly be regarded as representative and should, perhaps, be left out altogether. Before delving into the analysis of these sources, two caveats should be made. First, with the exception of the register of the Bishop of Coventry and Lichfield, recording both the dates of the deaths and new appointments, episcopal registers recorded, in the vast majority of cases, only the dates of new appointments. On the basis of the available data, it is possible to estimate that an average period between the death of a previous clerk and the appointment of a new one is about two months. Therefore, the dates of deaths of appointees' predecessors are assumed to have taken place two months before the appointments. Second, court rolls do not record the dates of tenants' death; instead, heriots were collected in the next available court after tenant deaths. Hence, it is impossible to reconstruct the monthly seasonality of tenants' deaths. Instead, I have attempted to sketch a seasonal trend, on the basis of three four-month periods.

As Figure 7.2 clearly demonstrates, famine mortality peaked between July and October 1316, with August being possibly the most lethal month. This undoubtedly reflects the exhaustion of the available grain resources and the peak of crop prices immediately before the harvest. But does this trend agree with the seasonal distribution of mortality in 'normal' years? A close analysis of court rolls from non-famine years (1310/1–1314/5, 1317/8–1320/1) indi-

[110] Ó Gráda, *Ireland Before and After the Famine*, pp. 107–08.

[111] Dyson, 'Famine in Berar, 1896–97 and 1899–1900, pp. 107–08.

[112] Dyson, 'Demography of South Asian Famines II', pp. 281–82.

cates that the peak in heriot payments was between January and May rather than during the 'hungry gap'. Conversely, it was precisely during the same 'hungry gap' period and early autumn that the levels of heriot payments were at their lowest. Using the same methodology as for the famine years, it is possible to establish that in non-famine years, mortality peaked between late autumn and early spring, mimicking the seasonality of the early modern period. In other words, the trends in mortality seasonality in 1315–1316 did not conform to those of non-famine years. The most likely explanation for this conundrum is the fact that the reality in 1315–1316 was truly unique, as far as the catastrophic shortage of food and the abnormally high prices go. As we have seen in Chapter 4, in no other year, famine or non-famine, did grain prices soar so high, to the point that very few could afford buying food. True, grain prices would always go up towards the harvest time, but their rise would never be so aggressive: between *c.* 1300 and 1349, summer prices in non-famine years were, on average, no more than 20–25 per cent higher than their previous September levels.[113] But even in the years of shortage, the rise was not nearly as aggressive as during the Great Famine. This implies that despite the perpetuated concept of the 'hungry gap', there always was enough available and affordable food during the summer time. Conversely, in non-famine years, the winter–early spring peak in mortality levels is most likely caused by cold weather, long nights, and airborne infections. In colder temperatures, people had to waste more energy to maintain body heat, and this fact would compromise the immune system of weaker individuals and expose them to various pathogens and diseases. This was especially true in the case of infants and elderly people. In the case of infants, this was exacerbated by a seasonality of birth peaking in late winter and early spring, as a result of a seasonality of conception peaking in late spring and early summer, as the evidence from the early modern period shows.[114] The mortality of elderly and frail people can be inferred from the account rolls of Clyst Gabriel Hospital (Devon), a unique source recording the names of local lodgers, consisting of sick, disabled, and old clerks. In the course of 1316–1317, four out of the total fifteen clerks died, with each inmate dying in October, January, February, and April, but none in the summer.[115] The 1316–1317 patterns mirror closely the mortality patterns in non-famine years in the period

[113] See above, Chapter 4.

[114] Wrigley and Schofield, *Population History of England*, pp. 288–89.

[115] Exeter Cathedral Archives (ExCA), 5212.

1313–1348: 39 per cent of all deaths occurred between January and March.[116] If a seasonality of conception and births in the early fourteenth-century was not different from that in the early modern period, and if the mortality patterns of the Clyst Gabriel inmates are reflective of the seasonality patterns of elderly people elsewhere, then it appears that Figure 7.2 based on the seasonality of heriots, is by no means representative of all age groups. It is possible that contrary to non-famine years, when a seasonality of mortality was uniform across different age groups, the first year of the famine witnessed a divergence in seasonality, with adults succumbing in the summer and infants (and possibly elderly) dying in the winter and early spring.

Intriguingly, the second year of the famine did not replicate the seasonality pattern of 1315–1316. Here, the high season of mortality lasted from late autumn into early spring, resembling, to a certain degree, the trend of non-famine years. Although mortality levels were still alarmingly high between autumn 1316 and summer 1317, there was a graduate seasonal decline in mortality levels in the aftermath of the 1316 harvest, despite its poor performance. During the 'hungry gap' of 1317, mortality levels reached their lowest point, despite the prevalence of shortage and high prices. This may indicate that frailest individuals may have been weeded out during the first year and the first half of the second year of the famine, while fitter individuals were dying in smaller proportions over the course of the second year of the famine. But one may also argue that it was the improved weather in the aftermath of the 1316 harvest that kept humans in somewhat healthier state. As we have seen in Chapter 2, the torrential rain had stopped, and the following months seem to have been much milder than the two previous years. With drier and milder weather, hungry humans were not as weakened and susceptible to pathogens and diseases as they were during the previous year. Nevertheless, the first half of 1316–1317 (that is, September 1316–February 1317) undoubtedly witnessed very high mortality levels. As we have seen in Table 7.4, more people seem to have died during the second year of the famine than during the first.

Decline in Fertility Rates

Famine demography cannot be fully appreciated unless death rates are juxtaposed against marriage and fertility rates. Unfortunately, just as with death rates indicated by heriot payments, marriage and fertility rates, revealed by

[116] Orme, 'Mortality in Fourteenth-Century Exeter', p. 201.

merchets (marriage dues payable to the lord), present many methodological and interpretational challenges. Just as with heriots, not every series of court rolls record merchets. Those that do, do not, with rare exceptions, distinguish between first marriages and remarriages. Fortunately, however, those few court rolls that do provide some sort of clue as to whether merchets were paid for the first marriages or remarriages indicate that the vast majority of those fines were concerned with the former rather than with the latter.[117] Also, as some studies have clearly shown, there is a chronic under-recording of the total number of marriages, either because of lack of scrutiny of manorial officials or because of evasion to pay merchets. Indeed, there are enough examples of tenants being fined for marrying without the lord's licence. To make things even worse, however, merchets were far more likely to have been paid by the daughters of more substantial villeins than by their poorer counterparts.[118] Therefore, merchets, just as heriots, have their clear limitations and should be treated with caution. Nevertheless, they may still provide an indirect guideline into general trends of marriage and fertility patterns before, during, and after the famine years.

As Figure 7.3 demonstrates, during the famine years, the level of merchets has declined by approximately 50 per cent (but taking into account regional variations of the crisis, the figures have been readjusted in Appendix 6 to 41 per cent for 1315–1316 and 42 per cent for 1316–1317). If merchets are indeed a firm indication of nuptial trends, then their decline indicates a pronounced impact of the crisis on marriages and, consequently, fertility and birth rates. As the figure shows, although the figures were remarkably depressed in 1315–1316 and 1316–1317, they also were in decline in both 1314–1315 and 1317–1318, namely, a year before and after the famine. The decline in merchet payments during the famine are fairly straightforward to explain. First, poverty and reduced food security negatively affected the ability of potential grooms to create and provide for their brides and future children. Conversely, potential brides would choose to stick to their parents' and siblings' homes rather than entrust their fate to their suitors. Furthermore, some were undoubtedly unable to offer good dowries for their marriages. Second, as some nutritional studies have shown, continuous malnourishment depresses the reproduction functioning of both men and women. It lowers the hormone leptin, thus decreasing

117 The clearest indicate of remarriages is in the 'widow' designation of brides (*X relicta/uxor quondam Y*).

118 Poos, Razi, and Smith, 'The Population History of Medieval English Villages', pp. 316–20; Smith, 'Further Models of Medieval Marriage'.

libido by reducing testosterone levels of men and causing amenorrhea (absence of a menstrual period) among women.[119] There are numerous instances when couples reduced their sexual activity or stopped maintaining sexual relationships altogether during historical famine episodes. The Great Irish Famine, the Soviet famine of 1918–1922, and World War II starvation are just a few of many such examples.[120] Cormac Ó Gráda also cites men's migration in search of food and work as another factor for the reduction in sexual activity within a starving population. Although this may well be the case also during the Great European Famine, its extent may not be that great, owing, on the one hand, to the restrictive mechanisms of manorial society, which imposes migration limitations on unfree tenants, and also peasant conservativism and reluctance to leave their communities and families, on the other. The impact of the Great Famine on migration needs to be quantified and analysed before any definitive conclusions can be reached. Also, it should be borne in mind that medieval marriage was more about property than sex.

A more surprising trend, however, is the decline in merchet levels in 1314–1315 and 1317–1318, namely, before and after the famine. In the former instance, the levels declined from about 1.16 (Michaelmas 1313–Michaelmas 1314) to about 0.91 (Michaelmas 1314–Michaelmas 1315). One possibility is the impact of weather. As some studies have demonstrated, insufficient sunlight tends to affect negatively one's mood, by reducing the brain's serotonin levels and increasing depressive inclinations and other varieties of emotional disorders. In particular, serotonin levels tend to be low during the cold and rainy months.[121] The mechanisms of libido are largely triggered by neurotransmitters, first and foremost dopamine and serotonin. The decline in serotonin levels may have a strong impact on the depression of one's sex drive.[122] Is it possible, then, that the incessant torrential rain and excessively cold winters of 1314–1315 provided exactly the right climatic scenario for depressing libido and, consequently, nuptial levels?

[119] Ó Gráda, *Famine: Short History*, pp. 106–07.

[120] Ó Gráda, *Famine: Short History*, p. 106; Sorokin, *Golod kak Faktor*, pp. 177–96; the English translation is *Hunger as a Factor in Human Affairs*.

[121] For instance, see Praschak-Rieder and others, 'Seasonal Variation in Human Brain Serotonin Transporter Binding'. For a somewhat speculative application of this view on the historical context of seventeenth- and eighteenth-century European urban environment, see Akkerman, 'Winter Cities and Mood Disorder'.

[122] Dragoa and others, 'Acute Low Doses of Melatonin'.

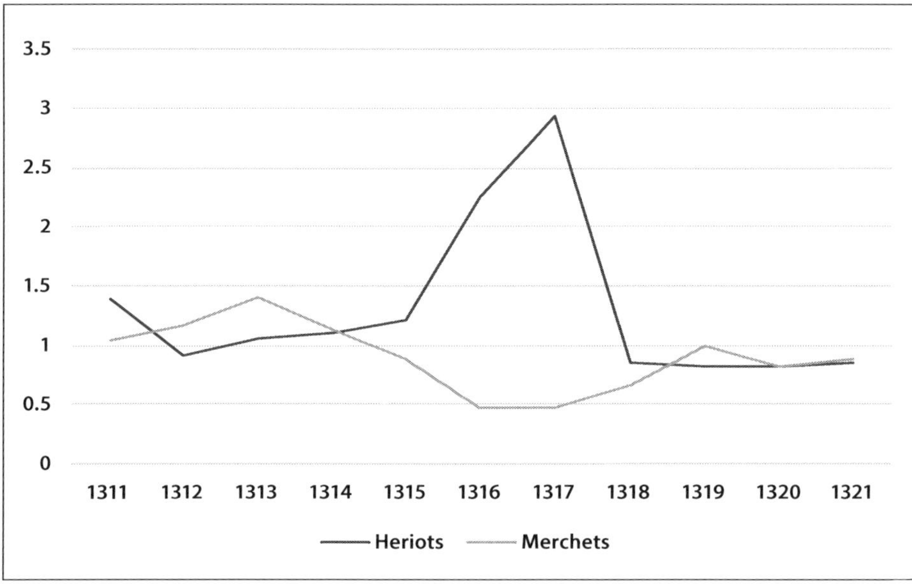

Figure 7.3. Heriots (forty-seven manors) and merchets (nineteen manors), indexed on 1311–1315 and 1318–1321. *Source*: As in Table 7.1; BAHPS, 346233–346239; 350357–350360 (Halesowen, the Worcestershire-Shropshire border); LinRO, KR 42–104 (Alsacoe, Manley and Corringham, all in Lincolnshire); University of Chicago Library, Special Collections, Bacon Rolls 119–20 (Hinderclay, Suffolk); BL, Add Ch 63433 (Rickinghall, Suffolk); Suffolk Record Office (the Lowestoft Branch), HA12/C2/3–7 (South Elmham, Suffolk); Hertfordshire Archives and Local Studies (Henceforth, HertALS), 65498–65518 (Norton, Hertfordshire); HertALS, 40702 (Bramfield, Hertfordshire); HertALS, DE/B/2355/M/111 (Hexton, Hertfordshire); HertALS, DE/B/2067/B/M1 (Ridge, Hertfordshire); HertALS, 40700 (Sandridge); and HantsALS, 11M59/B1/66–76 (Winchester Pipe Rolls).

Note: The respective figures indexed on 1311–1315 and 1318–1321 (1311–1315 and 1318–1321 = 1.0), appearing on the vertical axis, are as follows: (1) heriots: 1.0 = 195, and (2) merchets: 1.0 = 54.

In the course of 1317–1318, the merchet levels still stood below the non-famine figures, amounting to about 0.64. Obviously, one way to explain it is by considering the overall population losses during the famine. If England lost, as it is argued below, about 15–20 per cent of her population during the starvation period, then the corresponding figures for potential grooms and brides may have, with certain variations, declined accordingly. But given the fact that the remarkably low figure of 0.64 does not reflect the expected figure of 0.80–0.85, additional factors should be sought. One such possible factor is a time lag between the end of famine and the relative economic recovery of households. Many famine survivors were undoubtedly devastated around the time of the 1317 harvest, which, again, was anything but abundant (with composite net crop yields standing at about 26 per cent below average).[123] It may have required yet another year to pass for the situation improved considerably, around the harvest of 1318, which was, by all means, bountiful, with composite net crop yields reaching about 19 per cent above average.[124] Another possible factor is impact of malnourishment on human physiology. Those who survived the famine may have done it with great difficulties and not without physiological sacrifices and compromises. Long-term malnourishment leads to atrophy, a physiological disorder, by which organs and tissues of the body waste away, causing disastrous weight loss and deformation. The most ill-famed example is concentration camp survivors during World War II, whose physical and emotional state was studied in the immediate aftermath of the war.[125] After the end of the famine stage of the Leningrad Siege, the city abounded in dystrophic civilians for at least a few months.[126] In such frail states, both men and women were temporarily unfit for sexual reproduction, with reduced libido and the inability to bear children. Although the period of weight recovery via re-nourishment varies from case to case, it requires at least three months to reach pre-famine body mass, as the famous Minnesota Starvation Experiment of 1944–1945 has shown. The Minnesota Starvation Experiment was a clinical study conducted at the University of Minnesota and involved thirty-six male volunteers; it studied the physiological and psychological impact of a long-term dietary deprivation and the process of nutritional recovery process.[127] It is important to bear in mind, however, that the Minnesota experiment was conducted in a scientifically-controlled manner, with more than sufficient quantities of food available for re-nourishment. The poor harvest of 1317 was unlikely to be capable of feeding all in a sufficient manner. Therefore, there is no reason to suppose that the process of re-nourishment and recovery of the famine victims in 1317–1318 was in any way as fast as the Minnesota experiment. Also, a close analysis of court rolls indicates that a certain proportion of post-1317 merchets were paid for remarriages rather than first marriages. In

[123] See above, Chapter 2.

[124] Calculated from Campbell, *Three Centuries of English Crop Yields*.

[125] Eitinger, *Concentration Camp Survivors in Norway and Israel*, esp. pp. 57–68.

[126] Kirschenbaum, *The Legacy of the Siege of Leningrad*, pp. 257–58.

[127] The results of the experiment were published by Keys and others, *The Biology of Human Starvation*. On the process of body mass recovery after under-nourishment, see also, Martins and others, 'Long-Lasting Effects of Undernutrition'.

the aftermath of the famine, we would expect a substantial rise in the number of marriages involving one or two partners remarrying because of the previous death of his/her spouse. For instance, at Langham (Norfolk), William Heymes paid a merchet for his first marriage on the court of 29 October 1314, only to pay another one for remarriage on 11 February 1318.[128] Unlike first marriages, remarriages were not invariably associated with pregnancies. Still, in the context of population decline, remarriages were the most (and in, effect, the only) efficient method of demographic recovery. Remarriages were a commonplace phenomenon after various mortality crises. Examples include the Great Irish Famine and the Ethiopian Famine of 1983–1985.[129]

In contrast with the immediate aftermath of the famine, before and during the famine the vast majority of merchets stood for first marriages. In the context of the pre-Black Death reality, where extra-familial employment opportunities were few, where life expectancy was disturbingly low, and where kinship ties meant everything, the vast majority of first pregnancies must have followed first marriages right away. In this sense, merchets provide an indirect measure of first pregnancies. It is unfortunate that with the exception of 'childwite' and 'leyrwyte' (fines for delivering a child from outside of wedlock or committing adultery and bearing from outside of wedlock), manorial court rolls (or indeed any other contemporary documents) do not record the actual number of births. Moreover, given the disastrous proportions of malnourishment and physical debilitation during the famine, it is plausible that a certain proportion resulted in stillborn births or in newborn babies perishing shortly after birth. Therefore, for the famine years, merchets may represent a good proxy for first marriages and possibly first pregnancies, but not necessarily successful births rates. According to the same logic, the births resulting from first marriages rates may have been proportionally smaller than nuptial rates. To put it differently: if merchet/nuptial levels stood at 0.50 per cent of their non-famine levels, the corresponding figures for birth rates may have been, say, 0.45 (given the inability/unwillingness to bear), or lower still, if we account for miscarriages, stillbirths, or outright neonatal deaths. Moreover, merchets, just as heriots, account for only a small proportion of the total population of England. Thus, it is unclear if they reflect changes in nuptial patterns across other social strata. Is it possible, for instance, that worse-off tenants, such as non-agricultural rustics and town

[128] RHM, Langham Court Rolls 1314–1315 and 1317–1318.

[129] Guinnane, *The Vanishing Irish*, pp. 99–101; Rahmato, *Famine and Survival Strategies*, pp. 193–94.

dwellers, being more vulnerable to the famine than servile and customary tenants, experienced an even higher decline in nuptiality? Would that mean that the actual decline was even greater than 50 per cent? The lack of comparable data for other social echelons does not, alas, permit any educated estimates to be made, even though it is highly probable that these figures were higher than those reflected in recorded merchets.

It would be instructive to compare these figures to other famines and subsistence crises. During the Great Irish Famine, there was a decline in birth rates by about 25 per cent, compared to some ten years before the famine. At the height of the Bangladeshi famine in 1975, birth rates declined by about 38 per cent in relation to the period of 1966–1974.[130] In Nazi-occupied Greece, the figures declined by 45 per cent.[131] None of these statistics are, however, comparable with the figures from besieged Leningrad. Here, birth levels in 1942 declined by some 75 per cent, with about 40 per cent of all birth being premature and a further 25 per cent of babies born stillborn or perishing within a month. As Cormac Ó Gráda has stated, the Leningrad figures were 'unparalleled in history'.[132] As numerous studies have shown, intrauterine malnourishment, closely related to maternal starvation, can have a detrimental effect on pregnancy, miscarriage, and stillbirth rates. During the crisis of the 1590s, stillbirth rates at the parish of St Botolph without Aldgate, London, stood at close to 100 per 1000 live births, compared with the period of 1609–1623, when the figures dropped to 40.[133] In the course of the Great Leap Forward Famine in China (1958–1961) and the Bangladeshi famine of 1974–1975, a proportion of miscarriages and stillbirths went up a great deal.[134] It should be pointed out that in non-famine circumstances, stillbirths do not, as a rule, delay the subsequent pregnancy, in spite of the emotional hardships suffered by mothers and their partners.[135] Sadly, however, as one study of the Bangladeshi

[130] Boyle and Ó Gráda, 'Fertility Trends'.

[131] Hionidou, *Famine and Death in Occupied Greece*, pp. 178–90.

[132] Ó Gráda, *Famine: Short History*, p. 107.

[133] Forbes, *Chronicle from Aldgate*, pp. 63–65. On the topic of stillbirths and miscarriages in early modern England, consult Woods, 'The Measurement of Historical Trends'; Galley, 'The Stillbirth Rate in Early Modern England'.

[134] Cai and Feng, 'Famine, Social Disruption, and Involuntary Fetal Loss'; Hernández-Julián, Mansour, and Peters, 'Effects of Intrauterine Malnutrition'.

[135] Lamb, 'The Impact of Previous Perinatal Loss'; Campbell-Jackson and Horsch, 'The Psychological Impact of Stillbirth'.

famine of 1974–1975 has shown, women who were pregnant during the famine experienced a slightly higher probability of miscarriages and stillbirths in post-famine years.[136]

Decline in Sexuality: Evidence from Fornication Cases

As mentioned, the childwite and leyrwyte fines do not objectively reflect the trends in famine births. Rather, they can be taken to reflect changes in illegitimate births and adulterous relationships. As Richard Smith and Zvi Razi have shown, in their respective studies of two Suffolk manors (Redgrave and Hinderclay) and Halesowen (on the Worcestershire-Shropshire border), the ratio between merchets and childwite/leyrwyte fines varied from place to place in the pre-Black Death period. On the two Suffolk manors, for each merchet there were approximately 0.33 childwite fines within the period 1259–1293, while at Halesowen, the ratio was higher, standing at about 1.0:0.53 between 1270 and 1348.[137] Although the ratio between merchets as indicators of first marriage/first pregnancy and childwite fines as indicators of the rates of adultery and illegitimate births could in theory be a good measure, it is not without drawbacks. First, it does account for a temporal gap between marriages and hypothetical first births. Second, and more importantly, as Richard Smith has himself acknowledged at a later point, this juxtaposition is problematic, since childwite fines were more likely to be paid by poorer villein ladies, while merchets were collected from higher status villein females.[138] In other words, comparing childwites with merchets is a bit like comparing apples and oranges. Nevertheless, and keeping in mind all these caveats, Table 7.5 records the sheer numbers of merchets paid *in a previous year* (and most likely resulting in legitimate births *in a present year*) and those of childwite/leyrwite fines paid *in a present year*, as well as the ratio between the two, on six manors (Horsham St Faith (Norfolk), Rickinghall, Redgrave, Hinderclay, South Elmham (all in Suffolk), and Halesowen (on the Worcestershire-Shropshire border)). This juxtaposition, with all its methodological and conceptual pitfalls, may still be a valid

[136] Hernández-Julián, Mansour, and Peters, 'Effects of Intrauterine Malnutrition', pp. 1789–90. To be sure, the figure of 2 per cent cited by the authors is perhaps too meagre to have any impact on demographic recovery in post-famine communities.

[137] Smith, 'English Peasant Life-Cycles', pp. 454–61; Razi, *Life, Marriage and Death*, pp. 64–71.

[138] Poos and Smith, '"Legal Windows onto Historical Populations?"', p. 150.

Table 7.5. Ratio between a previous year's marriages and a present year's illegitimate births on six manors, 1312–1321.

Year	Marriages (previous year)	Illegitimate births (current year)	Ratio
1311–1312	19	14	0.74
1312–1313	23	11	0.48
1313–1314	26	11	0.42
1314–1315	16	8	0.50
1315–1316	8	3	0.38
1316–1317	7	2	0.29
1317–1318	9	1	0.11
1318–1319	18	5	0.28
1319–1320	19	10	0.53
1320–1321	15	8	0.53
Non-famine years, average	18	8.8	0.48
Famine years	8	1.5	0.20

Source: NRO, NRS 19505 (Horsham St Faith, Norfolk); University of Chicago Library (UCL), Special Collections, Bacon Rolls 119–20 (Hinderclay, Suffolk); UCL, Bacon Roll 11–14 (Redgrave); BL, Add Ch 63433 (Rickinghall, Norfolk); Suffolk Record Office (the Lowestoft Branch), HA12/C2/3–7 (South Elmham, Suffolk); BAHPS, 346233–346239, 350357–350360 (Halesowen, the Worcestershire-Shropshire border).

exercise, as long as we acknowledge that it establishes the ratio between legitimate births by better-off young villein ladies and illegitimate births by villein girls of a lower socio-economic standing rather than the general ratio between legitimate and illegitimate births across all villein strata.

Since manorial court rolls do not record legitimate births, I had to make an educated guess that most fresh brides became pregnant either shortly before or shortly after the payment of merchets and, thus, most first births happened approximately nine–twelve months after marriages. Within our manorial sample, based on six manors for the period 1311–1321, the ratio between the previous year's marriages and the present year's illegitimate births stood at 0.48 for non-famine years, declining to the unprecedented low of 0.20 during the famine years. In particular, in the course of 1317–1318, for each legitimate child, only 0.11 illegitimate children were born. This, again, indicates the weakened libido associated with famine and deprivation, reaching its peak during 1316–1317. But it also reflects the socio-economic depression within local communities. As both Smith and Razi have convincingly shown, most fines for illegitimate bearing were paid by young ladies from lower socio-economic strata, either for material support of their partners or in hopes of marrying their partners.[139] It appears that the unprecedentedly low figures, in both absolute and relative terms, may hint that there were fewer men willing to enter into illicit sexual relationships because of their financial inability to support their partners and their offspring. Moreover, fewer women could expect to convince their partners to marry them after a period of unmarried relationships, for the same reason.

How Many People Did England Lose?

Despite the fact that manorial court rolls and other related documents can shed much light on famine demography during the crisis, they do not provide any indication of how many people actually died during the famine. Both merchets and heriots are very indirect measures of population growth and decline, as they provide very partial and relative, rather than absolute, figures. In that period, there were no parish registers, let alone censuses, and even some best population estimates remain conjectural. In their estimates, scholars often rely upon manorial surveys (rentals, terriers, and extents), which record the names of local tenants. However, there are very few manors for which such surveys survive in a fair number, and, in most cases, there tend to be one or two surviving surveys in a hundred years. In the course of my research, I was unable to identify a single manor for which at least two surveys survive for any two points immediately preceding and following the famine (say, *c.* 1305–1315 and *c.* 1317–1327). This creates a serious methodological challenge, which renders such an exercise virtually impossible. There are, nevertheless, two methods that have some potential to arrive at some very crude estimates of population loss, but these still need to be treated with much caution.

One method is to use the lists of tithing penny payments, recorded in some court rolls. One advantage is that these tithing payment lists include not only tenants, but, in fact, all male adults aged twelve and over, living within a given parish community and held responsible for maintaining law and order there. We get some rough ideas of population change if we look at the cumulative totals on an annual basis. Indeed, such method has been utilized by Jan Titow,

[139] Smith, 'English Peasant Life-Cycles', pp. 458–59; Razi, *Life, Marriage and Death*, pp. 64–71.

Table 7.6. Cumulative totals of adult males paying tithing penny at eleven English manors.

Place and years	**Before famine**	**After famine**	**Ratio between pre- and post-famine population**
Great Waltham, Essex (1306 and 1319)	320	254	0.79
High Easter, Essex (1306 and 1319)	288	248	0.86
Chatham Hall, Essex (1308 and 1318)	63	61	0.97
Margaret Roding, Essex (1316 and 1318)	28	24	0.86
Wymondham, Norfolk (1314 and 1319)	60	55	0.92
Chesham Bois, Buckinghamshire (1315 and 1325)	122	104	0.85
Great Horwood, Buckinghamshire (1316 and 1318)	108	92	0.85
Longbridge Deverill, Wiltshire (1311–1315 and 1316–1320)	89	73	0.82
Monkton Deverill, Wiltshire (1311–1315 and 1316–1320)	29	31	1.07
Taunton, Somerset (1315 and 1319)	1,338	1,202	0.90
Halesowen, Worcestershire/Shropshire (1311–1315 and 1321–1325)	485	412	0.85
Total	**1,107**	**942**	**0.85**

Source: Poos, *A Rural Society after the Black Death*, pp. 106–07 (Great Waltham, High Easter, and Chatham Hall); University College Archives, Oxford, UC ER/CR1/1–4 (Margaret Roding); NRO, NRS 18475 and 18481 (Wymondham); Centre for Buckinghamshire Studies (CBS), D/BASM/18/2–12 and Marshall, *A View of Medieval Chesham*, pp. 17–19 (Chesham Bois); The New College, Oxford, MS 3913 (Great Horwood); Longleat House, MS 9660 (Longbridge Deverill and Monkton Deverill); Titow, 'Some Evidence of the Thirteenth Century Population Increase', p. 224 (Taunton); Razi, *Life, Marriage and Death*, pp. 31 and 40 (Halesowen).

Zvi Razi, Larry Poos, and some other scholars. On the basis of their analysis of tithing payments, these scholars were able to estimate that the population seems to have declined by some 10–15 per cent during the famine.[140] Together with these published estimates, I was able to identify only eleven manors that recorded tithing penny payments in the early fourteenth century (Table 7.6). As the table shows, these lists indicate that the number of adult males within those tithings declined by approximately 15 per cent during the famine.

[140] Poos, *A Rural Society after the Black Death*, pp. 106–07; Titow, 'Some Evidence of the Thirteenth Century Population Increase', p. 224; Razi, *Life, Marriage and Death*, pp. 31 and 40.

Although the tithing lists are by far the single more reliable statistical source on population movements, even they are not without their own limitations. First, these tithing penny lists do not record mortality, but tithing penny payments. Some individuals may not appear on some lists on account of non-payment of tithing penny, while some other individuals may have disappeared from the same lists altogether because they have moved elsewhere. Indeed, as Poos and Smith have convincingly shown, tithing penny lists do not necessarily record all male residents within the same tithing.[141] Second, the same lists capture neither women nor children. Third, these eleven manors are not necessarily indicative of the situation as a whole. There is nothing to suggest that the situation on Essex or Somerset manors reflected that in war-torn and impoverished Cumberland or Durham, or that in Cornwall, where the effects of famine seem to have been much less devastating than elsewhere. Finally, there is nothing to suggest that we can extrapolate these figures on urban settlements. Taken together, these facts point out that we should take these lists with a grain of salt. All we can say that they *do* reflect excess mortality, but they *do not* quantify it. The figure of 15 per cent relates to the decline in the number of tithing penny payers, not the proportion of dead individuals. It is possible that the former closely reflects the latter, but even if we are to accept this possibility, we still need to acknowledge that the population loss in the area of 15 per cent relates only to these eleven manors.

This leads us to the second possible method of estimating population losses. This method conjoins the levels of excess mortality and reduced fertility, estimated earlier in this chapter, with estimated levels of annual population growth on the eve of the famine (for detailed calculations, see Appendix 6). As we have seen, I have estimated that the 'national' mortality levels stood, respectively, in 1315–1316 and 1316–1317, at about 2.94 and 4.26 their non-famine levels, while the fertility levels may have stood at about 0.59 and 0.58 their non-famine levels in the same years (Appendix 6, Table 7.2, and Figure 7.3). I have assumed the annual population growth rates *c.* 1311–1315 to have been in the area of 0.5 per cent.[142] For the purpose of the present exercise, I have assumed death rates *c.* 1311–1315 to have been about 30/1000, while birth rates to have stood at about 35/1000. To estimate the respective death and birth rates during

[141] Poos and Smith, 'Shades Still on the Window', p. 410.

[142] This estimate differs from Broadberry and others, *British Economic Growth,* p. 31, where it has been assumed that the annual population growth rates between 1270/9 and 1300/9 was 0.23 per cent because of the subsistence crisis of 1290. However, this figure is inapplicable for the period 1311–1315, characterized by a series of bountiful harvests. Elsewhere (p. 20), how-

the famine, we need to multiply the normal years' death rates by the respective factors of 2.94 and 4.26 (rendering 88/1000 and 128/1000) and the normal years' birth rates, in both instances, by the factors of 0.59 and 0.58 (rendering 21/1000 and 20/1000). If the total population of England around the harvest of 1315 was in the area of 5.25 million, then we may estimate that during the first year of the famine (1315–1316), England lost about 462,000 people via deaths and gained about 110,250 people via births, while the total population declined to about 4,898,250 people. During the second year of the famine (1316–1317), there may have been about further 626,976 deaths and 97,965 births; by the end of the famine (around the harvest of 1317), the population would have stood at about 4,369,239 people. This reckoning suggests that between the harvests of 1315 and 1317, England may have lost as much as 17 per cent of her population. Obviously, this is a very crude estimate, which can easily be criticized as an exaggeration, but this may be the best (and probably the only) educated guess.

While this figure is indeed remarkable, it not an impossible one, judging by comparable statistics from other great famines. For instance, between 1845 and 1851, the population of Ireland declined from about 8.5 to 6.5 million people (about 24 per cent), with about one million accounting for deaths and missing births and a further million accounting for emigration.[143] Between 1931 and 1934, the population of Soviet Ukraine declined by about 17 per cent, from about 23.3 to about 19.4 million people (with about 2.4 million perishing in the *Holodomor* and the rest being lost to missing births and emigration).[144] During the Siege of Leningrad, about 750,000 out of about 3 million people (that is 25 per cent) died of hunger, but a large proportion of evacuees and reduced fertility rates meant that the population decline was much greater than 25 per cent: when the blockade was lifted, there were only about 546,000 inhabitants in the city.[145] If we to stick to the notion that the Great European Famine was the worst subsistence crisis of the Middle Ages, and one of the very worst famines in the history of Europe, then the figure of 17 per cent, even though surprising,

ever, Broadberry and others estimate that between 1279 and 1290, the total population rose from 4.43 to 4.75 million, rendering the annual population growth rates of 0.46 per cent, a figure much closer to my estimate of the annual growth rates between 1311 and 1315.

[143] Boyle and Ó Gráda, 'Fertility Trends', 543–62.

[144] Snyder, *Bloodlands*, pp. 42–46.

[145] Figures derive from Yeliseyeva and Gribova, eds, *Sankt-Peterburg* (in Russian); Barber and Dzeniskevich, eds, *Life and Death in Besieged Leningrad*, pp. 1 and 11.

is by no means an impossible one. It appears particularly credible if we turn back to the tithing payment lists, recording the decline of about 15 per cent in sheer numbers of adult males paying tithing pennies. These lists did not include children below the age of twelve, who were more prone to famine deaths than those above that age and who accounted for at least 25 per cent of the total population (Table 7.2).

Chapter 8

Fighting for Their Entitlements: Coping Strategies

Famine and Coping Strategies

In his *Poverty and Famines*, Amartya Sen listed four types of 'legal entitlements' to food: (1) trade-based entitlement; (2) production-based entitlement; (3) labour-based entitlement; and (4) inheritance/transfer-based entitlement.[1] In the circumstances of famine, the second type of entitlement becomes increasingly rare, as it becomes reserved for the highest social echelons, possessing sufficient landed resources to ensure direct access to food, without the interference of a third party. In the case of the vast majority of rustics, their parcels of land were too small to provide them with sufficient calories. The prevalence and relevance of the third type of exchange (labour-based entitlement) depends on the extent of the labour market and employment rates within a given economy. While manorial *famuli* could consider their hired labour as a blessing and salvation, they constituted a very tiny and, indeed, privileged group within the society, made up largely of family-employed/unemployed elements. Therefore, this type of entitlement could hardly work in early fourteenth-century England and other parts of the British Isles. The fourth type of entitlement (inheritance/transfer-based entitlement) could function in a scenario of the death of one's parent and one's entry onto his/her parent's parcel of land. All the same, in many instances, harvest failures and institutional food extraction would also imply that the same parcel of land could not ensure sufficient entitlement to food. 'Transfer-based entitlement' manifested itself in charitable activities asso-

[1] Sen, *Poverty and Famines*, pp. 1–8.

ciated primarily with religious institutions, whereby food was being transferred in a voluntary manner. However, with no single and centralized regulation of poor relief, the extent and success of these charitable works were far from being decisive, as we shall see below. This leaves us with the first type of entitlement, namely trade-based entitlement. Given the widespread poverty and abnormally high grain prices, few commoners would have sufficient cash to acquire sufficient food. In the context of the short-term shock of money depreciation in relation to calories, commoners had no alternative but to resort to another type of exchangeable capital: land. The remarkable spike in land transactions during the famine is a clear testimony to the importance of trade-based entitlements.

Apart from the four legal entitlements, Sen also touched upon 'illegal extra-entitlement transfers' of food (for instance, food theft and looting).[2] However, he did not dwell sufficiently on these channels, and this conceptual flaw has long been pointed out and criticized by several famine scholars, including Stephen Devereux and Olivier Rubin.[3] Famine is not only a bioecological, but also a socio-economic disaster; and hence, no discussion of famine can ignore the socio-economic impact of subsistence crises on the shift from legal to illegal entitlements. Under a considerable stress of hunger, starving humans are pushed back to their most basic survival instincts to secure their most basic need: food. These instincts force them to trespass and ignore the previously accepted boundaries of social and cultural behaviour and act in a self-centred and often unacceptable and potentially harmful manner. The single most apparent manifestation of anti-social behaviour, in conjunction with dearth and starvation, is a shift from legal entitlements to food to illegal extra-entitlement transfers of food. This implies that famine-hit societies will inevitably experience a drastic rise in crime rates, in particular those related to securing food or funds to buy food.

Illegal Entitlements: Crime Rates

Wherever legal channels of food security did not work, the starving population had to resort to illegal entitlements. For England, our information regarding the trends in crime rates on a comital level, rather than the manorial one, comes from gaol delivery rolls. These documents recorded, in a systematic and often detailed manner, the names and crimes of felons delivered to a local prison by

[2] Sen, *Poverty and Famines*, p. 45.

[3] Rubin, 'The Entitlement Approach'; Devereux, 'Sen's Entitlement Approach', pp. 66–90.

circuit justices. Prison deliveries occurred several times a year. A statute of 1299 established the practice of holding gaol delivery at least once a year, but no more than three times.[4] In 1330, another statute was issued commanding at least three deliveries a year.[5] In other words, circuit justices travelled through different counties several times a year, visiting prisons and trying felons incarcerated there. We are fortunate to have complementary information on the situation in Ireland, deriving from Justiciary Rolls, surviving in a calendared format, for 1314–1318.[6] The only glimpse into the situation in Wales is through the analysis of manorial court rolls from Dyffryn Clwyd (Denbighshire, north-east Wales).[7]

As Figure 8.1 demonstrates, crime accusation rates went up in England, the Lordship Ireland, and on the manor of Dyffryn Clwyd. In England in 1315–1316, crime rates were three times higher than the 1313–1314 level, while in the following year the figures were more than 400 per cent above the level of that year. The figures were still twice the average in 1317–1318, perhaps reflecting the time lag between the crimes during the later phase of the famine and the actual capture of criminals. Similarly, in Ireland, the 1317 figures were about 237 per cent their 1314–1315 level, while the 1317–1318 ones rose to about 330 per cent their 1314–1315 level. Here, the high levels of crime in 1317–1318 is explained by the fact that food crisis was not over before the end of the Bruce War in October of that year. At Dyffryn Clwyd, the rise was not as pronounced, with the 1315–1316 and 1316–1317 figures standing at about 63 and 130 per cent above the 1313–1314 ones. Nevertheless, this was a non-negligible rise, which should by no means be ignored. In all three regions, the relative share of food-related crimes went up during the famine. In England and the Lordship of Ireland, the figures rose from about 30 per cent in the pre-famine years to about 50 per cent during the famine, while at Dyffryn Clwyd, the respective figures rose from about 36 to 70 per cent.

These patterns resemble crime statistics from other historical famines. For instance, during the Great Irish Famine of 1845–1852, the figures (indexed on the year 1844) rose to about 2 in 1846, reaching 3.3 in 1847, remaining at about 2.3 in 1848 and 1849, and finally declining to 1.7 and 1.5 in 1850 and

[4] *The Statutes of the Realm*, ed. by Raithby, i, 129–30.

[5] *The Statutes of the Realm*, ed. by Raithby, i, 261.

[6] The National Archives of Ireland (NAI), KB 2/1–12.

[7] TNA, SC 2/215/71–76, 216/3.

1851, respectively.[8] Similarly, crime figures doubled, tripled, and quadrupled in the Madras and Bombay Presidencies (and indeed other parts of British India) during the Great Indian Famine of 1876–1878.[9] Perhaps the most extreme example would be the Soviet famine of 1918–1922, when, according to the official statistics, the levels of crimes in Moscow rose from the indexed figure of 100 in 1914 to about 4688 in 1918.[10] One major difference between the early fourteenth-century famine and some other famines with documented crime is the share of homicidal cases. Despite the apparent rise in the sheer number of persecuted crimes in general and violent nature of robberies in particular, the overall proportion of homicide charges during the Great Famine in England was relatively small: about 8.6 per cent of all crimes, a major decline from about 18 per cent for 1311–1315 and 1318–1321 (which closely agrees with 19 per cent for eight counties for the entire period of 1300–1348, studied by Barbara Hanawalt).[11] This is in a sharp contrast with the Great Irish Famine, when the total number of homicide instances in Ireland went up from about 110 a year in 1840–1843 to about 180 a year in 1846–1850.[12]

Before delving into the analysis of volume and trends in food-related crimes during the famine years, it is essential to clarify that although crime rates go up invariably during episodes of starvation, the relationship between hunger and crime is by no means simple and straightforward, as suggested by Barbara Hanawalt and Bruce Campbell.[13] It should be borne in mind that court rolls and gaol delivery report accusations rather than crimes, and it was at the hands of the local juries to decide whether the accusations were substantiated or not. Thus, some accusations of 'theft' could, in reality, be disguised instances of failure to repay debts by borrowers or distraints taken by creditors to force debtors to pay their debts. Or, alternatively, these 'thefts' could disguise a situation in which some poor relatives or members of clientage of better-off individuals, receiving handouts or payments of grain in 'normal' years, had to help themselves, in the face of the reluctance of their relatives or patrons to deliver these

[8] Ó Gráda, *Ireland: A New Economic History*, p. 203.

[9] Digby, *The Famine Campaign in Southern India*, I, 183 and 251–55.

[10] Sorokin, *Golod kak Faktor*, pp. 336–39.

[11] Hanawalt, *Crime and Conflict in English Communities*, p. 67.

[12] McMahon, '"A Violent Society"?', and *Homicide in Pre-Famine and Famine Ireland*, pp. 12–31.

[13] Hanawalt, *Crime and Conflict in English Communities*, pp. 251–52; Campbell, 'Nature as Historical Protagonist', pp. 291–92.

handouts during the famine. Also, the remarkable speed of the grain price increase would, as we shall see later in this chapter, prompt some creditors to demand the payment back and, in the case of non-payment, to accuse their borrowers of 'theft'. Finally, just as with the *Assize of Bread and Ale*, studied in Chapter 4, authorities may have been more anxious to impose a stricter control of law on local communities, in hopes of preventing a total break-down of social order. All these factors are very likely to have contributed to the increased numbers of accusations of crime, and food-related felonies in particular. With all these caveats, there is still no doubt that the spike in *crime accusation* during the famine years does reflect *the actual rise in crime levels* during those years. Disentangling 'real crimes' from 'accusations' can be extremely tricky and challenging, and such exercise cannot be undertaken within the framework of the present study. It may be sufficient to note that during the famine years, about 28 per cent of all accusations resulted in convictions and executions, and these figures were comparable with non-famine years, when the corresponding proportion was about 25 per cent.[14] In other words, a non-negligible proportion of suspected individuals may have been found not guilty because they were innocent. Therefore, to play on the safe side, I will use the term 'accused of x and y' rather than 'committed x and y' for the rest of the section.

Although all food crimes were related to the desperate attempts of accused individuals to secure their calorific intake in conjunction with omnipresent scarcity, the scale of crimes varied from case to case. It is essential to draw a distinction between small-scale thefts of crops for subsistence and large thefts of crops and livestock for monetary gain, or perhaps as distraint used to secure the payment of loan. The vast majority of cases involved small-scale thefts. Thus, on 26 June 1316, Isabel Ramsey was charged with breaking into the house of Agatha Cheteleman in Riseley (Bedfordshire) and stealing 1 peck (quarter of a bushel) of dredge.[15] On 22 Sep 1316, John Byng was accused of taking away a goose worth 6d from Sir Roger le Sauvage in Warden (Kent).[16] On 28 April 1316, Dauit ap Adda ap Gwillym of Dyffryn Clwyd (Denbighshire) was accused of stealing 1 bushel (*hopa*) of oats from the barn of Gronw ap Gwillym, of the same.[17] Overall, the average corn theft amounted to 0.9 quarters of crops, while the average livestock robbery involved driving away 1.24 livestock

[14] Hanawalt, *Crime and Conflict in English Communities*, p. 59.

[15] TNA, JUST 3/1/1, membr. 15^{r}.

[16] TNA, JUST 3/27/3, membr. 5^{r}.

[17] TNA, SC 2/215/75, membr. 17.

units.[18] This reflects that the majority of food-related thefts were survival felonies rather than robberies for profit. This, however, does not mean that large-scale thefts were unheard of. In some instances, felons were indeed daring. Thus, Adam le Chapman de Hawenes was accused of driving away two oxen and fifty-two sheep from several locations in Bedfordshire, for which he was delivered to Bedford Prison on 26 June 1316.[19] On 23 February 1318, Adam le Feure and his unknown companions were accused of stealing 10 crannocks (quarters) of wheat and maslin (together, worth 40s) stored in Kilmallock Priory (Limerick).[20] Obviously, grains and livestock were the most obvious target of theft or robbery during the famine because of their ability to be readily converted into food, without the need to be converted into cash first. Salt was another sought-after commodity. Thus on 4 August 1317, Philip Camayl and his accomplices were delivered before the justiciar at Waterford and accused of stealing, *inter alia*, 1 crannock of salt (worth 10s) from the cellar of Andrew Gerard.[21] On the same day, another felon was tried at Wateford for robbing 9.5 crannocks of salt (at 10s per crannock) and carting it to Carrick-on-Suir.[22] In many instances, however, securing comestibles through illegal appropriation could be challenging, on account of the omnipresent dearth of food. In those cases, felons had to resort to stealing non-comestible objects, which could be converted into cash and then into calories. These objects included primarily wool, textiles, and various household articles, including kitchenware, tapestry, and agricultural utensils. It should be borne in mind that converting these articles into cash could be difficult for two reasons. First, few people would have any spare cash to be invested in anything but food because of the abysmal dearth of grain. Second, to avoid detection, thieves would have to market stolen goods outside the community in which the thefts occurred. This would imply additional travel and search costs. Therefore, appropriating cash appeared to be a much less attractive prospect for felons.

[18] 'Livestock units' are calculated following the methodology proposed by Campbell, on the basis on weighting each animal group according to its relative financial value (horses = 1.0; oxen and mature cattle = 1.2; immature cattle = 0.8; sheep and swine = 0.1). See, Campbell, *Seigniorial Agriculture*, pp. 104–06. Poultry units are calculated, following my methodology outlined in Slavin, 'Late-Medieval Goose Farming', p. 6.

[19] TNA, JUST 3/1/1, membr. 13r.

[20] NAI, KB 2/10, p. 20.

[21] NAI, KB 2/8, p. 89.

[22] NAI, KB 2/8, pp. 89–90.

Cash felonies could be associated with either burglaries or violent robberies, but, because of ubiquitous poverty, felons often targeted people of means, chiefly merchants. For instance, on 27 October 1316, Robert le Ropere and his accomplice Richard de Lomherst were delivered to Canterbury gaol for robbing 10s from William Durant *mercator*.[23] In some cases, the assaults on merchants resulted in homicide. On 26 June 1316, Henry le Ros and Robert le Ran, both of Chellington (Bedfordshire), were delivered to Bedford for murdering an unknown merchant in the fields of Pakenham (in the same county) and seizing his cash, worth 5 marks.[24] These robberies of merchants are to be seen in a wider context of the formation of crime gangs, in the famine context. These bands operated primarily on highways, in open fields, and in woods, looking for and attacking passers-by. As far as the available evidence can tell, these gangs were small in size, consisting usually of a number of people, in many instances, family members. For instance, during the famine years, Geoffrey le Fishere and William his son, both of Sharnbrook (Bedfordshire), controlled a highway between Sharnbrook and Bletsoe (both villages located in a proximity to Bedford).[25] On 20 July 1317, Luke de Mallyng and his two sons were accused robbing and murdering a certain stranger merchant (*quendam mercatorem extraneum*) near Farningham (Kent).[26]

A similar situation prevailed in other parts of the British Isles. In the Lordship of Ireland, Thomas son of John de Rape, David son of William Brettenagh, and Philip Candelen were accused of robbing numerous merchants and better-off men passing between New Ross (Wexford) and Thomastown (Kilkenny).[27] It appears that Irish highways became a highly unsafe space for travel in the context of the ongoing war: in February 1316, people could not come to the King's Bench in Dublin 'because of the dangers of the ways and the ambushes of the enemies'.[28] In south Wales, during Llywelyn Bren's revolt (28 January–18 March 1316), local gangs were operating, too, although here available evidence is scarce. One entry indicates that a certain Llywellyn ap Ieuan and Ieuan ap Gwillym, robbers, were delivered to Cardiff gaol, under the super-

[23] TNA, JUST 3/27/3, membr. 8r.

[24] TNA, JUST 3/1/1, membr. 13r.

[25] TNA, JUST 3/1/2, membr. 5r.

[26] TNA, JUST 3/27/3, membr. 19r.

[27] NAI, KB 2/12, pp. 76–79.

[28] NLI, MS Harris 1, fol. 427 and Trinity College Dublin Library (TCDL), MS 579/1, fol. 191v (p. 398). The entry is also accessible, in English translation, on the *CIRCLE* website.

vision of no less than six guards.[29] It is tempting to interpret this gang violence not only as a surviving strategy of robbers but also as their resentment towards and scapegoating of merchants. There is no doubt that grain merchants, as we have seen, were perceived as profiteers, whose quasi-usurious activities contributed to the misery of starving communities.[30] Indeed, such anti-mercantile sentiment has been a commonplace social manifestation of many historical famines. For instance, during the Great Russian Famine of 1601–1603, runaway serfs and Cossacks formed highway bands and terrorized passing merchants, both Russian and foreign (especially in western Muscovy).[31] In the course of the French Famine of 1693–1694, Parisian citizens attacked local grain merchants, accusing them of hoarding food supplies.[32] During the Wollo Famine (1972–1974) and the Great Ethiopian Famine of 1983–1985, grain merchants were subject to scapegoating and social hatred, leading to numerous attacks and deaths. It has been estimated that the total number of grain merchants declined from about 25,000 to less than 5000 over the course of these ten years.[33] At the same time, however, the connection between famine and anti-merchant violence could be indirect: attacks on merchants were not unique to bad years, as gaol delivery rolls indicate. During the famine years, highway robbers were doing essentially what they would normally do in non-famine years. The only difference was in frequency and volume of attacks, with the famine years clearly standing out as exceptionally violent years.

Merchants were not the only target of hungry felons. There are numerous instances when lords and their properties were under attack. For instance, on 25 January 1317, William Frere de Staundon was delivered to Hertford gaol for the theft of 6 bushels of crops from a granary of the Priory of Holy Trinity, Aldgate, in Braughing (Hertfordshire).[34] A certain John Grunyng was accused of burglarizing a granary of Gilbert de Clare, Earl of Gloucester in East Peckham (Kent), and stealing 1 quarter of oats, sometime before his delivery to Canterbury gaol on 27 October 1315.[35] In the Lordship of Ireland, a number of granaries of local churches and monasteries were robbed, time and again,

[29] TNA, SC 6/1202/8.

[30] See above, Chapter 4.

[31] Dunning, *Russia's First Civil War*, pp. 105, 195–96, 222, 280, and 395.

[32] Ó Gráda, *Black '47 and Beyond*, p. 134.

[33] De Waal, *Famine Crimes*, p. 111.

[34] TNA, JUST 3/22/1, membr. 2ʳ.

[35] TNA, JUST 3/27/3, membr. 9ʳ.

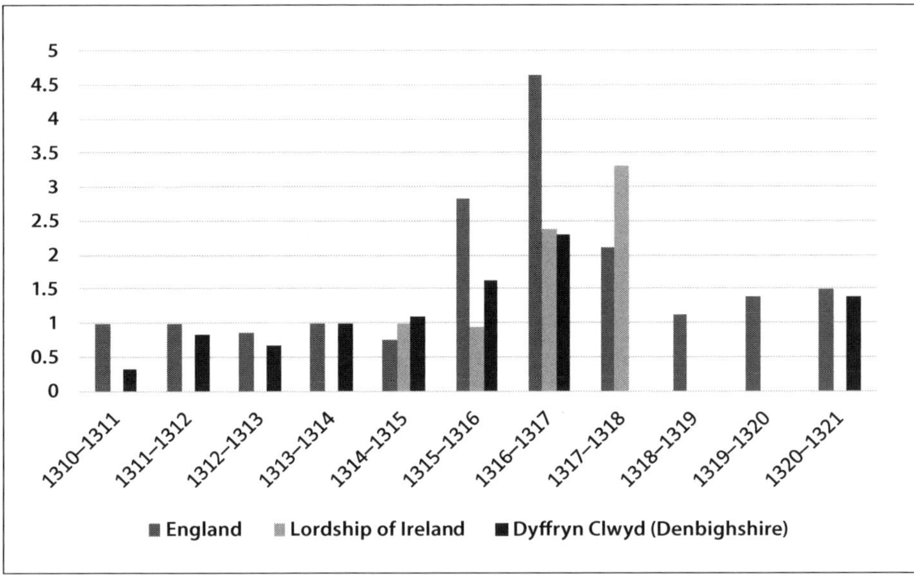

Figure 8.1. Crime Charges in England (fourteen counties), the Lordship of Ireland (nine counties), and Dyffryn Clwyd (Denbighshire, Wales), 1311–1321 (Indexed on 1313–1314 for England and Dyffryn Clwyd, and 1314–1315 for Ireland). *Source*: (a) England: TNA, JUST 3/1/1–3 (Bedfordshire); JUST 3/5/2–3 (Buckinghamshire); JUST 3/19/1–4 (Essex); JUST 3/222 (Herefordshire); JUST 3/23/2 and JUST 3/22/2 (Hertfordshire); JUST 3/26/2 and JUST 3/27/2–3 (Kent); JUST 3/40/2, JUST 3/41/1, and JUST 3/42/1 (Middlesex, including London); JUST 3/48 and JUST 3/49/1 (Norfolk); JUST 3/51/2, JUST 3/51/1, and JUST 3/51/3 (Northamptonshire); JUST 3/63/4 (Suffolk); JUST 3/112 (Surrey and Sussex); JUST 3/75 and JUST 3/76 (Yorkshire); and JUST 3/110 (various counties); (b) Ireland: NAI, KB 2/1–12; and (c) Dyffryn Clwyd: TNA, SC 2/215/71–76, 216/3.

Note: The fourteen English counties are Bedfordshire, Buckinghamshire, Essex, Herefordshire, Hertfordshire, Kent, Middlesex (including London), Norfolk, Northamptonshire, Somerset, Suffolk, Surrey, Sussex, and Yorkshire; the eight Irish counties are Cork, Dublin, Kildare, Kilkenny, Limerick, Louth, Meath, Tipperary, and Waterford.

by both local felons and ravaging marauders. Thus, the granaries of St Mary's Abbey, St Mary de Hogges Nunnery, and Abbey of St Thomas the Martyr (all in Dublin) were repeatedly plundered, leading to their gradual impoverishment by 1318.[36] In some instances, anti-seignorial felonies were initiated and committed by manorial officials. Thus, Thomas Kyng, the reeve of the manor of Newington (Oxfordshire), belonging to Canterbury Cathedral Priory, was

[36] NAI, KB 2/7, pp. 47–48; KB 2/8, pp. 7, and 23.

accused of misappropriation of one cart of grain, containing 4 bushels of grain, with the help of a local swineherd; the grain was later recovered and the reeve was fined at 4s on 27 November 1316.[37] Although there is no direct indication that Kyng was acting out of resentment against his lord, his corrupt administration would certainly be considered as the act of contempt towards the lord (*in contemptu domini*) — a common formula found in court rolls.

Although it may appear that starving commoners were united in their resentment against grain merchants and lords, they were far from being united as communities standing up against the challenges of the famine. On the contrary, it appears that the ties of community, friendship, and even kinship were breaking loose in the context of rising crime rates, with felons stealing from their own neighbours and, at times, relatives. Thus, on 14 November 1317, Thomas Wagger of an unspecified location in Cork, was accused of breaking into his neighbours' barns and stealing their grain.[38] On 2 April 1316, Thomas Pycot of Stoke Hammond (Buckinghamshire) was accused of breaking into the house of his relative William Pycot and taking away tunics and two silver brooches.[39] Perhaps the most unusual example comes from Birton (West Yorkshire), where on 21 April 1316 John, son of Adam of Bray, was accused of stealing 1 bushel of oats from John, his father.[40]

Although the vast majority of crimes (about 87 per cent) were committed by men, it is important to consider the place of women in the patterns of crime. About 37 per cent of all crimes committed by women consisted of larceny, primarily food thefts. This is somewhat higher, but still agrees with the figure of 32 per cent for eight counties for the entire period of 1300–1348, studied by Barbara Hanawalt. Within other sectors, too, female felonies seem to follow the general patterns of the pre-Black Death period. Thus, the proportion of homicide charges pressed against women stood at about 12 per cent during the famine, similar to 14 per cent calculated for the entire period of 1300–1348. Likewise, robbery charges accounted for about 7 per cent of all felonies committed by women, an insignificantly different figure from 6 per cent in 1300–1348. It is within the burglary sector that we see the main difference. During the famine, burglary charges represented only 15 per cent of all crimes committed by women, as opposed to about 29 per cent of all crimes committed

[37] CCA, U15/26/59.

[38] NAI, KB 2/12, p. 55.

[39] TNA, JUST 3/5/2, membr. 7r.

[40] *Court Rolls of the Manor of Wakefield*, ed. by Lister, IV, 95.

by women in 1300–1348.[41] The remaining 29 per cent consisted of primarily non-violent crimes, including harbouring criminals, receiving and concealing stolen goods, and assisting other felons. In some instances, women acted in teams, often made up of female relatives. Thus, on 18 March 1317, Godeleva ate Knokk de Goutherst and her daughter, both of Goudhurst (Kent), were found guilty of stealing wool and mutton from Roger Taketon, their neighbour.[42] On 7 December 1317, Alicia Britoun and her sister Joan were accused of breaking into houses and taking crops at Harlington (Bedfordshire).[43] The fact that most felonies committed by women, regardless of types of crime, were connected to food may have something to do with their domestic role of preparing food for their households — a daily task shared by women across virtually all socio-economic strata.[44]

The last two examples reflect the fact that, in some cases, crime carried a pronouncedly familial character, with a number of close relatives engaging in collaborative felonies. For instance, on 25 January 1317, Ralph Godlok, his wife, and daughter were charged with coin falsification at Ware (Hertfordshire).[45] On 14 November 1317, John Bay, Johana his wife, and Felicia their daughter were accused of stealing half a crannock of oats from Philip le Bret and five beehives from two tenants at an unspecified location in Cork.[46] On 2 November 1316, Richard son of Roger le Messager the chaplain, his four brothers, and five additional men were accused of a series of burglaries at Little Gravenhurst (Bedfordshire).[47] Roger le Messager was by no means the only clerk accused of felony: for instance, on 26 June 1316, Richard Godenok, clerk of Sandy (Bedfordshire), was charged with breaking into the granary of Robert de Hendelowe, of the same parish, and taking 1 bushel of dredge.[48] John de Chadeleshunte, chaplain, was caught and delivered to Newgate prison in London on 17 March 1316 for committing robbery at Kensworth (Hertfordshire).[49]

[41] Hanawalt, *Crime and Conflict in English Communities*, p. 118.

[42] TNA, JUST 3/27/3, membr. 15ʳ.

[43] TNA, JUST 3/1/3, member. 1ʳ.

[44] Bennett, *Women in the Medieval English Countryside*, p. 186; Bardsley, *Women's Roles in the Middle Ages*, p. 61.

[45] TNA, JUST 3/22/1, membr. 3ʳ.

[46] NAI, KB 2/12, p. 54.

[47] TNA, JUST 3/1/2, membr. 9ʳ.

[48] TNA, JUST 3/1/1, membr. 14ʳ.

[49] TNA, JUST 3/40/2, membr. 35ʳ.

A month later, Thomas Folger de Freton, clerk from Kent, was caught in Farringdon (London) with three stolen liturgical books, possibly to be sold or exchanged for food.[50] Although it may appear somewhat surprising that the terrible starvation of the famine years also prompted some clerks to secure foodstuffs through illegal channels, the clerical misbehaviour was by no means unique to the famine years. As numerous entries from gaol delivery rolls indicate, clerical crime was not unheard of both before and after the famine.

Food Riots?

In his *Popular Protest in Late Medieval English Towns*, Sam Cohn Jr. spoke about a 'wave of insurrection' in English towns between 1312 and 1318. In particular, he noted the Bristol uprising of 1313–1316, as well as more minor risings at Winchester, Canterbury, King's Lynn (Norfolk), Ravenser Odd (Yorkshire), and Nottingham. Cohn was correct in stressing that these revolts had very little or nothing to do with the Great Famine, because they were initiated and led by the urban elite against royal impositions rather than by urban paupers against the dearth and high prices.[51] William Jordan, in his monograph on the Great Famine, while explaining the lack of food riots during the famine, stated that '[b]y and large, riots were an urban, not a rural, form of protest'.[52] Buchanan Sharp, however, conceded that there may be *some* evidence for 'disorder of food...in the first two decades of the fourteenth century', states that 'the evidence, if it exists, has not yet been discovered'.[53] For Sharp, the earliest example of what he considers to be a food riot are urban movements that shook Bristol, King's Lynn, Thetford, Boston, and some unidentified ports in Kent in 1347.[54] Indeed, no chronicle mentioned any mass movement expressing their frustration about the omnipresent dearth and hunger during the Great Famine years. But does it mean that there were no such movements? In order to deal with this question, it is essential to define 'food riots' first. According to E. P. Thompson's dictum, food riots were rooted at the heart of the idea of 'moral economy', which was a popular and communal consensus that 'fair prices' were more beneficial than 'free-market prices'. Food riots were peaceful

50 TNA, JUST 3/40/2, membr. 37r.

51 Cohn, *Popular Protest*, pp. 144–58.

52 Jordan, *The Great Famine*, p. 113.

53 Sharp, *Famine and Scarcity*, p. 57.

54 Sharp, *Famine and Scarcity*, pp. 56–57.

rather than violent actions, based on this principle, and their main aim was to secure and regulate the same 'fair prices'.[55] Thompson's own work was dealing with a specific chronological, geographic, and cultural context: eighteenth-century England, which although saw a number of food shortages, did not suffer any substantial subsistence crisis (unlike Ireland). Food riots are by no means uniform phenomena, and their attributes vary a great deal from setting to setting. Certainly, the Great Famine of the early fourteenth century was radically different from the eighteenth-century shortages and, thus, Thompson's view of food riots in particular and moral economy in general may not necessarily hold true there.

Crudely oversimplifying and generalizing, I propose, for the purpose of the following discussion, to define food riots as collective actions consisting of masses protesting against food shortage, caused by the failure of local governments, and aiming to ensure public access to food supply, either through verbal or physical means (such as the use of power to gain access to food stocks, or/and attacks of officials in charge of local food policies). Although food riots are usually perceived as a purely urban phenomenon in both scholarly literature and in political discourse, as Ray Bush has recently pointed out, food riots may involve both urban and rural protests, as the example of the 2007–2008 crisis, when food prices spiked, has shown. In particular, Bush stressed the need to investigate the nature and peculiarities of rural protests in conjunction with food crises.[56] The Great Famine presents us with an opportunity to delve into this important question.

Unlike felonies, committed by individuals or small groups of people, food riots carry, by their definition, a pronouncedly collective character, often involving entire communities, or at least a few dozen people. Although such mass movements appear in neither manorial court rolls nor in gaol delivery rolls, they pop up in royal writs from the famine period. In these writs, Edward II addressed local royal officials and demanded them to take appropriate action against 'rebels' involved in collective attacks on local lords' properties. Unfortunately, in the vast majority of cases, the writs do not date the attacks themselves, and, therefore, it is unclear what the average time lag was between the attacks and the writs. However, the number of such writs reached its peak between 1315 and 1319, averaging at thirteen per year — in contrast with only four per year between 1311–1314 and 1320–1321. It is possible that the high number of

[55] Thompson, 'The Moral Economy of the English Crowd'.

[56] Bush, 'Food Riots'.

writs in 1318 and 1319 reflect some outstanding cases from the famine years. A few examples would illustrate the dynamics of these food riots. On 25 July 1317, Edward II issued a commission of *Oyer and Terminer* on the complaint of Nigel de Kenynton that a crowd of thirty-four people, 'with others', reaped and carried away his crops growing at Shottisham (Suffolk).[57] On 28 May 1316, a crowd of at least twenty-nine people invaded the Abbot of Hailes' manor of Great Wormington (Gloucestershire), broke down a stone bridge there, made dikes across the road (so that no goods could be carted), placed his cattle in arable fields to trample down and consume crops growing there, and attacked local manorial officers.[58] On 1 September 1316, an unspecified number of people of Great Doddington (Northamptonshire) assaulted the holdings of William de Thurleby, a parson of the local church, overthrew carts with crop tithes, took their contents away, drove away cattle (ten cows and four bullocks), and imprisoned a local servant. Because of the hasty nature of the attack, some of parson's grain had to be left behind, to perish.[59] As these examples illustrate, the attacks carried both scavenging and destructive character; the aim was not only to gain but also to damage. The assaults and imprisonment of local officers undoubtedly reflect locals' resentment not only towards their lords but also towards their officials, namely, their own neighbours. As the writs indicate, both men and women and both laymen and local clergy took part in these assaults. For instance, at some point in 1316, a crowd of at least twenty-one people, which included two women and a vicar, plundered four Lincolnshire manors of William de Welleby (Welby, Swinstead, Stamford, and Ancaster).[60] Moreover, manorial officers, in many instances, targets and victims of attackers, could also be the latter's collaborators and leaders. Thus, at some point in late 1315, the parishioners of Downham Market (Norfolk), headed by William de Bedynham, bailiff of the Abbot of Ramsey's manor there, reaped and carried away crops of Catherine Maister-Williamessoster (*sic*) and attacked her and her servant.[61]

Although there is no doubt that these riots carried a collective character, the question whether we can regard them as 'food riots' still remains open. It is true that these attacks consisted of some characteristic aspects of food riots', namely, the plundering of food storages, the forceful appropriation of food

[57] *CPR, 1313–17*, p. 581.

[58] *CPR, 1313–17*, p. 500.

[59] *CPR, 1313–17*, pp. 590–91.

[60] *CPR, 1313–18*, pp. 682 and 684.

[61] *CPR, 1313–17*, p. 431.

supplies, and assaults on local owners of food supplies, namely, landlords and their officers. At the same time, however, one central element of food riots is missing here: protests over high prices. As we have seen, the attacks were aimed at local producers, primarily lords, and their officials, who had nothing to do with grain price regulation. After all, grain prices were determined in accordance with annual crop yields and the current availability of crop stocks, and they were enforced by local sellers and market officers, not by lords.[62] This was in sharp contrast with urban food riots, where the popular demands to reduce grain prices and stop crop exports stood at the centre. Thus, during the 1328–1329 crisis in northern and central Italy, the hungry populace of Siena gathered together on the city *piazza* exclaiming 'Mercy! Mercy!', begging local civic authorities to cut down grain prices.[63] During the 1347 food riots in English towns, local crowds prevented civic authorities from exporting grain to the Continent and sold the appropriated grain at less than market price.[64] The Great Famine 'food riots' saw nothing of this kind. Do, therefore, these rural riots constitute food riots? One way to answer this question is to bear in mind Ray Bush's view that food riots are not exclusively an *urban* phenomenon and that *rural* food riots carry their unique attributes, which are yet to be properly studied. The anti-seignorial attacks during the famine years may be a manifestation of this still under-studied and under-appreciated phenomenon, when masses of starving rustics, devastated by the hardships of the famine, vented to their frustration with local food policies commanded by their lords, many of whom, as we have seen, hoarded grain stocks, allowing for prices to go up and, thus, depriving commoners of their calorific entitlement.

The Fate of Felons and Rioters

As stated above, the spike in crime levels during the Great Famine mirrors the situation during other historical famines. In particular, rising crime rates filled local prisons during the Great Irish Famine, primarily with thieves and burglars. Apart from ensuring their short-term calorific entitlement through illegal transfer channels, some Irish felons also hoped to secure access to prisoners' rations.[65]

[62] Davis, *Medieval Market Morality*, pp. 224–25.

[63] Cohn, *Popular Protest*, pp. 144–58.

[64] Sharp, *Famine and Scarcity*, pp. 56–87.

[65] Ó Gráda, *Ireland: New Economic History*, pp. 202–04; Ó Gráda, *Black '47 and Beyond*, pp. 188–91.

This is one fundamental difference between the two famines: unlike the mid-nineteenth-century starving Irish, early fourteenth-century felons could not have possibly hoped that prison meals would save them from starvation. Late medieval prisons, situated primarily in urban castles and other fortified structures, did not practise free food distribution among inmates. As a rule, there was little food available at local prisons, and felons were expected to pay for it, as well as for their 'accommodation' there. In some instances, this could be compensated by food deliveries by kinsmen and friends.[66] This, certainly, implied various logistical challenges, especially in those instances when felons came from villages rather than the same town in which they were incarcerated. Relying on the support of hungry kinsmen must have been particularly difficult during the famine times, given not only food scarcity but also various obstacles in transferring food from the countryside into towns. As we shall see later in this chapter, some religious institutions would come to the rescue, too, but the amounts of food they could allocate might not always be sufficient to feed the exhausted prisoners, especially given their increasing number.

Unlike modern prisons, medieval gaols did not form penal institutions for long-term seclusion and isolation. As a rule, felons would spend no more than a year in prison before either an indictment or an acquittal would be pronounced by the jury. During the sojourn, felons were subjugated to harsh imprisonment conditions. The excessively high food prices meant that food was deficient and inaccessible to many, and the spike in crime rates meant that cells were overcrowded. Indeed, as both gaol delivery rolls and coroners' rolls indicate, there was a spike in deaths among inmates during the famine years. For instance, between 1310–1311 and 1320–1321, the number of dead inmates in London prisons varied from zero to two a year in non-famine years; conversely, the figures stood at twelve and thirteen in 1315–1316 and 1316–1317, respectively.[67] Likewise, no less than twenty-three thieves died at Northampton castle prison in the summer of 1316.[68] At Wallingford Castle prison (Berkshire), twenty-eight people died between 24 June and 15 August 1316.[69] Furthermore, the share of convictions increased during the famine years, standing at about 26 per cent — in contrast with 18 per cent in non-famine years. Most convicts

[66] Geltner, *The Medieval Prison*, pp. 4, 15, 60, 73, 76, 80, and 84; Slavin, *Bread and Ale*, pp. 175–79.

[67] TNA, JUST 3/40/2, 3/41/1, and 3/42/1.

[68] TNA, JUST 2/108B. This is also noted in Hanawalt, *Crime and Conflict*, p. 253.

[69] Berkshire Record Office (BerksRO), W/JCm/1.

would be executed by hanging.[70] Some prisoners who refused to accept a jury trial were returned to the prison and put on a special severe diet and sojourn conditions to force them to plead. Two particular conditions are mentioned in the rolls: *ad dietam* (torture by penal starvation) and *peine forte et dure* (torture by putting a board on a prisoner's body and piling weights on it until he/she consented to answer to the charge).

To make things even worse, there is no discernible pattern of correlation between the severity of crime and the juridical outcome. A suspected homicide or large-scale theft could result in acquittal, while a petty theft of poultry worth a few pence could result in hanging. Thus, Thomas Rossel, suspected of murdering two women in Marston (Bedfordshire) and taking their clothes and other possessions sometime before his arrest on 14 June 1317, was acquitted.[71] William son of William Boys the miller of Hoggeston (Buckinghamshire) managed to get away free with stealing 5 quarters of wheat sometime before his delivery to Aylesbury gaol on 2 April 1316.[72] The fact that Thomas Rossel and William, son of William Boys were acquitted reinforces the impression that *accusation rates* should not be equated with *crime rates*. Conversely, John de Kereseye was convicted and hanged for just stealing six hens worth 12d from his neighbour at Sholden (Kent), shortly after his delivery to Canterbury Castle on 30 March 1317.[73] Thomas ate Heye from Essex was found guilty and hanged for merely stealing one bacon from the Hospital of St Mary without Bishopsgate (London), sometime before the Easter of 1316.[74] In other words, while resorting to crime may have been a risky, but inevitable choice, gaol delivery and incarceration were anything but the desired outcome for many desperate and hungry felons.

Post-Harvest Gleaning

Another, and by far less risky and more socially acceptable strategy undertaken by some tenants was to glean. Gleaning is the practice of gathering crop leftovers from post-harvest fields. As various manorial by-laws indicate, this activity was reserved for the poorer and frailer elements within community, while

[70] Similar statistics are mentioned in Hanawalt, *Crime and Conflict*, p. 58.

[71] TNA, JUST 3/1/2, membr. 2^{v}.

[72] TNA, JUST 3/5/2, membr. 6^{r}.

[73] TNA, JUST 3/27/3, membr. 14^{r}.

[74] TNA, JUST 3/40/2, membr. 37^{v}.

better-off and able-bodied members, capable of reaping, were not allowed to glean.[75] Obviously, the temptation for gleaning was strong, and, as court rolls reveal, tenants were caught and fined for gleaning from both the demesne fields and their neighbours' parcels. There can be little doubt that the ubiquitous scarcity of food made the prospects of gleaning all the more tempting. This is indeed reflected in some court rolls, recording unusually high numbers of individuals fined for gleaning. It should be noted, however, that unlike the amercements for breaking the *Assize of Bread and Ale* or for food regrating, gleaning fines were recorded in a highly irregular manner. Because of that, it is impossible to systematically quantify the trends in gleaning in normal years and during the famine. As a result, the discussion is based on anecdotal and fragmented rather than quantifiable and comprehensive evidence.

The most detailed information on gleaning activities comes from the court rolls of Horsham St Faith and Wymondham, both in Norfolk. As these records indicate, in non-famine years, the number of gleaning amercements stood anywhere between nil and four; during the 1315 harvest, the respective numbers on the two manors went up to nine and eighteen. During the 1316 harvest, seven individuals were caught and amerced for gleaning at Wymondham.[76] Who were these individuals? The Horsham St Faith and Wymondham records may be supplemented with cursory and fragmentary evidence from a further ten manors in East Anglia.[77] As court roll entries hint, many gleaners seem to have come from the poorest individuals within the community. Since these individuals would have been considered to be too poor to be amerced by leet courts, their names were not recorded. But what about these individuals who were considered to be well-off enough to be amerced? Some individuals appear to have come from better-off strata within their communities, judging by their involvement in the credit market and the food trade. For instance, John Rakke and John Minne of Horsham St Faith both appear as creditors on several occasions around the famine time.[78] Similarly, at Heacham, an unnamed wife of Thomas Hellebulle (alongside her husband) and Matilda Deynes acted as lenders in the 1310s.[79]

[75] Ault, 'By-Laws of Gleaning', pp. 210–13; Jordan, 'The Gleaners'.

[76] NRO, NRS 19505; NRO, NRS 18475 and 18481.

[77] The manors are: Bressingham, East Raynham, Gressenhall, Heacham, Hindolveston, Langham, Sedgeford and Tivetshall (Norfolk), South Elmham (Suffolk), and Balsham (Cambridgeshire).

[78] NRO, NRS 19505.

[79] NRO, Lest DA/7–8.

At Wymondham, Emma le Mayoun and Isabel Poey appear on several brewing amercement lists in the same decade.[80] These, however, were exceptions, and the majority of individuals amerced for gleaning seem to have come from the bottom of the socio-economic pyramid. This view is supported by the fact that of 162 individuals from the twelve East Anglian manors amerced during the famine years, only 15 (that is, 12 per cent) appear as land buyers, creditors, or food traders in the 1310s. Furthermore, at least two individuals are described as non-residents (*manentes in domo*): Auxilia Gille, hosted by Robert Ive of Horsham, and Agnes Ede, staying with Matilda Lank, of the same.[81] Outsiders were not allowed to glean, and if they were, their hosts were held responsible for their actions.[82] Two women from Wymondham, Isabel Poey and Margaret Poey, seemingly relatives, were described as 'fugitives' who did not pay their amercements.[83] Overall, it appears that gleaning was predominantly a female activity: of 162 individuals from the twelve East Anglian manors, 128 gleaners (79 per cent) were women. Just as with food-related crimes, the preponderance of women in post-harvest gleaning may have something to do with their domestic role of preparing food for their households.

Some entries provide a glimpse into the nature and scale of gleaning. Usually, gleaning could earn gatherers a few grain sheaves. For instance, John Rake of Horsham collected three barley sheaves (about 0.6 bushels) in the aftermath of the 1315 harvest — which was fewer than compared with the six wheat sheaves (1.2 bushels) he had collected in a much more bountiful harvest of 1314. Simon Garbote, of the same manor, collected four sheaves of grain (0.8 bushels) in the wake of the 1314 harvest.[84] Most entries, however, do not specify the quantities and merely record that amerced individuals 'gleaned wrongfully' (*glenavat male*); some entries add that gleaners 'did damage to the harvest' (*fecit dampnum in autumpno*), and in some instances, gleaners were described as 'acting stealthily' (*furtim, furtando*). Indeed, just as regrating, so gleaning, too, carried a certain degree of social stigmatization: gleaners were perceived as potential sheaf thieves.[85] Undoubtedly, individuals like John Rake, known also as a regrater, contributed to strengthening this perception among his fellow parish-

[80] NRO, NRS 18475 and 18481.

[81] NRO, NRS 19505.

[82] Ault, 'By-Laws of Gleaning', p. 213.

[83] NRO, NRS 18481.

[84] NRO, NRS 19505.

[85] Ault, 'By-Laws of Gleaning', p. 215.

ioners. There are numerous instances when gleaners gathered harvest leftovers from their neighbours' parcels rather than the demesne fields. Thus, John Bolex, from the parish of Norton, within the manor of Wymondham, was amerced for collecting grain sheaves from his neighbours (*blada vicinorum*) in the wake of the 1315 harvest.[86] Likewise, Simon Bosse of Balsham (Cambridgeshire) gathered harvest leftovers 'of various people' (*blada diversorum hominum*) in the aftermath of the 1316 harvest.[87]

Just as with food trade amercements, gleaners were fined rather lightly. The size of fines stood between 1d and 6d per person, with 3d being the most frequent figure. In some instances, some individuals were condoned altogether. Some of these pardons may have been extended on the account of one's poverty. For instance, neither Cecilia, wife of Richard de Barkesdale, nor Ralph Pilewet, both of Heacham, appears in any court roll entries between 1311 and 1321. Olivia Pybel was amerced as a regrater on several occasions in the 1310s. As we have seen in Chapter 4, regraters could be both poor and better-off individuals, and it is possible that Olivia Pybel was counted among the former. All the same, the correlation between condonation and poverty should not be taken for granted. For instance, one Heacham woman, designated as 'wife of Thomas Hellebulle', was married to Thomas Hellebulle, who acted as both creditor and debtor during the famine.

Could we interpret gleaning amercements as licences to gather harvest leftovers? As we have seen in Chapter 4, the *Assize of Bread and Ale* fines seem to have been, in many instances, licence payments to bake and brew rather than actual amercements. The payment of 3d or 6d would have eventually returned much higher profits for bakers and brewers. But was it as profitable to be amerced in relation to the volume of grains gleaned? Some entries may provide a partial answer. At Horsham, John Rakke was fined at 3d for gleaning three barley sheaves worth 1.5d, in the aftermath of the 1315 harvest. That the size of fines tended to exceed the value of collected grain is likewise hinted in another Horsham court roll entry, where Simon Garbote was amerced of 6d for gathering four wheat sheaves in the wake of the 1314 harvest, when grain prices were considerably lower than during the next three years.[88] On the other hand, for those who were condoned or were fined at lower rates (say, 1d or 2d), gleaning may well have been more profitable than for John Rakke, Simon Garbote, and their likes.

[86] NRO, NRS 18481.

[87] LMA, ACC/1876/MR/02/001, membr. 24v.

[88] NRO, NRS 19505.

There can be little doubt that the spike in gleaning amercement reflects the desperate efforts of starving individuals to secure some food. But it may also indicate, just as in the case of the *Assize of Bread and Ale* and, as we shall see in the next section, the suit of mill, the fear of local manorial authorities of the potential break-down of law and order. This fear may have prompted local manorial authorities to monitor the situation more closely than ever and impose fines for breaking the suit of mill more stringently, even in those instances when amerced individuals might have been only *suspected* of gleaning. By doing so, manorial authorities would have discouraged any illicit gleaning, on the one hand, and reassured their communities that law and order were kept, notwithstanding the widespread distress, on the other.

Breaking the Suit of Mill

Another illicit, but quasi-acceptable and quasi-decriminalized activity liable to fine rather than criminal prosecution, was breaking the suit of mill (*secta molendini*). This was a manorial custom expecting tenants to mill their harvest produce at their lord's mill. To do so, tenants had to pay their grinding fees known as *multure*, usually varying between one-twelfth and one-twentieth of the amounts of grains brought to be milled. Once the grains had been ground, the tenants would retain the rest in flour. Our information regarding grinding of tenants' grain produce comes from the 'mill issue' (*exitus molendini*) rubric (often absorbed into the grain issue section) of manorial accounts and court roll entries containing lists of tenants fined for breaking the suit of mill. The *exitus molendini* rubric records both the types and quantities of each grain paid in *multure* and the financial profit from their sales. The quantities of grain varied from year to year, reflecting the annual fluctuations in the quantities of tenant grain milled at lords' mills (Figure 8.2). As expected, the famine years saw the lowest levels, standing at 37, 41, and 26 per cent below average in 1315, 1316, and 1317, respectively.

There can be little doubt that this decline in the quantities and profits from *multure* reflect the low crop yields within both the demesne and tenancy sectors, discussed at length in Chapter 2. The 1315–1316 figures reflect, to a large degree, the demesne yields and tithe receipts, while the 1317 ones were lower than both the former and the latter. But it could hardly be the only factor here. We should also account for the reluctance of local tenants to grind their meagre grain supplies at the lords' mills, on account of their poverty and starvation. As various court rolls indicate, the sheer number of amercements related to suit

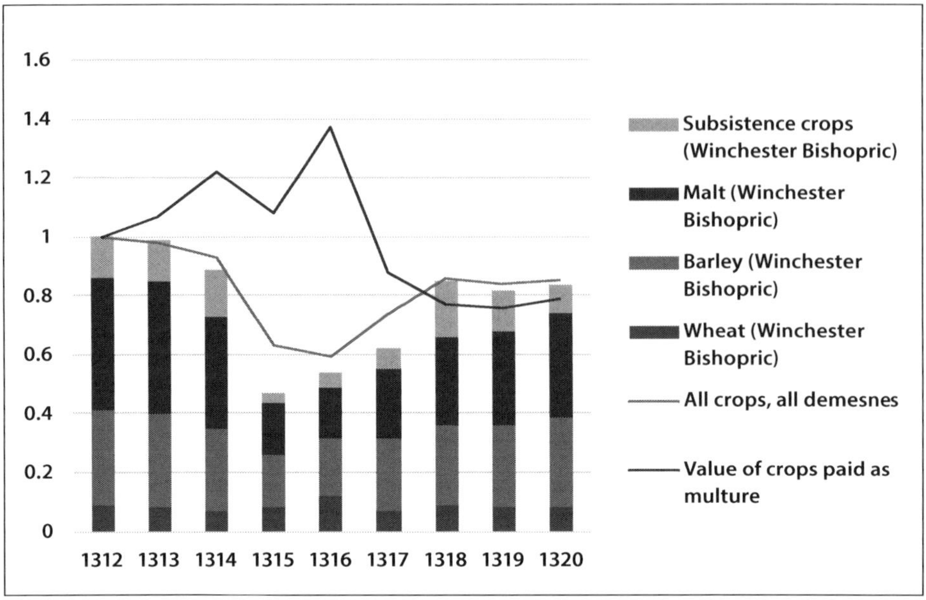

Figure 8.2. Amounts of tenant grain paid as *multure* and their financial equivalent on thirty-three demesnes, 1311/2–1319/20 (indexed on 1311–1312). *Source*: Canterbury Cathedral manors (thirteen manors): CCA, DCc/BR/Appledore/14–21; DCc/BR/Chartham/8–14; DCc/BR/Copton/7–12; DCc/BR/Eastry/36–48; DCc/BR/Great Chart/26–32; DCc/BR/Hollingbourne/25–29; DCc/BR/Ickham/19–24; DCc/BR/Loose/28–34; DCc/BR/Meopham/57–65; DCc/BR/Merstham/18–26; DCc/BR/Middleton/25–32; DCc/BR/Orpington/23–20; and DCc/BR/Westwell/10–17. Winchester Bishopric manors (12 manors): HantsALS, 11M59/B1/67–75. Byfleet (Surrey): TNA, SC 6/1011/2. Feering (Essex): WAM 25626–36. Hinderclay (Suffolk): University of Chicago Library, Special Collections (UCL), Bacon Rolls 444–49. Langley (Hertfordshire): TNA, SC 6/866/18–25. Longbridge Deverill (Wiltshire): LH, 9644, 9655, 9645, 10688, 9646–48, 9656. Wootton St Lawrence (Hampshire): HantsALS, DC/J1/11–15. Cheddington (Buckinghamshire): Merton College Archives (MCA), 5541–50.

Note: The respective figures indexed on 1311–1312 (1311–1312 = 1.0), appearing on the vertical axis, are as follows: (1) all crops (wheat+barley+malt+subsistence crops) on the Winchester Bishopric estates (twelve demesnes): 1.0 = 604 quarters; (2) all crops, all thirty-three demesnes: 1.0 = 1169 quarters; and (3) value of all crops (all thirty-three demesnes) paid as *multure*: 1.0 = £206 6s.

of mill seems to have risen considerably during the famine years. Unlike the amercements for breaking the *Assize of Bread and Ale*, or food regrating, and similarly to gleaning fines, manorial court rolls tended to record the instances of breaking of the suit of mill in a highly irregular manner. Thus, in many instances court rolls contain no entries related to suit of mill fines. Because of this fact, it is virtually impossible to quantify the changing trends in the breaking of the suit of mill in a systematic manner over space and time. Hence, the following discussion will be based on clearly under-recorded and fragmented evidence.

Despite under-recording, however, there is enough evidence to support the claim that the number of suit of mill amercements went up considerably during the famine years. Again, just as with the *Assize of Bread and Ale*, studied in Chapter 4, and with crime accusations, analysed earlier in this chapter, the increased number of mill amercements during the famine years may reflect not only the actual amercements but also the fact that local authorities were more anxious to impose a stricter control of law on their communities to prevent a total break-down of social order. Again, this stood in a sharp contrast with non-famine years, when breaking the suit of mill may not have been considered an offence serious enough to break down social order, and, thus, tended to be under-recorded in local court rolls. It was this combination of the increasing incidence of the breaking the suit of mill, on the one hand, and a sterner monitoring of local tenants by manorial authorities, on the other, that explains a drastic rise in the numbers of milling amercements. For instance, at Langham (Norfolk), thirty tenants were fined in 1315–1316, in contrast with eighteen in 1304–1305, ten in 1322–1323, and six in 1324–1325.[89] Likewise, on 5 April 1317, thirty-seven parishioners of Hunstanton (Norfolk) were fined, in contrast with twenty-three people amerced on 21 August 1311.[90] Similarly, at Isleworth (Middlesex) and Gressenhall (Norfolk), thirty-eight and fourteen tenants were fined, respectively, on 19 December 1315 and 4 October 1316, for breaking the suit of mill obligation.[91] At the manor of Wakefield (West Yorkshire), no less than twenty-one instances of breaking the suit of mill are recorded between 1315 and 1317. By contrast, very few or no suit of mill fines are recorded in non-famine years on the same manors. Similarly, at Coltishall (Norfolk), there are hardly any references to breaking the suit of mill between 1309 and 1326, except 2 October 1316, when four cases are reported in a single court.[92] Similar trends can be spotted on other manors, including Balsham (Norfolk) and Chesham Bois (Buckinghamshire).[93] Again, the non-famine fig-

[89] RHM, Langham Court Rolls, 1304–1305, 1322–1323, 1315–1316, and 1324–1325.

[90] NRO, ING 227; NRO, ING 43.

[91] London Metropolitan Archives (LMA), Acc 1379/2; NRO, ING 41.

[92] King's College Archives (KCA), Col/363, membr. 9.

[93] LMA, ACC/1876/MR/02/001; Centre for Buckinghamshire Studies (CBS), D/BASM/18/2–12.

ures may well be under-recorded. But there can be little doubt that the dire circumstances of the famine, when every bushel mattered, forced some tenants to ignore the centuries-old custom and to avoid, as much as possible, losing a fraction of their meagre harvests to *multure* payments.

Largely speaking, there were three types of accusations in conjunction with breaking the suit of mill: (1) not observing the suit of mill, (2) using portable hand mills, and (3) not milling at the lord's mill (and milling elsewhere). Thus, at Wakefield, sixteen tenants were amerced for grinding their grain elsewhere, and a further five people were fined for using their hand mills.[94] Although hand mills may not have been as numerous as watermills or windmills, many peasant households still owed a quern, and, as such, they still represented a non-negligible share within the English milling industry. It has been estimated that *c.* 1300, about 20 per cent of all grain grinding was done by hand mills.[95] Certainly, they were the single most affordable alternative for peasant households. Given the low amercements for using hand mills, usually varying from 1d to 6d per case, such choice made a lot of economic sense. Some tenants, however, did not get away that lightly: thus, one Thomas of the Wode of Rastrick (West Yorkshire) had to pay 2s for using a hand mill, contrary to the suit of mill.[96] Gender-wise, the vast majority of amercements were paid by male tenants (about 75 per cent). Still, women do appear in milling lists, indicating that at least some hand mills were operated by women, as John Langdon suggested.[97]

Even though breaking the suit of mill could undoubtedly be beneficial for some poorer tenants, does it mean that those who chose not to do so were at loss? The analysis of the composition of crops at seignorial mills may provide an answer to this question. Although many accounts do not specify the types of grains paid in *multure*, merely referring to them as 'mill-mixture' (*tolcorn*, *mixtura*, *multura*, etc.), the Winchester Bishopric Pipe rolls meticulously distinguish between different kinds of multure crops. A close reading of the relevant entries, related to the composition of milled crops, is indeed revealing. As Figure 8.2 indicates, the famine years saw an increase in the share of wheat from around 9 to 17 per cent, at the expense of 'subsistence' crops (maslin, rye, and occasionally oatmeal), whose figures declined from about 15 to 9 per cent. Similarly, the proportion of malt declined from about 41 to 35 per cent. These

[94] *Court Rolls of the Manor of Wakefield*, ed. by Lister, IV, 23, 33, 41, 45, 121, 191, and 193.

[95] Langdon, *Mills in the Medieval Economy*, p. 126.

[96] *Court Rolls of the Manor of Wakefield*, ed. by Lister, IV, 191.

[97] Langdon, *Mills in the Medieval Economy*, pp. 19 and 129.

shifts do not necessarily reflect short-term changes in peasant crop cultivation, but rather short-term coping strategies. The dire economic reality of the famine years undoubtedly prompted many peasants to switch to the consumption of inferior crops. The optimal strategy would be to substitute wheat, the most premium grain, with cheaper subsistence crops. To maximize their profits, the tenants would have to sell wheat in a ground, rather than raw, form. The value of flour is higher than that of its unprocessed equivalent, and this fact was crucial, especially during the famine years, when the high prices and the demand for processed or half-processed food would imply unprecedented profits for flour sellers. Although there are no, to my best knowledge, price data on flour from the fourteenth century, the evidence from the fifteenth and sixteenth centuries suggests that the ratio between raw wheat and wheat flour was in the area of 1:0:1.75, *by weight*.[98] A thorough calculation reveals that a hypothetical tenant made a profit of about 50–75 per cent for each quarter of wheat he/she brought to a seignorial mill and paid a *multure* fee for, depending on the quality of flour that the remaining wheat was converted into (see Appendix 7). Over the course of the 1315–1316 accounting year, this would have amounted to about 6s 10.5d if milled into bran-less flour (*flora*), incurring 85 per cent extraction rates (that is, 15 per cent weight loss), or about 10s 6 1/2d if retained as coarse *farina*, made up of both flour and bran. These sums could have purchased between 1.25 and 2 quarters of oats, depending on a milling scenario. If this interpretation is correct, could this be one of many survival strategies of tenants in face of the dire straits of the crisis? It is possible that this strategy was undertaken by better-off tenants, who actually had sufficient quantities to be milled and who could withstand paying the *multure* fees. The increased incidence of the suit of mill amercements, on the other hand, may refer to poorer elements, who would have attempted to save every pound of grain from being paid as *multure*. Obviously, hand mills were by far not as prolific as seignorial mills and, thus, not always a readily available option. Apart from their physical capacity, wind- and watermills had two clear advantages over hand mills: their operational speed and power. This was especially true in the climatic context of 1314–1316, when the high winds and strong river tides undoubtedly boosted up the power and increased the grinding speed.

There is, however, another interpretational possibility. As John Langdon has convincingly shown, in regions south of the Humber, private hand mills of tenants were seen as commercial competition to seignorial mills, and lords

[98] Calculated from Thorold Rogers, *A History of Agriculture and Prices*. Also, see Appendix 7.

did their best, therefore, to curtail their proliferation. This was in contrast with the situation in the northern counties, where tenant hand mills were either very few or virtually non-existent because of more extractive milling customs.[99] Because of the high prices and, consequently, the value of ground crops, *multure* became an exceptionally profitable source of income for lords, whose interest was to assure that tenants respected the suit of mill and ground their grains exclusively at their mills. This may have prompted the local manorial authorities to monitor the situation more closely than ever and impose fines for breaking the suit of mill more stringently. As Figure 8.2 reveals, in spite of low *multure* quantities, the profits from manorial mills, expressed in the financial equivalent of crops collected as *multure* payments, rose during the famine years. The figures reached their peak in 1316, when the revenues stood at almost 40 per cent above the 1311–1314 levels, despite the total receipts in grain standing at 40 per cent below the pre-famine levels. As a rule, the vast majority of *multure* grains — usually in the area of 90–95 per cent — were eventually sold, while the remainder was usually allocated to local *famuli*. Any source of marketable grain, which could potentially be hoarded, was undoubtedly welcomed with open arms by lords during the famine years, and *multure* payments were undoubtedly no exception.

To sum up, during the famine years, suit of mill cases increased partly because more tenants were evading tolls as a survival strategy. But then again, this strategy was not unique to the famine years — we witness accusations of breaching the suit of mill in non-famine years too, although in smaller proportions. However, the number of individuals breaking the suit of mill rose not only because more tenants resorted to this strategy. It was also because lords were tightening up enforcement in order to maximize revenue from mill tolls during the famine years.

Trade-Based Entitlement: Land Transfers

Resorting to illegal entitlements may not have always been the most attractive option for starving commoners, who, for the most part, sought legal ways to secure their food entitlement. One such channel was the land market, which allowed them to sell or transfer a parcel of land in exchange for some cash, which would be converted into calories. The land market was one of key features of the English economy in the later Middle Ages, with both freeholders

[99] Langdon, 'Lordship and Peasant Consumerism'.

and unfree tenants being actively engaged in buying, selling, exchanging, accumulating, and losing land parcels. This would not have been possible without the encouragement of local landlords. The impact of this seignorial encouragement has long been debated by historians. For Ambrose Raftis and his Toronto School, this was a sign of lords' goodwill, on the one hand, and the spirit of peasant individualism and entrepreneurialism, on the other, boosting what Raftis called 'peasant economic development'.[100] Bruce Campbell's view is far less optimistic. For him, the uncontrolled land market and subsequent morcellation (subdivision) of land units constituted hindrance, rather than progress, because it allowed for the population to grow virtually unchecked and for structural poverty to become widespread.[101] Indeed, as we have seen in the previous chapter, the vast majority of English rustic households held meagre parcels of land, barely sufficient to sustain them in good years. The disastrous harvest failures of 1315 and 1316 prompted many desperate and hungry smallholders to offer their meagre parcels for sale to secure some calories, thus transforming their small holdings into even smaller units of land or, in some instances, losing them altogether. Indeed, some entries indicate that land-plots changed hands because of the crisis. Thus, at Hindolveston (Norfolk), Joan daughter of John West surrendered eighteen perches of arable land, 'because of the great famine' (*pro magna fame*).[102] Thus, land market activity had reached an unprecedentedly mad level during the crisis years of 1315–1316 and 1316–1317, as is indicated by the surge of transactions within both the customary and freehold land sectors (Figure 8.3).

Figure 8.3 is based on two sets of documentation. The customary land transactions are entered into court rolls, while freehold transactions were entered into 'feet of fines', namely final concords recording the conveyance of free land.[103] As the figure demonstrates, the surge in distress sales of land was especially pronounced within the customary land sector. The average level of *inter-vivos* (that is, among living parties, to be distinguished from *post-mortem*) land transactions increased three-fold in 1315–1316 and two-and-a-half fold in 1316–1317, before falling back to slightly above-average levels in 1317–1318. The figures varied from case to case. At Hinderclay (Suffolk), Hindolveston, and Heacham (both in Norfolk), the transaction levels increased between four-

[100] Raftis, *Peasant Economic Development*.

[101] Campbell, 'The Agrarian Problem', pp. 50–55, and *The Great Transition*, pp. 182–98.

[102] NRO, DCN 60/19/3; Hudson, 'The Prior of Norwich's Manor of Hindolveston', p. 211.

[103] Yates, 'The Market in Freehold Land', pp. 580–81.

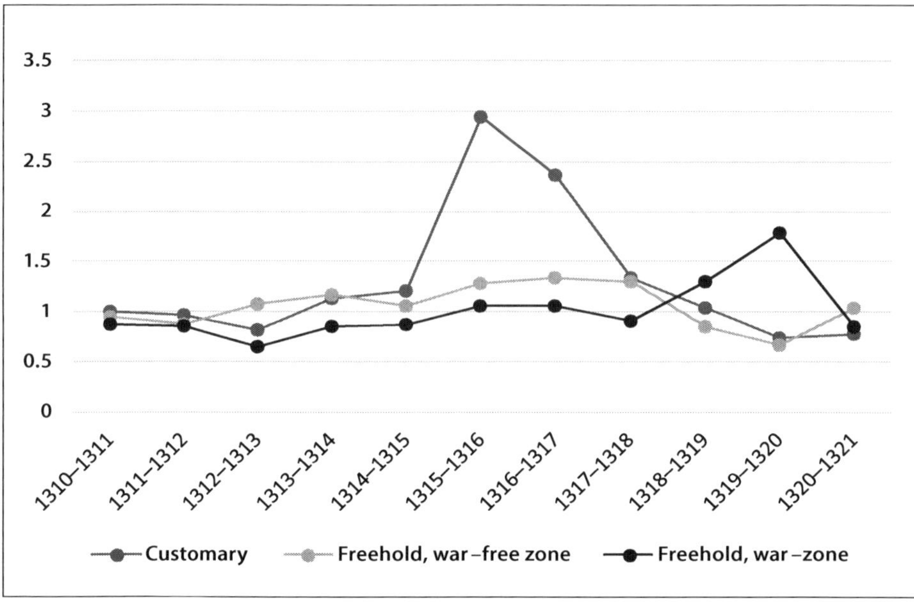

Figure 8.3. Levels of land transactions (customary and freehold) in England, 1311–1321 (indexed on non-famine years). *Source*: (a) Court rolls: NRO, NRS 19505 (Horsham St Faith, Norfolk); Wisbech and Fenland Museum, WK/2 (Walsoken, Norfolk); NRO, WAR/3–11 (Bressingham, Norfolk); NRO, Lest DA/6–8 (Heacham, Norfolk); NRO, DCN 60/19/3 (Hindolveston, Norfolk); NRO, DCN 60/32/9 (Sedgeford, Norfolk); RHM, Langham Court Rolls (Langham, Norfolk); UCL, Bacon Roll 11–14 (Redgrave, Suffolk); UCL, Bacon Rolls 119–20 (Hinderclay, Suffolk); BL Add Ch 63422–32 (Rickinghall, Suffolk); SuffRO (Lowestoft Branch), HA12/C2/3 (South Elmham, Suffolk); LMA, ACC/1876/MR/02/001, membr. 10–39 (Balsham, Cambridgeshire); CUL, EDC 7/4A (Sutton-in-the-Isle, Cambridgeshire); BoL, Gough Camb 1, fols 73r–108v (Chesterton, Cambridgeshire) and NorthantsRO, Montagu Collection, Box X364A (Brigstock, Northamptonshire); (b) Feet of fines: Slavin, Feet of Fines Database (see Appendix 8).

Notes: Customary transactions are indexed numbers of average transactions per court deriving from court rolls of fifteen manors; freehold transactions are indexed numbers of total transactions recorded in each foot of fines. The respective figures indexed on non-famine years (1310/1–14/5 and 1317/8–1320/1 = 1.0), appearing on the vertical axis, are as follows: (1) customary transactions: 1.0 = 276 transactions a year on all fifteen manors; (2) freehold 'war-free zone' transactions: 1.0 = 489 transactions a year; and (3) freehold 'war-zone' transactions: 1.0 = 60 transactions a year.

and five-fold during the famine years. At Horsham St Faith (Norfolk), on the other hand, the transactions increased by only two-fold and one-and-a-half fold in 1315–1316 and 1316–1317, respectively.[104] The surge was not nearly

[104] NRO, NRS 19505.

as dramatic within the freehold sector. Here, because of the remarkable geographic and chronological survival of the feet of fines, we get a national coverage. In particular, it is instructive to distinguish between the 'war-free zone' (south of the Humber) and the 'war zone' (north of the Humber), as discussed in Chapter 6. Within the 'war-free zone', the figures did not increase by more than 30 per cent during the famine years. The main exceptions were the metropolitan counties of Middlesex, Bedfordshire, and Surrey, where the figures rose by 100 per cent. Similarly, in Herefordshire and Derbyshire, the rise was even more remarkable, with the volume of transactions going up by about 140 per cent. Elsewhere, the surge was much less pronounced. An altogether different picture emerges within the 'war zone'. Contrary to the southern counties, the land market seems to have been depressed here. In Lancashire, the famine years recorded only five and four transactions each: in contrast with the average twelve annual transactions in non-famine years. In Cumberland, no transactions were recorded in 1315–1316. In Northumberland, in both years the number of transactions stood at three, in contrast with four annual transactions on average, in non-famine years. Intriguingly, the northern figures rose pronouncedly in 1318–1319 and 1319–1320, namely, after the end of the famine in the south, standing at 1.3 and 1.8 of their non-famine average. This has undoubtedly to do with the fact that the food crisis in the north persevered for a much longer period than in the south because of the destructive effects of the Anglo-Scottish warfare and the cattle pestilence, as we have seen in Chapter 6.

Why did the levels of freehold land transactions not rise as aggressively as within the customary sector? One possible explanation is that an average free tenant household held, as a rule, smaller parcels than his/her unfree counterpart. As Bruce Campbell has shown, around *c.* 1300, 'free tenants, in fact, had become the new poor' because of the ease with which free land could be transferred and because of the usually lower rents, resulting in a faster morcellation of freeholdings. In other words, free tenants, with their meagre holdings, had very little or next to nothing to offer. Liquidating one or two acres of land may mean the total liquidation of one's assets. But also, freeholders could use their land as collateral for a loan, without sacrificing it in exchange for cash and thus, unlike customary tenants, they did not have to face the dilemma between selling up or hanging on to their land. Therefore, many freeholders may have chosen to take a chance and hold on to a few acres of land they had, in hopes to make it through the crisis without sacrificing their most precious asset, for the sake of their family and heirs.

Indeed, within both sectors, the vast majority of transactions involved very small plots of land, most of which amounted to less than an acre. At Chesterton

(Cambridgeshire), the mean size of land-plot sold during the famine years was 0.83 acres.[105] At Bressingham (Norfolk), the comparable figure was half an acre.[106] Many transactions involved even smaller units. Thus, at Raynham (Norfolk), Roger Hynyne surrendered a small piece of land, 28 feet in length and 24 feet in breadth, to Margery Drew.[107] At Walsoken (Norfolk), Adam Wade surrendered a small cottage containing just four perches to Adam le Hunte.[108] At Gillingham (Dorset) and Horsham St Faith (Norfolk), most land transactions recorded at famine courts regarded either small pieces of land (*pecie terre*) or crofts.[109] The small size of the surrenders is hardly surprising, given the widespread poverty and, thus, unaffordability of large parcels of land. Also, it appears that poor tenants were ready to sacrifice *some* of their land, but not *all* of their land — again, perhaps thinking about their moral obligation to preserve their land for their heirs, whether or not they themselves survived the famine.

The grim reality of the famine changed the absolute number of land transfers, but it hardly brought about any discernible change in the break-down of transactions. Of all *inter-vivos* land transactions during the famine years, the proportion of familial transfers on fourteen East Anglian manors and a Northamptonshire manor of Brigstock accounted for only about 15 per cent. This was a figure largely comparable with the situation in 1311–1315 (14 per cent).[110] This was in contrast with the situation at pre-Black Death Halesowen (on the Worcestershire-Shropshire border), where the proportion of intra-familial transfers accounted for a much larger share: about 63 per cent.[111] Likewise, on the manors of the Bishop of Winchester (located primarily in southern counties), in the pre-Black Death era, inter-familial land transfers accounted for 46 per cent of the total.[112] Indeed, as the work of both Richard Smith and Phillipp Schofield has shown, the nature of the land market in East Anglia was much less family-oriented in the pre-Black Death era than else-

[105] BoL, Gough Camb 1, fols 89^r–99^r.

[106] NRO, WAR/7–9.

[107] RHM, Raynham Court Rolls, Edward II.

[108] Wisbech and Fenland Museum (WFM), WK/2.

[109] DHC, D/FAN, 10 Edward II; NRO, NRS 19505.

[110] The East Anglian manors in question are: Horsham St Faith, Walsoken, Bressingham, Heacham, Hindolveston, Sedgeford, Langham (Norfolk); Redgrave, Hinderclay, Rickinghall, South Elmham (Suffolk); and Balsham, Sutton-in-the-Isle, Chesterton (Cambridgeshire).

[111] Razi, 'Family, Land and the Village Community', pp. 4–5.

[112] Page, 'The Peasant Land Market', pp. 62–64.

where in the country.[113] It appears that in East Anglia, the inheritance practice of partability and possibly the geographic mobility of local tenants encouraged the formation of what Razi referred to as 'low kin density', in sharp contrast with the situation elsewhere.[114] If this hypothesis of Razi, backed up by the figures above, is correct, then it appears that in the case of East Anglia there were comparatively few individuals to support their starving relatives. This, however, does not mean that such help was not extended when close relatives were living nearby.

Although intra-familial *inter-vivos* land transfers accounted for only 15 per cent of all transactions, there are more than enough examples to suggest that in many cases these intra-familial transfers were guided by purely altruistic motives, deriving from strong sense of family and kinship, rather than profit. At times, this implied self-sacrifice for the sake of spouses, children, and siblings. For instances, at Chesterton (Cambridgeshire), William Pittock transferred five and a half acres of land to his three sons, before dying sometime in the summer of 1317.[115] Similarly, Walter Reyne, of the same manor, handed over two units of one acre each to his two sons sometime in 1316. After that, Reyne's name disappeared from court records, possibly indicating that he did not make it through the famine. In any event, his name does not appear under any heriot rubric, implying that he may have died landless.[116] Some individuals were lucky to have enjoyed support of multiple relatives. Thus, Peter Turkeby of Horsham received three parcels of land from his father William, his brother John, and from another kin called William, totalling two and a half acres and a messuage.[117] Transfers among siblings were not unheard of too. At Walsoken (Norfolk), Stephen son of Gregory assigned two parcels of land to his siblings Walter and Agnes.[118] At Hindolveston (Norfolk), Adam Imagyne transferred three rods to his sister Isabel.[119] In some instances, siblings would yield landplots to their siblings shortly after their receipt. Thus, Adam son of Hervey le Lethe of Brigstock (Northamptonshire) transferred one and a quarter acres

[113] Smith, 'Some Issues concerning Families', pp. 19–21; Schofield, 'Social Economy of the Medieval Village', p. 42.

[114] Razi, 'Myth of the Immutable English Family', pp. 15–18.

[115] BoL, Gough Camb 1, fol. 93^{v}.

[116] BoL, Gough Camb 1, fol. 89^{v}.

[117] NRO, NRS 19505.

[118] WFM, WK/2.

[119] NRO, DCN 60/19/3 (12 July 1316).

of land to his sister Alicia, having just received them from his mother Isabel, on 24 August 1316.[120] It is possible that Alicia was unmarried and thus potentially more vulnerable to famine, and this fact promoted Adam to act out of selfless fraternal motives. The care for unmarried daughters is reflected in a transfer of two plots of land, less than one acre each, by Andrew atte Lathe of Walsoken (Norfolk) to his two daughters, Catherine and Agnes.[121] The fact that Catherine and Agnes were mentioned alone in these transactions, and that there is no court roll entry indicating their spouses, may hint that they were, in fact, unmarried. While parents took care of their children, children did their best to help out their ageing parents too. Thus, at Chesterton, Henry son of Robert de Stapilford transferred half an acre to Sarah, his mother.[122] Similarly, Robert son of Robert Faber of the same manor gave an acre to his mother Alice.[123]

As noted above, some individuals seem to have lost everything or nearly everything, either via surrendering plots to their relatives because of their sense of kinship responsibility or through transferring their parcels to non-related parties because of the practical need to make ends meet. Thus, at Horsham St Faith (Norfolk), Thomas Picock and Maud Herring sold their respective plots on four occasions, but bought none, during the famine years.[124] Similarly, at Chesterton (Cambridgeshire), Jordana Parys transferred land no less than six times, without acquiring any parcel even once.[125] After the famine, these tenants disappear from records. Conversely, some better-off tenants amassed land at the expense of their impoverished neighbours. Thus, at Chesterton (Cambridgeshire), Robert son of John de Marbilthorp and Margaret, his sister, managed to accumulate six acres of land during the famine period.[126] At Bressingham (Norfolk) John Joye, managed to acquire five and a half acres of land between May and November 1316 alone. At Horsham St Faith (Norfolk), William Croberd and William Lanke, clerk, bought land on four occasions and sold only once each.[127] At Hinderclay (Suffolk), Henry Breton purchased six plots and sold none dur-

[120] NorthantsRO, Montagu Collection, Box X364A/29.

[121] WFM, WK/2.

[122] BoL, Gough Camb 1, fol. 93^{r}.

[123] BoL, Gough Camb 1, fol. 89^{v}.

[124] NRO, NRS 19505.

[125] BoL, Gough Camb 1, fols 95^{r}–96^{v}.

[126] BoL, Gough Camb 1, fols 89^{r}–89^{v}.

[127] NRO, NRS 19505.

ing the famine.[128] At Hindolveston (Norfolk), Adam Carpenter amassed about four and a half acres of land in twelve transactions from his neighbours during the famine, selling nothing.[129] More examples can be cited.

In his study of the Great Famine, William Chester Jordan conceded that the phenomenon of crisis land sales contributed to the process of stratification and 'kulakization' within villages, whereby some better-off tenants abused the situation to accumulate much land.[130] While there is no doubt that the abnormally hyperactive land market during the famine years contributed to the widespread impoverishment of tenants and exacerbated the problem of land morcellation, there is very little to suggest that it created much stratification, let alone kulakization. As we have seen, in the vast majority of cases, an average transaction involved about half an acre of land (and in many cases, much less than that) and rarely exceeded one acre'. Moreover, in most cases, an individual or household acquired land in one or two transactions, and there were very few cases of *per capita* land accumulation. But even such individuals as Adam Carpenter of Hindolveston, Henry Breton of Hinderclay, or John Joye of Bressingham, all buying land in multiple transactions, did not manage to acquire more than five or six acres overall. While these distress purchases undoubtedly made these individuals richer than their neighbours, it would be incorrect to label them as 'kulaks'. Even if some of these rose to socio-economic prominence, there were very few of these individuals. Although the land market has always been highly active in the context of the fast demographic growth in the thirteenth and the early fourteenth century, it was not nearly as active as during the famine, and it is clear that there was much less land available for sale in non-famine years. Throughout his career at Hindolveston, Adam Carpenter appears as a buyer in forty land transactions (twelve of which took place during the famine), but these forty transactions amounted to about only fourteen acres in total. Can, then, Adam Carpenter, who was clearly a rare bird, be called a 'kulak'?

One last question relates to the profitability of exchange of land for calories. Unfortunately, direct information on the market price of land-plots is very rare. After all, land transactions were business conducted between two tenants, and, thus, the prices paid by buyers to sellers were not recorded into

[128] Schofield, 'Social Economy', p. 50 n. 47.

[129] NRO, DCN 60/19/3. The land market activities of Adam Carpenter have been noted by Hudson in his 'Prior of Norwich's Manor', pp. 212–13 and subsequently analysed by Jordan, *The Great Famine*, p. 103.

[130] Jordan, *The Great Famine*, pp. 103–04.

court rolls. There are, however, references to acquisition of land-plots by lords, and there is no reason to suppose that these values were considerably different from those paid by tenants. Roughly speaking, an average acre of arable in East Anglia could be valued, *c.* 1300, anywhere between 15s and 60s.[131] Let us stick with the approximate average of 37s and bear in mind that an average size of a plot sold during the famine stood at half an acre. Therefore, selling a plot of land would have endowed a seller with, say, 18s 6d. If sold in the summer shortly before the harvest, during the so-called 'hungry gap' period, this sum could have been converted into about 0.85 quarters (about 6.8 bushels) of wheat (assuming the peak price of 22s per quarter) — an amount just sufficient to last one individual for one year, but certainly insufficient to serve a family of four or five hungry mouths for more than two months. Still, this figure would have been certainly above a composite crop yield per the same unit of land. As we have estimated in Chapter 3, around 1300, half an acre of sown land would have yielded, net of seed and tithe, about 3.9 bushels of composite crops, while during the famine years, the comparable figures would have been considerably lower (about 2.7 in 1315, 2.2 in 1316, and 3.2 in 1317).[132] The obvious disadvantage of such exchange was that unless bought back, the sold land-plot would have remained in the hands of a buyer, and the conversion of land into food calories would have been a one-off enterprise. In other words, a seller would be deprived of his/her sold plot's fruits for years to come. Therefore, while winning the desperately needed calories in the short-term, one would lose them in the long run. But then again, what other alternatives would starving and needy tenants have in those dire straits of famine?

'Fuzzy Entitlements': Credit Extension

An alternative coping strategy was to seek credit. The extension of credit was closely linked to land transfers. As Phillipp Schofield's work on late thirteenth-century Hinderclay has shown, land buyers often acted as creditors, while land sellers often featured as lenders.[133] In the context of the famine, the credit market helped villagers to achieve the same goals as the land market: to secure some cash or crops to be converted into food. It was not until recently that students of subsistence crises began paying attention to the role of credit mechanisms as

[131] Campbell, 'The Agrarian Problem', p. 52 n. 153.

[132] See above, Table 3.10.

[133] Schofield, 'Dearth, Debt'.

a mitigating factor in those crises. To a large degree, it was thanks to the work of Stephen Devereaux, who drew our attention to the existence of 'hidden', or to use his language, 'fuzzy entitlements', based upon household and community networks and revolving around common property rights, claims, and responsibilities.[134] Seeking credit on a local level as a coping strategy against environmental shocks and potential famine risks has been studied by Vanhaute and Lambrecht in their work on two subsistence crises in Flanders, in the 1740s and the 1840s, showing that, in both instances, subsistence crises did not unfold into famines thanks to efficient village community networks and mechanisms, credit being one of them.[135] Most recently, Tyler Goodspeed has convincingly shown that during the Great Potato Famine of 1845–1852, even worse-off regions with available credit suffered smaller population losses and augmented their animal stocks to a larger degree than better-off regions, where microfinance funds were largely unavailable.[136] Today, the informal credit market on a village level still plays a substantial role in reducing disaster risk in some parts of Africa, including northern Nigeria.[137] Likewise, in early fourteenth-century England (and indeed other parts of the British Isles), rural credit was conducted almost exclusively on a small-scale and a local level. It would be impossible to calculate interest rates, as these are entirely hidden from court rolls, because any explicit financial profit would put lenders at risk of being ostracized as usurers. Indeed, according to Catholic theology and the Canon Law, usury was a grave sin, an unnatural and immoral behaviour, punishable by the fires of Hell.[138] However, given that the Canon Law allowed some form of compensation to be paid for lenders (in effect, interest rates in disguise), especially in the context of a delay in payment, we may assume that *some* sort of profit must have been made by lenders.[139]

As Chris Briggs has shown in his comprehensive study of rural credit in late medieval England, the credit market was omnipresent and practised by virtually any tenant who could afford extending loans.[140] Likewise, Phillipp Schofield demonstrated various channels of access to credit available to pre-Black Death

[134] Devereux, 'Sen's Entitlement Approach', pp. 66–89.

[135] Vanhaute and Lambrecht, 'Famine, Exchange Networks and the Village Community'.

[136] Goodspeed, *Famine and Finance*.

[137] Udry, 'Credit Markets in Northern Nigeria'.

[138] Munro, 'The Medieval Origins of the Financial Revolution'.

[139] Briggs, *Credit and Village Society*, pp. 76–77.

[140] Briggs, *Credit and Village Society*.

peasants.[141] At the same time, both authors argued that there were periods of short-term credit crunch in conjunction with bad harvest years. Briggs also noted, on the basis of the analysis of five manors, that although the agrarian crisis seems to have brought about a decline in the availability of credit, the situation may have varied from place to place.[142] Briggs' and Schofield's findings and conclusions can be supplemented and tested by additional information from both court rolls and the so-called *Condempnaciones Register* of Dean and Chapter of York Cathedral. The latter is a register of debt recognizances and other business in the Court of Audience of York Cathedral, covering the years of 1316–1321 and 1326–1330 and recording some 484 debt recognizances for the period between 12 January 1316 and 28 September 1317 alone.[143] When it comes to court rolls, the relevant entries contain debt plaints between lenders (plaintiffs) and borrowers (defendants) related to non-payments, as entered into court rolls. Unfortunately, the vast majority of debt plaint entries do not contain any information beyond the names of parties, and this fact creates a series of challenges. First, we are largely left to speculate about the time lag between the date of transaction and the first legal action. Second, there are very few instances when the entries provide detailed information about the size of debts. Both problems can be solved by the *Condempnaciones Register*, whose entries specify, in many instances, the dates of recognizance and payment, as well as the size of debts. For the famine period, there are 228 recognizance entries recording the duration of a loan term, rendering the median figure of 107 days.[144] To this we may add twenty-one recognizance entries from four East Anglian courts, rendering the median figure of eighty-nine days.[145] Taken together, the *Condempnaciones Register* and the court rolls render a median figure of 106 days. This figure is higher than the median figure of sixty-one days reckoned by Brigg, on the basis of thirteen cases from five courts, but because of the thick coverage of the *Condempnaciones Register,* I prefer to stick to the fig-

[141] Schofield, 'Access to Credit'.

[142] Schofield, 'Dearth, Debt', pp. 9–17; Briggs, *Credit and Village Society*, pp. 179–93.

[143] York Minster Library, M2/la. I am grateful to Prof. Robert Swanson of the University of Birmingham, who has drawn my attention to this unique source and kindly shared its digital photographs.

[144] In the case of repayment in instalments, the latest date is used.

[145] The manors in question are: Langham (Norfolk), Sedgeford (Norfolk), South Elmham (Suffolk), and Ingoldmells (Lincolnshire).

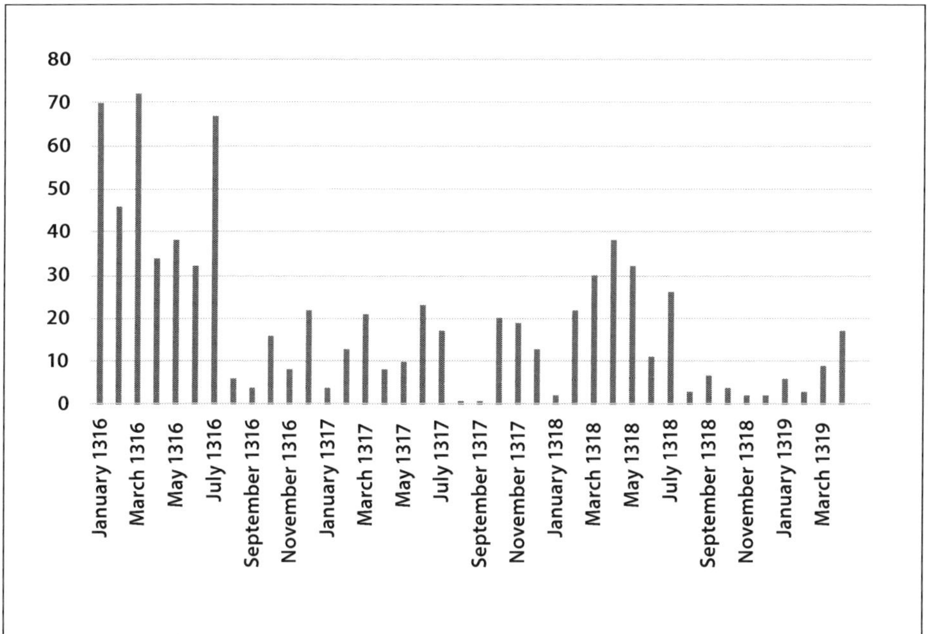

Figure 8.4. Total number of new debts registered each month into the *Condempnaciones Register* of York Dean and Chapter, 1316–1319. *Source*: York Minster Library, M2/la, fols 1^{r}–27^{v}.

ure of 106 days for the purpose of the present exercise.[146] As Briggs has shown, the duration between the due date and plaint initiation was in the area of 269 days. Adding this figure to 106 days of loan terms, we arrive at the estimate that the length between the date of initial transaction and the first action was approximately 375 days (or, twelve and a half months). If this estimate is not far from reality, then we may assume that if a new plaint was initiated in a court from December 1315, the actual loan had been extended sometime around November 1314. Therefore, in search for credit extended during the famine years, we need to examine court rolls from 1316–1317 and 1317–1318 rather than from 1315–1316 and 1316–1317. Obviously, this is a rough and educated guess, given that some larger-scale loans were to be repaid in several instalments, often covering a few years.[147] Still, such estimate is not to be dismissed.

According to Figure 8.4, the period between January and July 1316 saw the peak in the sheer number of new debts registered into the *Condempnaciones*

[146] Briggs, *Credit and Village Society*, p. 71.

[147] Schofield, 'Dealing in Crisis', pp. 267–69.

Register. The volume of new debts fell considerably afterwards, and, although fluctuating from month to month, it never reached that level again. This suggests that the credit market was hectic during the first year of the famine, but it was exhausted in the course of the second year and in the aftermath of the famine, before gaining ground again in the first half of 1318, only to shrink again considerably in August 1318. It is unfortunate that the register starts in January 1316, so there is no indication of the situation in the first few months of the famine — or, indeed, of the few years preceding it. But it is clear that the unusually high volume of new debts recorded between January and July 1316 indicates that, because of the dire circumstances of the crisis, there was an abnormally high demand for credit and there were enough people able and willing to extend credit, notwithstanding the same crisis. This, in turn, suggests that Briggs' pessimistic assessment of the 'credit crunch' during the famine years may hold true only as far as the second year of the crisis is concerned. Why, then, did the volume of new debts fell so drastically after July 1316? There are several possibilities. One possibility is that, as Schofield has suggested, the new (and cheap) opportunities of the land market prompted creditors to divert their capital from the credit sector and invest in land.[148] The second possibility is that the second back-to-back harvest failure of 1316 may have strongly discouraged potential creditors from extending loans, both because of the impoverishment and potential default of their debtors and because of their increasing need to hold on to their funds to make their own ends meet. Another possibility is that very few borrowers could afford to borrow, given all the stress and risks deriving from the crisis and impoverishment. The figures of 1316–1317 indicate that credit was still being extended, but its low volume reflects that there were few people who were either willing to lend or able to borrow. As Schofield has most recently argued, if credit was still being extended during the famine, then it must have been reserved to the 'particular and privileged groups', reflecting the 'particularized trust' discussed in Chapter 4, in conjunction with market and trade.[149]

The contrast between the first and the second years of the famine is reflected not only in the sheer volume of new debts but also in the length of loan terms. The median length of a loan term was in decline between January and August 1316, from 199 to 52 days. Between September 1316 and July 1317, the figures varied from month to month, but standing, in most instances, well above 100. Is it a mere coincidence? One possibility is that local creditors were will-

[148] Schofield, 'Dearth, Debt', pp. 9–17.

[149] Schofield, 'Dealing in Crisis', p. 268.

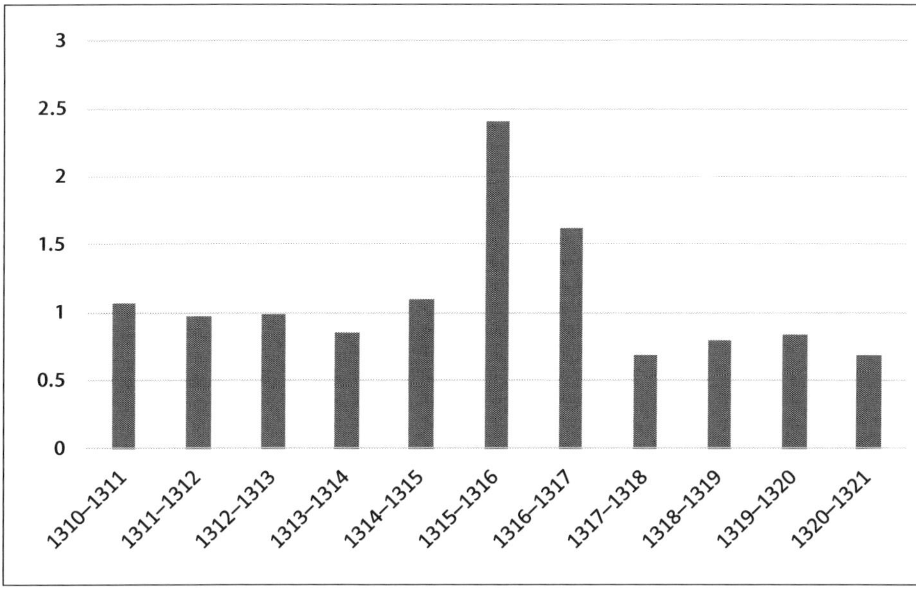

Figure 8.5. New debt plaints on eleven manors, indexed on 1310/1–1314/5. *Source*: Briggs, *Credit and Village Society*, pp. 181–86 (Oakington, Great Horwood and Balsham); Phillipp Schofield, 'Social Economy of the Medieval Village', p. 48 (Hinderclay); NRO, NRS 19505 (Horsham St Faith); NRO, Lest DA/6–8 (Heacham); NRO, DCN 60/19/3 (Hindolveston); NRO, DCN 60/32/9 (Sedgeford); RHM, Langham Court Rolls (Langham); SuffRO (Lowestoft Branch), HA12/C2/3 (South Elmham); and BL Add Ch 63422–32 (Rickinghall).

Note: The figures indexed on non-famine years (1310/1–14/5 = 1.0), appearing on the vertical axis, represent the total numbers of new debt plains registered each year on all eleven manors: 1.0 = 89 new debt plains a year

ing to help their friends and neighbours, but because of the challenging circumstances, they were willing to extend credit for shorter periods of time than usual. In other words, creditors needed their money back sooner rather than later because of very real hazards related to the omnipresent food dearth and the fast rise in prices. But apart from these hazards, creditors had a good reason to be concerned about debtors' inability or unwillingness to repay their debts on time. This fact prompted many creditors to seek repayment, as is indicated in a sudden surge in the number of new debt plaints in 1315–1316 and 1316–1317 (Figure 8.5), as well as in new recognizances. As the figure indicates, the volume of new debt plaints rose to 2.4- and 1.6-fold their pre-famine levels in 1315–1316 and 1316–1317, respectively. This rise reflects the fact that the debtors were not only extending new debts but were also keen to take legal action in order to collect old and existing debts. As noted above, in non-famine years, the average duration between the due date and plaint initiation was in the area of 375 days. Very few cases from the famine years indicate the duration between the two dates, but those that do show that creditors were not, in many instances, willing to wait for such a long time and appealed to the court to implement recoveries of debt much faster. For instance, at Heacham (Norfolk), the date of first litigation between Geoffrey Gosse and Helen Porteman, of a debt of 12s, was 19 January 1316. On 23 March 1316, Gosse took further action by complaining to the court that the recovery has not yet been implemented. The parties opted for concordance, whereby Porteman pledged to repay the debt no later than by Michaelmas the same year.[150] Similarly, at Langham (Norfolk), on 28 May 1316, Augustine de Elmham initiated a plaint against his debtor, Adam de Grymesby, in order to recover the debt of 6 bushels of barley. Less than two months later (20 July 1316), he sought the court's assistance to implement the recovery of the debt.[151] More examples can be added; indeed, the famine years' court rolls contain an unusually high number of complaints regarding the failure to implement debt recovery. Unfortunately, these cases do not show the due date for the repayment of the original debt, but, given the unusually short time lag between the initial transaction and the first litigation, one may conclude that creditors could not wait as long as they would in normal years. It appears, therefore, that the rise in debt cases in 1315–1316 and 1316–1317 reflects the rush to collect older and existing debts rather than only the availability of new credit. After the famine, and especially in 1317–1318, there seems to have been a slump in debt litigation. This reflects the credit crunch of the second year of the famine, when fewer and fewer creditors were willing or able to extend credit because of the gloomy economic reality, on the one hand, and because of the non-repayment of earlier debts, on the other.

Very few manorial court entries (fifteen recognizances in total, all related to the East Anglian manors of Langham and South Elmham) specify the nature and scale of loans, but this shortcoming can be compensated by the abundance of information in the York *Condempnaciones Register*. Out of 484 loans, 153 (32 per cent) were expressed in crops (that is, A owed x bushels of grain to B). This figure is not very different from the proportion of crop-related loans recorded in the court rolls: here, six loans out of fifteen (40 per cent) were related to crops. The proportion of crop-related loans recorded into court rolls

[150] NRO, Lest DA/7.

[151] RHM, Langham Court Rolls, Roll 1316–1317.

in non-famine years was considerably smaller, standing at just 11 per cent (five loans out of forty-two). The composition of crop-related debts varied across regions: while, in East Anglia, the vast majority of recorded crop loans consisted of barley, the composition of crop loans in Yorkshire was more even (about 20 per cent wheat, 40 per cent barley, and 34 per cent oats, with more minor crops occupying the rest). This reflects the differences in agricultural geography between the two regions: while barley was a local speciality of East Anglian manors, this crop played a much less significant role elsewhere. In the case of East Anglia, it also reflects the fact that during the famine, a lot of barley had been diverted from brewing into baking, as we have seen earlier.[152] The size of crop-related loans also varied across regions. Thus, the size of an average barley loan was 1.25 quarters (worth 15s 3d and 11s 10d in 1315–1316 and 1316–1317, respectively) on the two East Anglian manors, while an average crop loan in Yorkshire stood at 2.4 quarters (worth 24s 10d and 23s 2d, respectively). The median cash loan stood at 20s at Langham and South Elmham, while in Yorkshire, the median figure was 10s 3d. This agrees with Phillipp Schofield's view that the pre-Black Death credit market was dominated by larger debts, those of 5s and above.[153]

Did grain-related loans mean repayment in the same commodity? In dealing with the issue, Chris Briggs suggested that loans in kind may have been advance sales rather than loans, but most debts in kind still involved money at some point.[154] In other words, Party A would have paid Party B a certain amount of money, while Party B would have committed to deliver a certain amount of crops or other goods in exchange for that amount, on a date agreed between the two. As far as our evidence can tell, this may have been the case with most grain loans. Out of 484 loans entered into the York *Condempnaciones Register* during the famine, 66 (or, 14 per cent) can be positively identified as advance sales — either on the basis of price specification or by the formula *ex causa emptionis et venditionis*.[155] In some cases, some entries indicate that the creditor had already paid his/her debtor for a certain commodity. For example, on 14 May 1316, Michael de Northampton extended an unspecified amount of

[152] Campbell, 'Matching Supply to Demand', p. 833.

[153] Schofield, 'Dealing in Crisis', p. 258.

[154] Briggs, 'Money and Rural Credit'.

[155] This formula is discussed in the so-called Glanvil (aka, the *Tractatus de legibus et consuetudinibus regni Anglie*), Bk 10, Ch. 14. As of May 2017, the best printed edition of 'Glanvil' remains *De legibus et consuetudinibus regni Angliae*, ed. by Woodbine.

money to Robert de Fudaythorp for barley, 'which he purchased from him' (*quod ab eodem emit*).[156] In some instances, creditors were very clear that they expected to receive crops of good quality, described by the formula 'good, clean and marketable' (*bonum, mundum et pacabile*).[157] This was especially crucial in the context of the torrential rain and poor storage, which, as we have seen above, rendered some harvested grains corrupt, inedible, and un-marketable.[158]

Unfortunately, manorial court entries are not as specific as the *Condempnaciones Register*, and here we can determine the nature of loans on a case-by-case basis. In some instances, it was clear that crop-related loans were deferred payments. Thus, on 12 July 1316, Simon de Dallinge, a minor Norfolk lord, claimed that he was yet to be repaid £6 for an unspecified amount of grain that he had sold to three tenants from Hindolveston.[159] The facts that the first transaction leg consisted of grain deliverance, that the same grain was delivered by a lord to tenants (and not the other way around), and that only the monetary value of the debt was specified all point to the loan being extended in grain. Other entries are, however, more ambiguous, and it is often difficult to establish whether these were advance sales or actual loans in kind that were to be repaid with the same type of commodity. Thus, on 23 September 1316, Walter Man of Horsham St Faith (Norfolk) acknowledged that he owed to William Schade 1 quarter of barley to be repaid in three instalments (4 bushels on 6 October, 2 bushels on 31 October, and a further 2 bushels on 24 December).[160] While it is possible that Walter Man acknowledged that he failed to supply William Schade with barley, as he had promised to in return for an advance payment, it is equally possible that Schade had extended a loan in barley to Man, while Man failed to repay Schade with the same commodity, due to the dire circumstances of the famine. The latter interpretation is strengthened by the fact that the *Schade* v. *Man* entry does not indicate any agreed price or financial equivalent of the 4 bushels of barley. As Chris Briggs has shown, although moneyless credit transactions were extremely rare, they nevertheless were not unheard of.[161]

[156] York Minster Library, M2/la, fol. 7^{r}.

[157] York Minster Library, M2/la, fol. 8^{v}.

[158] See above, Chapter 4, p. 148 and Chapter 5, pp. 186–89.

[159] NRO, DCN 60/19/3.

[160] NRO, NRS 19505.

[161] Briggs, 'Money and Rural Credit', pp. 134–40.

In other words, it appears that grain loans could, at least in theory, entail two possible scenarios: (1) advance purchases, whereby a creditor extended a borrower advance *cash*, in return for his promise to deliver *grain* on a certain date; or (2) what Briggs refers to as 'exigency credit', whereby a creditor extended a borrower advance *grain*, in return for his promise to deliver *the same kind and quantity of grain* on a certain date.[162] In either scenario, the dire circumstances of the famine made repayment extremely difficult, if not impossible, because of the declining purchasing power of money, the chronic scarcity of grain, and the aggressive price rise. If, say, in early September 1315, a creditor agreed to extend a borrower 8s for 1 quarter of wheat to be delivered around Christmas, the former would be at loss if he managed to secure the same amount of grain, whose financial equivalent would have risen to around 12s, that is, about 50 per cent higher than the agreed sum, by the repayment date. Also in the case of grain-for-grain loans, debtors would be at considerable loss because of the rising financial value of grains. That repaying grain loans was challenging in the context of high prices is reflected in a rise in a relative proportion of cases related to grain debts compared to non-famine years. This, again, does not imply that grain credit was flourishing — especially during the second year of the famine. Rather, it indicates the concern of creditors to secure the repayment of debts, on the one hand, and the inability or unwillingness of debtors to repay the same debts, when the economic reality of the famine years was too adverse, to the point of being life-threatening, on the other.

Who were the creditors and debtors? Obviously, the majority of people involved in credit transactions (about 91 per cent), both as lenders and borrowers, were male tenants. Despite their insignificant proportion, women and their role in the late medieval rural credit market should not be dismissed. Judging by their name designation, most female creditors were widows. Although most female creditors appear only in one or two transactions recording modest loans, there were some examples of better-off women extending rather generous credit. Take the example of Agnes atte Hallogate of Langham (Norfolk). This woman was an active creditor in the 1310s. During the famine she extended several loans, including an advance loan of 4s for 1 stone of wool and another advance loan of 4.25 quarters of barley.[163] Alice de Morston, of the same manor, lent 10s to William ate Cros on 20 July 1316.[164] The name of Agnes le Shipman (of an

[162] Briggs, *Credit and Village Society*, pp. 47–50.

[163] RHM, Langham Court Rolls, Rolls 1315–1316 and 1316–1317.

[164] RHM, Langham Court Rolls, Roll 1316–1317.

unspecified parish) appears thrice in the *Condempnaciones Register* during the famine years; the total sum of her loans amounted to 7s 11d and 3.5 quarters of grain that she either paid in advance or lent.[165] In some instances, married women appear in debt plaints together with their husbands, suggesting that they were unable to operate independently of the latter. At the same time, there is evidence that some married women were involved in credit extension, independently of their husbands. For instance, at Heacham, on 29 January 1317, Isolde, wife of Thomas Torpere, took legal action against Edmund Shepherd (*bercarius*).[166] Isolde's husband, Thomas Torpere, appears as a creditor in several plaints in the 1310s, but neither of these included the name of Edmund Shepherd. He was certainly alive by the time Isolde was extending credit to Shepherd. This fact indicates that Isolde and Thomas may have acted independently. If this interpretation is correct, then it appears that the situation at Heacham may have been somewhat different from other East Anglian manors studied by Briggs, where married women seem to not have had freedom to act independently of their husbands.[167]

Another group included clergymen, whose active involvement in the credit market has been highlighted by Pamela Nightingale.[168] The proportion and composition of clerical creditors varied from place to place. Thus, they represented 16 per cent of all creditors and 8 per cent of all debtors entered into the York *Condempnaciones Register*. At Heacham (Norfolk), they accounted for only 10 out of the total 352 debt cases during the famine period. At Horsham St Faith (Norfolk), not a single clerk was involved in credit transactions. In the vast majority of cases (89 per cent of all transactions recorded in the *Condemptiones Register*), clergymen appear in credit transactions involving laymen rather than their fellow churchmen. Of all men designated with a religious title, 'clerks', with no further designation, accounted for about one third. Chaplains represented about 15 per cent of all clergy; rural vicars and rectors made, together, a further 10 per cent; masters represented about 23 per cent; and the remainder was made of York Minster obedientiaries and vicars. That clergymen were more prominent as creditors than debtors is hardly surprising: a good proportion of them (especially vicars, rectors, and masters, let alone monastic officials) were better-off individuals, with fair means. Some of them

165 York Minster Library, M2/la, fols 3^v and 10^v.

166 NRO, Lest DA/7.

167 Briggs, 'Empowered or Marginalized?', and *Credit and Village Society*, pp. 112–16.

168 Nightingale, 'The English Parochial Clergy'.

were of lordly status, as their designations 'domini' suggest. Vicars and rectors were often non-residents, and this fact did not allow them to be involved in direct business interaction with local parishioners.[169] Instead, these churchmen of higher status were usually associated with large-size loans, extended to better-off debtors, presumably substantial tenants and corn-mongers. Thus, Master Roger de Heselarton features as a creditor on four occasions between January 1316 and September 1317, extending credit for the total sum of some 70s.[170] On 8 October 1316, Master John de Walkefeld lent £7 6s 8d to a certain Philip son of Thomas Rosse de Staynsye.[171] Conversely, chaplains were often hired by beneficed clergy to perform their pastoral and liturgical duties and reside among their parishioners. These chaplains were assisted by local clerks. Because of their direct involvement in local communities, chaplains were also directly involved in business dealings with their parishioners. Unsurprisingly, the loans extended by chaplains to local parishioners were of a smaller size, reflecting the poverty of parishioners, as well as the modest economic status of chaplains. Because of that, chaplains were also more likely to appear as debtors than any other type of clergymen. In some instances, the same parish clergymen acted both as creditors and debtors. Thus, at Heacham (Norfolk), Robert son of John de Frausham, chaplain, appears as both as a creditor and a debtor of two parishioners, Geoffrey Pemond and Robert Felawe, in several plaints between April and August 1316.[172] Similarly, in the same vill, James de Ryngested, chaplain, developed a similar 'horizontal' reciprocal creditor-debtor relationship with William Canon, during or shortly before the famine.[173]

The cases of Robert son of John de Frausham and James de Ryngested are interesting ones, as they are related to a wider debate among scholars of late medieval credit. This 'reciprocal' (or 'horizontal') kind of indebtedness between two parties of roughly the same status and within the same community was first noted by Elaine Clark in her work on credit in fifteenth-century Writtle (Essex). Clark calculated that about 14 per cent of all debts were of a reciprocal character.[174] Conversely, Phillipp Schofield and (initially) Chris Briggs have argued for a pronouncedly 'vertical' character of creditor-debtor ties, marked

[169] Briggs, *Credit and Village Society*, p. 119.

[170] York Minster Library, M2/la, fols 8^{r}, 10^{r}, 15^{v}.

[171] York Minster Library, M2/la, fol. 11^{r}.

[172] NRO, Lest DA/7 (24 March 1316, 15 June 1316 and 24 August 1316).

[173] NRO, Lest DA/7 (21 December 1315).

[174] Clark, 'Medieval Debt Litigation', and 'Debt Litigation in a Late Medieval English Vill'.

by credit accumulation rather than mutually advantageous reciprocal indebtedness.[175] The famine-year court rolls reveal a contradictory and more complex picture. On the one hand, at Horsham St Faith, no two parties formed any reciprocal indebtedness relations during the famine years. In fact, during the entire period of 1311–1321, there was only one such instance.[176] An altogether different situation prevailed at Heacham, where reciprocal indebtedness was an omnipresent reality: during the famine period, 49 out of 160 cases (31 per cent) involved this kind of credit relationship, while during the non-famine years in the period 1311–1321, its proportion was higher still (94 out of 212, or 44 per cent). In other words, the situation varied from place to place, depending much on a local context. Why some places practised the 'reciprocal' indebtedness and others did not yet awaits thorough investigation.

Just as with land transfers, the majority of credit transactions were conducted between non-related parties. At Heacham, only 6 credit plaints (out of 160) are proven to have taken place between relatives, during the famine period. One case is particularly curious. On 24 April 1316, Alan Attebrigge Jr appeared in court as a plaintiff against his mother Matilda, defendant, over a debt case.[177] It is a shame there is no indication of the nature of the plaint, the nature of the loan/debt, and its circumstances. Was Alan in a desperate need to secure the repayment to avoid hunger? Were his economic circumstances so bad that he was willing to take legal action against his own mother and, thus, potentially expose her to the same hunger? Did the famine-related stress create tension between son and mother? Or, was it all about Matilda withholding Alan's legacy, as an executrix of his father's will?[178] The silence of the court rolls does not allow any further conclusion to be made.

The fact that the clergy constituted a non-negligible proportion of creditors leaves us with an ambiguous impression. On the one hand, it is possible that at least *some* clerks may have regarded credit extension as a charitable act, in the context of the famine. Indeed, although the economic and impersonal element did dominate credit and the credit market, it is essential to consider the other side of the coin: namely, the personal and non-materialistic motivations of creditors. The latter, whether clergy or laymen, thought not only of their profit

[175] Schofield, 'Dearth, Debt'; Briggs, 'Creditors and Debtors'.

[176] *Agnes Cruce* v. *Stenger Hakon with his wife Matilda*, at some point in late 1313 or early 1314 (NRO, NRS 19505, 10 October 1314).

[177] NRO, Lest DA/7 (24 March 1316).

[178] I am thankful to Dr Chris Briggs for this suggestion.

and interest but also about helping their parishioners, kinsmen, neighbours, and friends. The theological basis for this notion is found in several passages of the Old and New Testaments (for instance, Exodus 22:25–27, Matthew 10:8, and Luke 6:35), where the faithful are urged to lend to the needy without expecting anything in return. With no organized money-lending charities, such as the Italian *monte di pietà* and its derivatives, and before the appearance of community loan chests after the Black Death and the introduction of parish charitable endowments in the sixteenth century, loans of money and food in an hour of need could only be sought on an individual, interpersonal, and local level.[179] The issue of credit as a charitable act should be linked to the wider topic of poor relief in the context of famine. On the other hand, however, there is no evidence to support the idea that the majority of clerks were indeed driven by purely altruistic and man-loving motives rather than by the same impersonal and profit-seeking thoughts. Indeed, as we have seen in Chapter 3, during the famine, local tithe owners conducted business as usual, selling the vast majority of tithe receipts for market price rather than handing out or selling them for discounted prices to their parishioners. This impression is strengthened further by the fact that, as we shall see in the next section, charitable institutions did relatively little to save the situation and support the starving masses.

Famine Relief: Religious Institutions

Although land transfer and credit extension seem to have played a non-negligible role in helping some individuals to make ends meet in the very difficult circumstances of the famine, these mechanisms of help were largely inaccessible to marginal elements of the society, made up of poor tenants, landless rustics, urban paupers, vagabonds, the sick, and the imprisoned. It is here that food offerings and alms from local religious institutions came to the rescue of a small proportion of these individuals, who were left outside of various networks of the village community. These religious institutions acted in line with the theological precept of the Six Corporal Works of Mercy, which included: feeding the hungry and giving drink to the thirsty, welcoming the stranger, clothing the naked, visiting the sick, visiting the prisoner, and burying the dead.[180] The information about food and alms distribution by these charitable bodies is rather patchy, and it comes from annual rolls of either monastic officials (obedien-

[179] Dyer, 'Poverty and its Relief', p. 57; James, 'Charity Endowments'.

[180] Slavin, 'Feeding the Brethren', pp. 347–50.

tiaries), primarily almoners, or accounts of hospitals. The resources for distribution of cash alms and food could be obtained from different sources. In the case of monasteries (and in particular larger conventual houses), these would derive primarily from grain provisioned by their manors or purchased at local markets.[181] In the case of hospitals and alm-houses, these would be obtained normally by gathering of donations by locally appointed proctors. Often, donors would receive indulgences for a few weeks in exchange for their donations.[182]

There were two types of charitable activities associated with conventual houses. First, there was a weekly or daily distribution of food and alms either at monastic gates or local sub-kitchens, situated within monastic precincts. Second, there were festive charitable meals, usually running on anniversaries of local bishops or on major religious feasts. For instance, at Norwich Cathedral Priory, local almoners' accounts distinguish between wheat distributed among Norwich paupers on the anniversary of Bishop Herbert de Losinga, the founder of the cathedral (23 July), and other grains and legumes (rye, peas, and barley) distributed throughout the year.[183] Similarly, at Canterbury Cathedral Priory, the almoner was responsible to distribute wheat and rye among local needy on Archbishop Lanfranc's feast (28 May), in addition to offering grains, legumes, and alms throughout the year.[184] Similar charity meals were organized at Battle Abbey, Evesham Abbey, Durham Cathedral Priory, and many other houses.[185] As far as almshouses and hospitals are concerned, these were more 'inclusive' and closed institutions, catering for their residents, consisting of vaguely defined 'poor', mostly of old and physically disabled people (lame, crooked, blind, and bedridden).[186] In some instances, like the Great Hospital of Norwich and St John's Hospital, Winchester, local residents were meant to be fed on a daily basis. In some other instances, like Clyst Gabriel Hospital (outside Exeter), residents would receive weekly allowances and had to make their own food provision arrangements. The famine situation undoubtedly placed these charitable institutions in a difficult position. On the one hand, they had, more than ever, the moral and religious obligation to feed the poor, in

[181] Slavin, *Bread and Ale*, pp. 1–83.

[182] McIntosh, *Poor Relief in England, 1350–1600*, pp. 52–58.

[183] Slavin, 'Feeding the Brethren', p. 360.

[184] CCA, DCc/Almoner 1–26.

[185] *The Cellarers' Rolls of Battle Abbey*, ed. by Searle, p. 44; Dugdale, *Monasticon Anglicanum*, II, 29; *Extracts from the Account Rolls of the Abbey of Durham, ed. by* Fowler, pp. 484 and 531.

[186] McIntosh, *Poor Relief*, pp. 71–78.

Table 8.1. Disposal of Crops by Canterbury Cathedral Almoner, 1298/9–1324/5 (in quarters)

	Receipt					*Disbursement*			
Year	**Carry-overs from the previous year**	**Demesnes**	**Bought**	**Other**	**TOTAL**	**To the poor**	**Sold**	**Other**	**Carry-overs to the next year**
1298–1299	0.0	324.9	6.3	15.6	**346.8**	234.4	19.8	60.6	32.0
25.12.1299–29.9.1300	17.0	411.1	6.5	14.8	**449.4**	192.5	10.1	144.8	102.0
1316–1317	74.5	344.5	97.8	36.1	**552.9**	373.6	28.6	134.6	16.3
1317–1318	16.5	265.6	91.3	0.0	**373.4**	270.8	12.0	80.6	10.0
1318–1319	32.4	274.3	41.9	35.0	**383.5**	208.1	69.3	105.1	1.0
1324–1325	0.0	257.5	25.8	16.3	**299.5**	231.9	9.0	58.6	0.0

Source: CCA, DCc/Almoner 17, 18, 21, 20, 5, 23, 4, 6, and 26.

a desperate attempt to save at least some souls from the Third Horseman of the Apocalypse. But on the other hand, the abnormally high crop prices and the rising number of hungry mouths would require these institutions to increase their annual expenditure on food alms (in both relative and absolute terms) and thus potentially experience financial difficulties. Given the quasi-corporative nature of major conventual institutions, such as Canterbury Cathedral Priory or Westminster Abbey, any major diversion of annual funds towards feeding the poor would be at odds with their economic agenda. As available evidence indicates, these charitable institutions, both substantial and minor, attempted to solve this dilemma by a pragmatic compromise: they did increase the proportions of food expenses, but these increases were far from being sufficient to feed a great many starving people. Therefore, the efforts to reach out for and feed the starving were barely more than symbolic. Obviously, the levels of increase in food distribution varied from house to house, depending on the economic and financial abilities of these institutions and the impact of the crisis on the same institutions.

Unfortunately, the surviving data on charitable activities of religious institutions during the famine years are very patchy. In fact, the only set of almoners' accounts covering the famine period survives from Canterbury Cathedral Priory, and even these cover only the second year of the famine. Nevertheless,

Table 8.2. Composition of crops distributed among the poor by the Canterbury Cathedral almoner (in quarters), their approximate calorific equivalent, and numbers of people they were capable of feeding on a daily basis, 1298/9–1324/5.

Year	**Wheat**	**Rye**	**Legumes**	**Barley**	**TOTAL**	**Approximate calorific equivalent**	**Paupers per diem, if 1500 kcal per capita**	**Paupers per diem, if 1000 kcal per capita**
1298–1299	3.6	18.0	66.0	146.9	234.4	81,471,581	149	224
25.12.1299–29.9.1300	4.0	0.0	38.5	150.0	192.5	70,802,686	130	194
1316–1317	7.5	0.0	23.3	342.8	373.6	153,79,380	280	421
1317–1318	0.0	0.0	75.5	195.3	270.8	92,523,322	169	254
1318–1319	8.0	2.5	61.1	136.5	208.1	71,049,511	130	195
1324–1325	8.5	4.0	82.8	136.6	231.9	74,699,928	137	205

Source: CCA, DCc/Almoner 17, 18, 21, 20, 5, 23, 4, 6, and 26.

Notes: Since both barley and legumes were baked into bread, I have assumed their extraction rates to be 80 per cent (that is, 20 per cent calorific loss).

the Canterbury accounts provide an illuminating insight into charitable activities of this substantial conventual house. As Table 8.1 shows, the overall amounts of crops procured by the Priory almoner in 1316–1317 were well above average, amounting to 553 quarters, out of which 374 quarters (68 per cent) were distributed among the needy. In absolute terms, these figures were about 50 per cent higher than in non-famine years, when the annual quantities of crops reaching the almoner's granary stood at about 371 quarters, of which some 228 quarters (61 per cent) were allocated to the paupers. In other words, in 1316–1317, the volume of food served among Canterbury poor rose in both absolute and relative terms. To achieve that, the almoner had to increase the share of annual crop purchases and spend £41 12s 3d on 89.5 quarters of barley in 1316–1317. These may first come across as generous figures, both in terms of quantities of barley and amounts of money spent on it, all the more so, given the high prices of the famine years. However, as we shall see below, these figures were a drop in the ocean of need.

What was the composition of food alms distributed at the almoner's food kitchen in Canterbury? Barley was the single most prominent crop allocated to paupers' meals, and it accounted for 92 per cent of the total in 1316–1317 and for 67 per cent of the total in non-famine years, by volume (Table 8.2). Both during the famine and non-famine years, barley was mixed with legumes and baked into low-quality bread rather than brewed into ale.[187] This policy was certainly efficient, and particularly in the context of famine, when every calorie mattered and when the choice of baking over brewing could potentially save lives of starving masses. Other crops were distributed in much smaller quantities. Converting the sheer quantities of crops into their approximate calorific equivalent, we may estimate the numbers of paupers they were capable of feeding on a daily basis. If we were to assume that each day an average pauper was to receive the equivalent of 1500 kcal, then we may speculate that these quantities were capable of feeding some 280 paupers on a daily basis in 1316–1317. If, however, the daily meals were less generous, amounting to 1000 kcal, then the almoner's food kitchen could have fed as many as 421 paupers each day. In each scenario, it appears that during the famine, the almoner and his assistants were catering for twice as many people as in non-famine years. Yet, if these estimates are not too removed from reality, we may conclude that Canterbury Cathedral Priory, acting not only as a business organization but also as a charitable body, could provide food for only a very small proportion of the needy. It is unknown what proportion of the Canterbury population was on the verge of survival during the famine; but, using the 15–20 per cent mortality rates estimated in the previous chapter, we may speculate that the figures may have been in the area of 1350 to1800 people (with the total population of Canterbury being around 9000 on the eve of the famine). In other words, only about a quarter of the potentially needy people could be fed through the generosity of the Cathedral authorities.[188]

Who were these needy individuals that the almoner's food kitchen so empathetically catered for? Fortunately, there are several entries in the Canterbury almoners' accounts that indicate the nature and background of these people. These consisted of four main groups: (1) friars (Franciscans, Dominicans, and Carmelites), (2) widows, (3) infirm (*in lecto iacentes*, namely bedridden), and (4) paupers. The Canterbury evidence can be supplemented by granary accounts from Norwich Cathedral Priory (included in the Master of the Cellar

[187] This is indicated in CCA, DCc/Almoner 17, 18, 20, 5, 23, and 26.

[188] Campbell, 'Benchmarking Medieval Economic Development', p. 908.

Table 8.3. Allocation of wheat for *panis militum* and its distribution among the poor at Norwich Cathedral Priory, 1310/1–1320/1.

	In quarters					***In Loaves***		
Year	Total receipt by bakery	Baked into *panis militum*	To anchorites	To prisoners in Norwich Castle	To the poor on the Sunday mass	Approximate number of loaves for anchorites	Approximate number of loaves for prisoners	Approximate number of loaves distributed on the Sunday mass
1310–1311	948.4	403.1	2.5	8.0	10.0	536	1,716	2,145
1313–1314	1,054.1	551.8	2.5	26.0	8.5	536	5,577	1,823
1314–1315	1,045.9	530.5	2.5	21.3	6.0	536	4,558	1,287
1315–1316	1,132.8	603.5	2.5	33.3	7.0	536	7,132	1,502
1318–1319	911.3	417.8	0.0	18.8	8.0	0	4,022	1,716
1319–1320	1,057.0	570.0	0.0	23.5	10.0	0	5,041	2,145
1320–1321	1,132.9	616.0	0.0	26.0	10.0	0	5,577	2,145

Source: NRO, DCN 1/1/22–28.

accounts), studied by me elsewhere.[189] Each year, the local bakery would bake three kinds of bread of various degrees of fineness; about one half of all bakery wheat receipt was converted into the most inferior type of bread, called *panis militum* (knights', or servants', bread). The *panis militum* loaves were distributed among the workers of the Priory and among three groups of needy: Norwich anchorites, prisoners incarcerated at Norwich Castle prison, and local paupers. The paupers would be fed with *panis militum* on Sunday masses. As Table 8.3 demonstrates, just as in the case of Canterbury Cathedral Priory, the amounts of grain allocated for baking to the Norwich poor did not decline during the first year of the famine but, on the contrary, rose well above average in 1315–1316. Collectively, 42.8 quarters of wheat were converted into bread for the poor, in contrast with 30.6 quarters in non-famine years of the same decade. In particular, the number of loaves baked for Norwich Castle prisoners was

[189] Slavin, *Bread and Ale*, pp. 173–87.

pronounced, rising to about 7132 (in contrast with the annual average of 4415 loaves in non-famine years). This indicates that the Priory authorities were aware of the unusually high number of prisoners starving in the castle and made an effort to cater for as many of them as possible. Still, the quantities of wheat allocated for prisoners' bread did not rise nearly as pronouncedly as the number of felons delivered to Norwich Castle prison during that year. According to the 1315–1316 gaol delivery rolls for Norfolk, 139 new prisoners were handed in to the castle authorities during that year (compared with 33 in an average non-famine-year in that decade).[190] In other words, despite their good will, the Priory authorities were unable to feed every incarcerated individual in a sufficient manner. It is unclear what would have been an average allocation of loaves per prisoner because the duration of each felon's incarceration would vary. It is possible, all the same, to estimate that the amounts of wheat baked into bread to be distributed among the poor on the Sunday mass were enough to produce twenty-nine loaves each Sunday in 1315–1316. This figure was somewhat lower than that in non-famine years (approximately thirty-five weekly loaves).

The allocation of *panis militum* among the needy was the responsibility of the master of the cellar of Norwich Cathedral Priory. But it was the almoner who was in charge of a local food kitchen catering for the poor on a daily basis. Unfortunately, no almoner's account survives for the famine years; in fact, there are only three surviving accounts from the first half of the fourteenth century that indicate the annual distribution of food alms. According to these accounts, the almoner would allocate about 360 quarters of crops each year to be turned into rye loaves, legume pottage, and low-quality barley ale.[191] If these quantities were not significantly lower during the famine, then they were sufficient to feed some 320 paupers on a daily basis (assuming the allocation of 1500 kcal per capita). These estimates agree with those for the Canterbury Cathedral food kitchen, discussed above. However, it should be borne in mind that feeding some 320 (or about 280, as it was in the case of Canterbury Priory) paupers a day could hardly solve the issue of ubiquitous starvation. There can be little doubt that urban dwellers were faring in the crisis worse than their rural counterparts, and the proportion of hungry mouths in towns was alarmingly high. On the eve of the famine, the population of Norwich was in the area of 15,000 people, while that of Canterbury may have been about 9000 people.[192] But even

[190] TNA, JUST 3/48.

[191] NRO, DCN 1/6/9,12,13 (1310–1311, 1339–1340, and 1345–1346).

[192] Rutledge, 'Immigration and Population Growth'; Campbell, 'Benchmarking', p. 908.

if we account only for the proportion of town dwellers that were in a dire need, struggling on the verge of survival, it is unlikely that the food kitchens could feed but a small proportion of these hungry people. Moreover, it is possible that these food kitchens may have also attracted the inhabitants of nearby villages, thus creating even more pressure and tension among the starving paupers.

At Bolton Priory, a much less affluent institution, charitable activities mimic those of Canterbury and Norwich priories, albeit on a very small scale (Table 8.4). As the Bolton account book indicates, local officials managed to increase the proportions of grains allocated to the poor in 1316–1317. Just as at Canterbury Cathedral Priory, the Bolton authorities achieved that by purchasing unusually large quantities of grain (41.2 quarters of barley, in contrast with just a few quarters in 1311–1315), most of which (32.6 quarters) was distributed among the poor. In theory, these amounts were enough to feed about twenty-seven (if allowing 1500 kcal per capita per diem) or forty-one people (if allowing 1000 kcal per capita per diem). This was twice as much as in pre-famine years, despite that year's harvest failure and the disastrous state of the house's finances. Still, these amounts of crops could feed only a small number of hungry mouths, small even for such a thinly populated region as the Yorkshire Dales. Fortunately, Bolton was temporarily spared from Scottish attacks. The situation changed in 1318, when the priory became an easy target for recurrent Scottish raids, leading to its decline and the eventual dispersal of its community in 1325. This is reflected in the fact that, in 1320–1321, the priory was able to allocate only 3.7 quarters of grains (mostly beans), sufficient to feed only one or two persons at the rate of 1500 or 1000 kcal per diem.[193]

While Canterbury Cathedral Priory, the Norwich Cathedral priories, and Bolton Priory did manage to increase the sheer amounts of food to be distributed among the poor, some conventual houses could not do the same, because of various economic difficulties. Unlike Bolton Priory, Durham Cathedral Priory and rural estates were targeted by the Scots during the famine years; there were at least five attacks between 1312 and 1322. As a result, its bursar (official in charge of monastic finances) was able to allocate only £1 6s 8d to the almoner in 1316–1317, in contrast with £19 12s 8d in 1314–1315 and £18 3s 4d in 1317–1318.[194] Unfortunately, no almoners' accounts from Durham Priory survive for the famine period, and it is uncertain, therefore, if the local almoner was able to secure adequate amounts of grains, notwithstanding these cuts. Some conventual houses became impoverished without being involved

193 *The Bolton Priory Compotus*, ed. by Kershaw and Smith.

194 Durham Cathedral Archives, Bursar accounts 1314–1315, 1316–1317, and 1317–1318.

Table 8.4. Composition of crops distributed among the poor by Bolton Priory authorities (in quarters), their approximate calorific equivalent, and numbers of people they were capable of feeding on a daily basis, 1310/1–1320/1.

		To the poor								
Year	Total receipt, all grains	Wheat	Rye	Beans	Barley	Oats	All grains	Approximate calorific equivalent	Paupers per diem, if 1500 kcal per capita	Paupers per diem, if 1000 kcal per capita
1310–1311	3,559.8	0.3	0.0	29.0	3.0	0.0	32.3	4,881,269	9	13
1311–1312	4,093.7	2.5	4.5	22.0	1.5	0.0	30.5	6,795,128	12	19
1312–1313	3,856.5	Unspecified								
1313–1314	3,879.2	Unspecified								
1314–1315	3,473.8	4.2	2.8	23.1	3.1	6.0	39.2	9,943,099	18	27
1315–1316	2,292.7	0.4	2.5	19.0	8.1	0.0	30.0	7,180,276	13	20
1316–1317	2,259.8	1.0	0.0	3.8	32.6	0.0	37.4	14,910,463	27	41
1317–1318	2,318.4	4.4	0.0	24.5	5.2	0.0	34.1	7,446,540	14	20
1318–1319	3,311.8	2.8	1.0	22.5	3.4	0.0	29.6	6,048,907	11	17
1320–1321	2,041.0	0.6	0.0	2.8	0.3	0.0	3.7	765,634	1	2

Source: *The Bolton Priory Compotus*, ed. by Kershaw and Smith.

Notes: I have assumed that barley was baked rather than brewed, and thus I took its extraction rates to be 80 per cent (that is, 20 per cent calorific loss).

in the Anglo-Scottish conflict. One such example is St Werburgh Abbey in Chester. On 1 November 1315, this house was visited by Walter Langton, bishop of Coventry and Lichfield, who found the monastery in a deplorable and indebted state. The bishop instructed the brethren to cut all unnecessary expenses and demanded that all food leftovers should be turned into alms.[195] This request seems especially meaningful in the context of the famine, when the bishop insisted that even such a non-affluent house as St Werburgh Abbey should practise *some* form of food charity. Whether the Werburgh brethren took heed of the bishop's injunctions remains unknown, but the injunctions

[195] *The Register of Walter Langton*, ed. by Hughes, I, 106–07 (no. 809).

hint that even poor houses were expected to hand over some food to the poor during the famine years and that some of these houses did not do so without interference from above.

Hospitals and almshouses were another type of religious institution practising charitable activities related to food distribution during the famine years. Just as with almoners' accounts, there are frustratingly few sets of surviving hospital accounts from the late medieval period, and there are only (to my best knowledge) three sets of accounts partially covering the famine period: the one is for St John's Hospital, Winchester, the second for the Great Hospital of Norwich, and the third one for Clyst Gabriel, in Bishop's Clyst (near Exeter). In the first two instances, the respective accounts survive only for the first famine year, while in the case of Clyst Gabriel, only the 1316–1317 account survives.[196] This partial survival, nevertheless, provides an important insight into food policies of the hospital authorities during the crisis. As Tables 8.5 and 8.6 show, both Winchester and Norwich hospitals spent more than usual on foodstuffs during the famine. Each hospital, however, deployed a different strategy. Winchester Hospital increased its expenditure on both farinaceous and non-farinaceous products, with the grand totals rising from £59 6s in 1314–1315 to £77 4s in 1315–1316. Conversely, the Great Hospital of Norwich spent unusually large amounts of money on meat, while the outgoing sums on 'kitchen expenditure' (*expense coquine*) declined. Unfortunately, the stewards of the Norwich hospital do not indicate the nature of these 'kitchen diets', but the fact that they were being prepared in the hospital kitchen, rather than at a local bakery or brewery, implies that these were mostly non-farinaceous products (seafood, ready-made meat, fish), with occasional supplements of bread and ale.[197] To make things worse, the stewards' accounts do not record purchases of bread and ale, as they seem to have been the responsibility of a different officer. Therefore, it is impossible to estimate whether the Norwich hospital increased its expenditure on grains as well.

An altogether different situation prevailed at Clyst Gabriel, where local inmates were paid a food allowance (*putura*) rather than actual food. Here, the authorities would invariably allocate 9d per inmate each week (1.28d per diem), and this sum did not vary at all in the pre-Black Death period, in both normal and bad years.[198] At first these sums may appear meagre, especially in the con-

[196] HantsALS, W/H1/3; NRO, NCR 24A, 1315–1316 account; and ExCA, 5212.

[197] Rawcliffe, *Medicine for the Soul*, pp. 180–81.

[198] ExCA, 5212.

text of the high prices during the famine. In reality, however, these sums were good enough to buy almost 2.5 quarters of wheat in 1316–1317, far beyond the per capita requirement of about 1 quarter a year. It is unlikely that these food allowances would be supplemented by any additional food; in any event, the accounts do not indicate any additional food purchased or cooked for the inmates. As Nicholas Orme, who worked on the Clyst Gabriel accounts, suggested, the clergy may have agreed to hire a cook and pay his wages from their allowances.[199] This would have been more than doable, especially if both infirm priests and chantry priests (the latter receiving a weekly stipend of 18.5d) would do so collaboratively. Whatever the situation was, it is clear that Clyst Gabriel inmates were not starving during the famine. This impression is confirmed by the mortality seasonality of the Clyst inmates in that year: four infirm priests (out of fifteen) died, respectively, in October, January, February, and April, but not in the summertime, when the food prices were at their highest.[200] As we have seen in the previous chapter, the 1316–1317 trends in mortality of the Clyst priests mimic, to a certain degree, the seasonality of their mortality in non-famine years, when the majority of deaths would occur between January and March.

Although these three hospitals undoubtedly looked after their residents, consisting of old and disabled people, there is no evidence that they practised any charitable works outside their own communities. Many late medieval hospitals did not normally make handouts to non-residents, and there is evidence that they changed this rule during the famine. But even these that fed poor non-residents, such as the Great Hospital of Norwich, did so on a small scale. The Norwich hospital handed out food to thirteen paupers each day, and there is no evidence that it increased the levels of food handouts during the famine.[201] To qualify entering these hospitals, inmates had to be old or sick (or both), and in many instances, clerics; 'hungry' would qualify as sick. But even admitting new *qualified* inmates during the famine would hardly be doable: these institutions were always full and had strict residential capacity limits, which could not be expanded. When an inmate died, his place would have been taken right away by another sick inmate. The lifespan of an average inmate at Clyst Gabriel Hospital varied anywhere between a few months and a dozen years.[202]

199 Orme, 'A Medieval Almhouse for the Clergy', p. 6.

200 Orme, 'Mortality in Fourteenth-Century Exeter', p. 201.

201 Rawcliffe, *Medicine for the Soul*, p. 244.

202 Orme, 'A Medieval Almhouse for the Clergy', pp. 8–9.

Table 8.5. Food expenditure at St John's Hospital, Winchester, 1314/5–1319/20.

	Year	Meat	Fish	Bread and ale	Total
(a) In £ Sterling	1314–1315	22.1	11.3	26.2	59.6
	1315–1316	25.5	15.2	36.4	77.2
	1318–1319	16.9	9.9	15.1	42.0
	1319–1320	27.4	9.8	16.3	53.6
(b) As % of total food expenses	1314–1315	37.1%	18.9%	44.0%	100.0%
	1315–1316	33.1%	19.8%	47.2%	100.0%
	1318–1319	40.4%	23.7%	36.0%	100.0%
	1319–1320	51.2%	18.4%	30.5%	100.0%

Source: HantsALS, W/H1/2–6.

Famine Relief: Outside the Church

Notwithstanding a rather limited scale of poor relief, extended by religious institutions, they were still the main providers of the poor. Despite this fact, we should also look beyond these institutions when considering the topic of famine relief. As Chris Dyer has convincingly shown, various mechanisms and channels of poor relief were available within different social and institutional sectors of late medieval society.[203] Although Dyer's study is based primarily on post-1383 wills, some attributes of post-Black Death poor relief may have been in existence around the famine time. As we have seen above, the extension of credit can be considered as a charitable act amidst the grave reality of the famine, with at least *some* better-off creditors showing their good will and allocating funds to their worse-off parishioners, neighbours, friends, or relatives. This was especially true in the case of the first year of the famine. This idea of credit extension as a form of intra-village charity has been already brought up by Phillipp Schofield in his 2008 study of pre-Black Death Hinderclay (Suffolk).[204] Elsewhere, Schofield suggested that other, less obvious forms of poor relief were available within the parish, including grain sales for reduced prices and fine condonation.[205] In the same vein, we should attempt to dig for those often hidden, sparse, and unquantifiable forms of famine relief dur-

[203] Dyer, 'Poor Relief'.

[204] Schofield, 'Social Economy'.

[205] Schofield, 'Approaching Poverty'.

Table 8.6. Food expenditure at the Great Hospital, Norwich, 1315/6–1320/1.

	Year	'Kitchen diets'	Meat	Fish	Other	Total
(a) In £ Sterling	1315–1316	27	21	16	3	67
	1318–1319	33	1	14	8	56
	1319–1320	32	5	20	3	60
	1320–1321	36	15	15	7	73
(b) As % of total food expenses	1315–1316	40.6%	30.8%	23.6%	5.0%	100.0%
	1318–1319	58.9%	1.8%	25.0%	14.3%	100.0%
	1319–1320	53.3%	8.3%	33.3%	5.0%	100.0%
	1320–1321	49.3%	20.5%	20.5%	9.6%	100.0%

Source: NRO, NCR 24A.

ing the famine years. Thus, there are occasional hints about the reduction in grain prices. One such example has been discussed in Chapter 4 and is related to the purchase of 1 bushel of wheat for just 10d by a villein named William Waldrugge from Cuxham (Oxfordshire), with the permission of a local granary keeper.[206] Although it is possible that the granary keeper did not do so without the lord's authorization, it is equally possible that he, just as some other manorial officials, was ready to act contrary to his lords' (the Metron College scholars) will and, thus, provoke their ire by selling grains well below the market price level.

This, in turn, brings up the question: did manorial lords distribute any food among their tenants? As we have seen, monastic lords did indeed hand out some amounts of grain to the urban poor, which were, nevertheless, insufficient to make any difference. Was the situation any different in the countryside? A systematic survey of all the surviving famine years' manorial accounts reveals that lords and their officials were even less generous in distributing food among local tenants. There are, in fact, very few occasional references to grain alms. Thus, in 1315–1316, the officials of Downham (Suffolk) baked 2 bushels of rye and beans into bread for the poor.[207] At Littleport (Hampshire), 1 quarter of barley was given as a gift to a local winnower and an ewe-keeper, in the same

[206] 'Quia custos ei ita concessit' (MCA, 5840). That William Waldrugge was a local villein is mentioned in Harvey, *Medieval Oxfordshire Village*, p. 31.

[207] WSuffRO, 651/35/1.

year.[208] Around the Advent of 1316, 1 bushel of wheat was distributed among the poor of Overton (Wiltshire).[209] As some entries indicate, in some instances these food alms were offered 'at the lord's command' (*precepto domini*) rather than by the initiative of local officials.[210] Thus, according to the 1315–1316 account from Scratby (Norfolk), 3 quarters of barley were allocated to the lord's alms (*in elemosina domini*). Although some more examples may be added there, they all point out that such activities were irregular and confined to small quantities of crops. It is unlikely that the accounts under-recorded the actual amounts of grains given in alms or distributed by manorial officials among their relatives and friends: if they were smuggling food out, the auditors would have picked it up, while offenders would have fined by manorial courts. This is exemplified well in the case of Thomas Kyng, the reeve of the manor of Newington (Oxfordshire), who was fined at 4s for misappropriation of 4 bushels of grain on 27 November 1316.[211]

Just as manorial lords and their officials, the Crown, too, seems to have shown little sympathy towards the poor. As we have seen in Chapter 4, it did little to relieve the famine, apart from vague proclamations against speculation and hoarding, as well as securing insignificant quantities of imported grain, which ended up primarily with the armies rather than at local markets.[212] There is, however, an additional piece of information indicating Edward II's attitude to famine and famine relief. The information derives from the annual account books of the Royal Wardrobe, which recorded expenditures of various departments, including alms distribution. The person in charge of the distribution of alms was the royal almoner; during the famine years, this office was held by Sir Bernard de Kirkeby. As his accounts, recorded into the wardrobe books, indicate, the annual sums spent on alms were rather meagre, standing at around £389 in 1315–1316 and £382 in 1316–1317 — figures comparable with the £385 spent in 1317–1318.[213] It should be noted, however, that the vast majority of these sums were allocated to various friaries across the country rather than to paupers. For instance, on or shortly before 28 November 1315

[208] HantsALS, DC J/1/14.

[209] HantsALS, DC J/1/14.

[210] NRO, DCN 60/30/5.

[211] CCA, U15/26/59; see above, p. 379.

[212] See above, Chapter 4.

[213] TNA, E 101/376/7, fols 4^{r}–5^{v}; Society of Antiquaries of London (SAL) MS 120, fols 13–20 and MS 121, fols 3–7.

the almoner distributed 89s 8d among London friars to pray for the soul of Eleanor of Castile, Edward II's mother, on the anniversary of her death.[214] The Carmelites of Sheen (Surrey) received the princely contribution of £38 6s 6d to pray for the souls of Edward's ancestors.[215] It should be noted that the Friars could hope to receive alms not only from the royal almoner but also from various lay donors. For instance, John Dalderby, Bishop of Lincoln, encouraged people of his diocese to say prayers for those buried with the Franciscans at Grantham (Lincolnshire) and to donate to the friars 'during the harsh dearth' (*durante instanti caristia*).[216] Conversely, poor individuals received very small sums. Thus, between 13 November 1315 and 6 January 1316, the royal almoner expended a total of 87s 6d, distributed among 800 paupers on three occasions (that is, between 1d and 1.5d per capita). In the course of 1316–1317, the almoner distributed only 56s 1d among 198 sick (*infirmi*) and 240 paupers at various places.[217] These alms carried a symbolic rather than a practical character, for these sums were too meagre to support the hungry paupers, and it is unlikely that they were of any help at all. These entries stand in contrast with Buchanan Sharp's interpretation of the 'royal paternalism' of Edward II during the Great Famine.[218]

In other words, despite some occasional acts of charity, there was very little or nothing that most hungry paupers could hope for. The lack of organized poor relief outside of religious institutions (such as cathedrals, monasteries, or hospitals) implied that the hunger could only get worse and that there was no way to fight and overcome the crisis. As examples from later subsistence crises demonstrate, the lack or presence of such organized famine relief channels could be detrimental in aggravating or relieving such crises. For instance, the failed governmental response to the crisis in Ireland, both in 1740 and 1845–1847, ensured that subsistence crises turned into terrible famines in both instances.[219] Likewise, during the 1867–1869 crisis in Estonia and Finland, inadequate institutional responses of both the Czarist government and the local landlords caused harvest failures and food deficiency to transform into a harsh famine.[220]

214 TNA, E 101/376/7, fol. 5^r^.

215 TNA, E 101/376/7, fol. 5^v^.

216 Swanson, 'Indulgences for Prayers for the Dead', p. 198.

217 TNA, E 101/376/7, fols 4^r^–5^v^; SAL, MS 120, fols 13–20.

218 Sharp, 'Royal Paternalism', p. 628–47; Sharp, *Famine and Scarcity*, pp. 33–55.

219 Engler and others, 'The Irish Famine of 1740–1741'.

220 Lust, 'Providing for the Hungry?'; Myllyntaus, 'Summer Frost', p. 91.

Conversely, the food crises of 1740 and 1845–1847 in Flanders did not turn into famines, thanks to the ability of local communities to absorb the shocks of these crises via various mechanisms of famine relief.[221] More examples can be cited here.

Although it is tempting to accuse the royal government, the local landlords, and the lack of communal initiative, it is important to ask if the institutions had any means or resources to relieve the famine. With the total population of some 5–5.25 million people and the high proportion of landless people (about 37 per cent of the population consisted of landless rustics and cottagers, according to one estimate),[222] there may have been nearly two million people (and possibly more) suffering from severe malnourishment. As we have estimated in the previous chapter, the financial value of one million crop calories rose from about 10s 2d in an average non-famine year to 26s 2d in 1315–1316 and 26s 11d in 1316–1317 (see Appendix 5). To feed some two million hungry mouths each day at the very low per-capita allowance of 600 kcal (as it was during the Siege of Leningrad or during the Dutch Famine of 1944–1945) would have cost about £571,669 and £587,954 a year in 1315–1316 and 1316–1317, respectively. Around 1315, the total amount of the currency in circulation in England was in the area of between £1.2 and £1.5 million.[223] To feed all the hungry mouths, England would have had to recruit about 39–48 per cent of her available currency, which would have been anything but doable. Even to feed one-tenth of that figure would have been an impossible task. This is given the fact that the Crown was soaking in debts, owing well over £100,000 to various creditors on the eve of the famine, with a good share of its annual revenue being invested in financing and provisioning its armies and garrisons.[224]

Liturgical Responses

It would be appropriate to end the discussion with religious practices, namely with liturgical responses to the famine. As I have shown elsewhere, liturgy played an important role in collective responses to biological and economic crises in

[221] Van Haute and Lambrecht, 'Famine, Exchange Networks', 155–86.

[222] Broadberry and others, *British Economic Growth*, pp. 317–18.

[223] These estimates derive from Martin Allen, 'The First Sterling Area', pp. 86 and 91, on the assumption that England had about 82 per cent of all the currency within the Sterling Area (England, Wales, Ireland, Scotland, and the Isle of Man).

[224] Phillips, *Edward II*, p. 129.

late medieval period. Thus, during the outbreaks of sheep scab of 1279–1280 and cattle pestilence (most likely rinderpest) of 1319–1320, lay communities and men of religion alike resorted to special religious practices to ward off the diseases.[225] The reaction to the famine was similar. On 14 July 1315, Walter Reynolds, archbishop of Canterbury, instructed the prior and the brethren of Canterbury Cathedral Priory to ring the bells, organize solemn processions, sing litanies, and celebrate votive masses against the downpour and for the well-being of the king and his kingdom. In addition to taking part in these rituals, he urged both clergy and laity to pray, fast, and practise the Six Works of Mercy.[226] To encourage the laity to participate and do penance, Reynolds granted forty days of indulgence.[227] A year later, on 8 July 1316, when the effects of famine were too apparent to be ignored, Reynolds appealed to the bishops of his province for action once more.[228] Because of the bad weather and lack of food, as well as the dissention of the enemies, bemoaned the archbishop, people of the once bountiful country are now begging, tortured by the famine. To restore the peace within the realm, as well as to end the bad weather, Reynold commanded his suffragans to offer votive masses every Sunday and undertake barefoot processions each Wednesday and Friday in churches and market places. The archbishop expected the prayers and processions to be attended by lay crowds, in a humble and diligent manner.[229] Just as in 1315, Reynolds granted all the participants forty days of indulgence.[230] Reynolds's injunctions seem to have been taken seriously by the suffragans of his province: already on 15 July 1316 (a mere week after their proclamation), Gilbert Segrave, bishop of London, wrote to Roger Martival, bishop of Salisbury, instructing him of Reynolds's injunctions, while on 27 August, Martival, in turn, directed his archdeacons to publish the injunctions and inform him about their fulfilment.[231]

What was the nature of these processions and votive prayers? According to one London chronicle, local priests and monks were instructed, apparently by Gilbert Segrave, to process barefoot from their respective churches to the

[225] Slavin, 'Flogging a Dead Cow', pp. 120–22.

[226] CCA, Register I, fols 341^{v}–342^{r}.

[227] Wright, *The Church and the English Crown*, p. 353.

[228] Only the missive to the Bishop of London survives (CCA, Register I, fols 348^{v}–349^{r}).

[229] CCA, Register I, fols 348^{v}–349^{r}.

[230] Wright, *The Church and English Crown*, p. 353.

[231] *The Registers of Roger Martival*, vol. II: *The Register of Divers Letters*, ed. by Elrington, p. 83.

Church of Holy Trinity, carrying the consecrated Host and relics.[232] In theory, there was a wealth of votive masses that could be said in the context of the incessant torrential rain and the following famine: the mass for tempest repellence (*Missa ad repellendam tempestatem*), the mass for new harvests (*Missa de frugibus novis*), the mass for tithes (*Missa de decimis*), and the mass against famine (*Missa contra famem/pro fame*).[233] These masses are attested in numerous liturgical books in England; the mass against famine usually appears together with another votive prayer against pestilence.[234] It should be noted that these prayers were by no means a liturgical novelty; rather, these were centuries-old practices originating in Late Antiquity and the early Middle Ages and existing in a codified manner since at least the sixth century.[235] This is hardly surprising, given that the vast majority of liturgical practices, known and used in medieval Catholic rites, had been devised by the end of the first millennium and very few innovations had been introduced after the year 1000.

One interesting feature of those processions, votive prayers, and other liturgical practices is their polysemic character. Although performed in the context of bad weather and famine, these rites were aimed to achieve multiple goals. The vast majority of liturgical books (primarily, missals) do not have a separate rubric for masses against famine; rather, there is a common rubric for masses against famine *and* pestilence (*Missa contra pestilenciam*). Likewise, when Archbishop Reynolds addressed the bishops of his province, he instructed them to pray and organize processions not only to stop the downpour and relieve the famine but also to ward off human mortality, supress England's enemies (namely, the Scots), and bring peace to the realm. Hence, the anti-famine prayers should be regarded in the wider context of cognate votive prayers — just as with other late medieval disasters. For instance, when Ralph of Shrewsbury, bishop of Bath and Wells, instructed his archdeacons to undertake processions against the plague in August 1348, he stated very clearly that the processions would appeal for God's mercy in order to save His people from the pestilence and grant peace between Christian countries (in the context of the ongoing

[232] *Annales Paulini*, ed. by Stubbs, p. 278.

[233] *Corpus orationum*, ed. by Moeller, Clément, and Coppieters 't Wallant, pp. 211–30.

[234] For instance, *The Missal of Robert of Jumièges*, ed. by Wilson, p. 269.

[235] *Liber sacramentorum Romanae Ecclesiae*, ed. by Wilson, p. 260 (the Gelasian Sacramentary, *c.* 750); *The Leofric Missal*, ed. by Warren, p. 188; *Missale ad usum insignis Ecclesiæ Eboracensis*, ed. by Henderson, II, 171.

Anglo-French conflict).[236] Likewise, during the attack of the *pestilencia secunda* in 1361, John Thoresby, archbishop of York, urged his clergymen to engage in prayer and procession on account of 'the whirlwinds of war and...pestilences and other misfortunes'.[237] Similar polysemy is found in votive masses against the Infidel and for the liberation of the Holy Land, where liturgy was aimed not only at defeating the Muslims and freeing Jerusalem but also at subjugating heretics and rendering Christian rulers victorious and their realms peaceful.[238]

To the people who lived in that time, these liturgical practices represented an efficient mean to atone for their sins, to avert the ire of God, and to end the crisis. The illiteracy of masses and their very shallow and vague understanding of 'official' liturgy implied that they could participate in these rituals only in a passive manner, as attendants — as indeed indicated by Reynolds's injunctions. It was really clergy that were in charge of conducting and performing these various forms of intercessory liturgy, and as such, they were undoubtedly credited with possessing those arcane skills and knowledge that allowed them to act as intermediaries between the starving masses and their retributive God. This spiritual contribution of religious men has been duly acknowledged by the anonymous author of the *Vita Edwardi Secundi*, who stated that 'if the English church had not interceded for us, we should have perished long ago'.[239]

[236] *Black Death*, ed. and trans. by Horrox, p. 113.

[237] *Black Death*, ed. and trans. by Horrox, p. 119.

[238] Linder, *Raising Arms*, pp. 195–273.

[239] *Vita Edwardi Secundi*, ed. by Childs, pp. 110–11.

Epilogue

The End of the Crisis?

Between the Famine and the Plague: A Generation of Crisis

In the present study, considerable attention has been devoted to the question: What caused the Great Famine of 1315–1317? As we have seen, there is no uniform answer to that; the Great Famine, just as any other historical famine, was a very complex and eclectic phenomenon, with exogenous, endogenous, and demographic factors being interwoven one into another and acting as a three-headed hydra. But a no less important question to ask, while approaching the end of the study, is: What ended the famine? To say that it was a combination of population loss, the end of the torrential rain in the summer of 1316, and, consequently, the improved harvest of 1317 should not be too far from truth. Although the *net* harvest of 1317 was still mediocre at best, standing at about 26 per cent below average, the fact that a proportion of the population died in the course of the two preceding years implies that the 1317 harvest shortfall was not as devastating as it may have been otherwise. The following year's harvest was bountiful, with composite *net* crop yields reaching about 19 per cent above average. But did the economy and society of the British Isles return to their pre-famine *status quo*? When talking about famines, scholars tend to focus on famines themselves and their victims and often ignore the long-term impact on the survivors. It would be instructive to consider the fate of the famine's survivors and its long-term biological and socio-economic implications. Although the famine seems to have weeded out the poorer and frailer individuals, it also had an impact on the heath of its survivors.

The first half of the fourteenth century was arguably one the most troublesome periods in northern European history. Although the famine was over by

1317, and certainly by 1318, England and other parts of the British Isles were subsequently visited by several climatic and biological shocks with severe long-term repercussions. First, there was a cattle pestilence, which arrived in England around Easter of 1319. The disease spread throughout the British Isles, reaching Scotland shortly after September 1319, Wales by summer 1320, and Ireland in 1321. Thanks to the detailed nature of manorial accounts, recording cattle fatalities, it is possible to reconstruct that, in England and east Wales, about 62 per cent of bovids perished. The murrain was most likely caused by rinderpest, a viral disease with death rates approaching 100 per cent in infected animals. The disease was particularly devastating in female animals, killing about three quarters of all cows and heifers, owing to the fact that the immune system of lactating and pregnant animals was compromised by malnourishment and the abnormally damp and cold weather during the famine years. The morality rates of oxen, on the other hand, stood at about 50 per cent, which may be explained by their better resistance to pathogens because of stronger physiology and better diet, which included oats and legumes.[1]

The recovery of bovine stocks was a long and expensive process. As the manorial accounts indicate, it was not until the late 1330s that herds reached their pre-crisis levels. Because of their importance as draught animals, oxen were the first animals to be restocked; wherever such immediate replenishment was not possible, oxen were temporarily replaced with draught horses. Thanks to these steps, there is no evidence of depression within the arable sector until the Black Death, and this fact allowed the English population to grow anew. The dairy sector, however, remained depressed for some twenty years because of the comparatively slow recovery of cow stocks. The contraction within the dairy sector meant the English population was deprived of their most important source of some vital nutrients, including protein, calcium, and vitamin B_{12}. This fact seems to have had some harsh biological implications on humans, whether those who survived the famine or those born after.[2]

Although the cattle pestilence was the most serious blow, it was by no means the only one. The torrential rain returned, once more, in the autumn of 1320, pouring into the following year and ruining the harvest of 1321, when the composite *net* crop yields stood at approximately 34 per cent below average.[3] Although it did not create another famine, it did drive up crop prices a great

[1] Newfield, 'A Cattle Panzootic'; Slavin, 'The Great Bovine Pestilence'.

[2] Slavin and DeWitte, 'Between Famine and Death', pp. 37–60.

[3] Calculated from Campbell, *Three Centuries of English Crop Yields*.

deal, with a quarter of wheat, rye, barley, oats, and legumes selling, respectively, for 11s 7d, 9s 9d, 7s 9d, 4s 3d, and 9s 3d.[4] The downpour stopped at some point in late 1321 and was followed by a series of remarkably good harvests, in 1322–1327, 1329–1330, 1332–1338, 1340–1345, and 1347–1348, with crop yields standing well above their average.[5] This, however, does not mean that the situation was ideal all over the place. In the late 1320s and the early 1330s, parts of England experienced two clusters of exceptionally dry years (1325–1327 and 1330–1333), affecting the harvests of 1328 and 1331, but also causing human and livestock mortality, albeit on a local scale and of modest proportions. As David Stone has suggested, these may have been due to outbreaks of enteric infections, malaria, and winter and spring fevers, particularly among riverside communities.[6]

Possible Long-term Health Implications

Those recurrent disasters most certainly had harsh implications on the survivors of the famine and their offspring. A glimpse into the fast-growing body of palaeopathological evidence reveals some exciting insights. In 2005, Antoine and Hillson published the results of their study of dental layers of thirty individuals, known to have been born before the famine and buried at the East Smithfield Black Death Cemetery in London. On the basis of the analysis of enamel and dentine incremental structures, they built a detailed dental growth record of the early years of their lives and found signs of a pronounced growth disruption, whose chronology matches the famine years. These signs consisted primarily of hypoplastic surface defects and slow or disturbed growth, reflecting some severe malnourishment experienced by the survivors of the famine.[7] Similarly, the Hereford Cathedral cemetery evidence reveals clear signs of dental pathologies, including periodontal (gum) disease, alveolar atrophy (decrease in the size of the supportive tissues of an affected tooth), exfoliation of buccal (backside) roots, and heavy calculus deposits.[8] Malnourishment

[4] Munro, 'Revisions'; Farmer, 'Prices and Wages', in *The Agrarian History*, II, and in *The Agrarian History*, III

[5] Calculated from Campbell, *Three Centuries of English Crop Yields*.

[6] Stone, 'The Impact of Drought'.

[7] Antoine and Hillson, 'The Great Famine'; Antoine and others, 'Using Growth Structures', p. 65.

[8] Kacki, 'Influence de l'état sanitaire', pp. 277–85.

and decreased living standards may have also been responsible for caries, as is indicated in the mid-thirteenth-century burials at St Mary Spital (London), but also in the recently published analysis of the Black Death burials at Charterhouse Square (London).[9] Similar dental defects have been spotted on the post-famine cohort: linear enamel hypoplasia (horizontal lines on the enamel of an affected tooth) and periodontal disease seem to be prevalent in both pre- and post-famine generations.[10] This indicates that the nutritional crisis was not over with the end of the famine in 1317.

Recent cutting-edge advances in biological anthropology and palaeomicrobiology, and particularly in the field of stable isotopes, have a huge potential to take the available skeletal data to much greater heights. For instance, these advanced (and continuously advancing) methodologies are now capable of answering perhaps the single most burning question: What effect did medieval famine have on health and mortality? In particular, high-resolution isotopic analysis, when coupled with the available palaeopathological data, can determine whether people who survived the famine were at an elevated risk of mortality during the Black Death or were under 'normal' mortality conditions.[11] My own previous work with Sharon DeWitte on the impact of the Great Bovine Pestilence of 1319–1320 and the subsequent dairy shortage on the health of the English population, as evidenced by palaeopathological data from East Smithfield, shows that the post-famine cohorts (namely, those being *in utero* or born after 1317/8) tended to exhibit more pathological signs than the pre-famine cohorts.[12] Similar evidence comes from the Hereford Cathedral cemetery and the Charterhouse cemetery.[13] The presence of pathological signs, such as cribra orbitalia (lesions on orbital roofs) and porotic hyperostosis (lesions on cranial vault bones), is surely one indication of health in historical populations, but surely it is not the only one. As some recent research has shown, stable isotope analysis can be used not only to reconstruct skeletal health and diet but also physiological health, nutrition history, and geographic mobility patterns.[14]

[9] Pfizenmaier, ed., *Charterhouse Square*, pp. 132–33.

[10] DeWitte and Wood, 'Selectivity of Black Death Mortality'.

[11] I am grateful to Sharon DeWitte, who kindly shared with me her insights on this matter.

[12] Slavin and DeWitte, 'Between Famine and Death'.

[13] Pfizenmaier, ed., *Charterhouse Square*, pp. 132–33.

[14] Reitsema, 'Beyond Diet Reconstruction'.

Most recent scientific breakthroughs in the field of stable isotope can be exemplified in a high-resolution analysis of several skeletons from the recently excavated Charterhouse Square cemetery, London. Here, signs of malnourishment and health stress can be detected in several skeletons, including those numbered 159 (a male individual, aged 18–25, who died either during the first or second outbreak of the Black Death) and 188 (a male individual, aged 26–35, who died in the first outbreak of the plague). It has been established that Skeleton 159 was a native of eastern England, born either between 1323/24 and 1330/31 or 1336 and 1343. Although breastfed as an infant, the individual suffered a short-term malnourishment episode between two and four years of age. He also exhibited clear signs of enamel hypoplasia and periodontal disease.[15] Skeleton 188 was, most likely, a local man, born between 1313/14 and 1322/23, who experienced health stress between four and eight years of age. In other words, although malnourishment was particularly bad between 1315 and 1317, it was not only confined to those disastrous years, as the evidence from Black Death cemeteries reveals.

The already frail health of the famine survivors and the post-famine cohorts may have been compromised even further by air pollution. Most recently, a meticulous multidisciplinary and cutting-edge study of 2000 years of lead (Pb) deposit levels in ice-core layers of the Colle Gnifetti glacier (the Swiss-Italian Alps) has revealed that the atmospheric pollution in pre-Industrial Europe was considerably higher than previously thought, and it was mortality crises, first and foremost the Black Death, that brought lead production and, consequently, air pollution to their *minima*. This, in turn, may have had some far-reaching implications on the health of children and adults.[16]

Another promising venue is the still-nascent fields of palaeogenetics and palaeoepigenetics (namely, the study of changes to the genetic code in historical populations).[17] Here, some exciting work can be done to determine the epigenetic impact of prolonged malnourishment on (1) those growing up during the famine; (2) being *in utero* during the famine; (3) born into the famine; and (4) born within a generation after. The two most crucial periods when

[15] Pfizenmaier, ed., *Charterhouse Square*, p. 50.

[16] More and others, 'Next Generation Ice Core Technology'.

[17] I am grateful to Mr Jacob Spargo-Mabbs, my undergraduate student at Kent, who engaged me in thinking about the epigenetic impact of the Great Famine on post-famine generation and who has produced an undergraduate research seminar paper on the subject (Autumn 2015). My own insights here have been inspired in part by some of Mr Spargo-Mabbs' suggestions that he expressed in his seminar paper.

humans are vulnerable to environmentally-driven epigenetic alterations are *in utero* (especially during the first two trimesters of gestation) and during the slow growth period (SGP), between the ages of 8 and 12.[18] As recent research has shown, episodes of starvation and deprivation can have a detrimental effect on some significant epigenetic alternations. Thus, prenatal exposure to malnourishment during the Dutch *Hongerwinter* Famine of 1944–1945 and the Chinese Great Leap Forward Famine of 1958–1961 produced a trigger which caused schizophrenic and depression-related alterations within the epigenome. Also, it has been found that the same epigenetic alterations compromise the immune system in foetuses and children and make them weaker and more prone to various pathogens and diseases.[19] In the case of the Dutch famine, it has been found that prenatal exposure to famine increased metabolic and cardiovascular disease in the offspring, as well as the frequency of adiposity.[20] Similarly, some recent studies have suggested that the Great Irish Famine of 1845–1852 had a long-term impact on the mental health of Irish.[21] To make things even worse, however, famines do not only affect starving populations — they also have transgenerational effects. In other words, there is a chance that the same changes in genomic expression may be shared by the progenies of starving individuals for a couple of generations: as some studies in rats and humans have demonstrated, food deficiency can have a long-lasting impact on two subsequent generations.[22]

Although without collaboration with palaeomicrobiologists and palaeogeneticists, any suggestion regarding the possible epigenetic impact of the Great Famine would remain a mere hypothesis, but some suppositions can, nevertheless, be made. Given that the two most critical periods for epigenetic alterations are *in utero* and the SGP, then one may assume that those individuals born between *c.* 1304–1312 and early 1316–early 1318 were the most at-risk to be born with an inclination to mental disorders or/and a compromised immune system. Likewise, the same genetic traits could have been shared by the progenies of the same individuals, born anytime between the early 1320s and the

[18] Kaati and others, 'Transgenerational Response to Nutrition'.

[19] Susser and Stein, 'Timing in Prenatal Nutrition'; Susser and others, 'Schizophrenia after Prenatal Famine'; Heijmans and others, 'Persistent Epigenetic Differences'.

[20] Veenendaal and others, 'Transgenerational Effects'.

[21] Walsh, *Insanity, Power and Politics* and 'Gender and Insanity in Nineteenth Century Ireland'; Cox, *Negotiating Insanity*.

[22] Zamenhof, van Marthens, and Grauel, 'DNA (cell number) in Neonatal Brain'.

Black Death (and possibly later, if the parents survived the first outbreak of the plague). If this interpretation is correct, then the Great Famine may indeed have had a long-lasting impact on health of the pre-Black Death population, as is supported by palaeopathological evidence from the Black Death cemeteries, discussed above. The long-term deprivation during the famine years may have caused some epigenetic alterations in the generation of 'famine children' and potentially their offspring, making them prone to mental disorders and physical debilitation. This is in addition to the harsh health implications of the same generation due to the chronic deficiency of milk and other dairy products because of the tragic losses of bovids in 1319–1320 and the slow recovery of dairy cattle in the 1320s and 1330s. It was the same generation from which at least half succumbed to the first wave of what seems to have been the single deadliest pandemic in known history: the Black Death. Is it coincidental that the plague, or at least its first outbreak in the British Isles (1348–1350), was so devastating not only because of its biological force but also because of the biological weakness of its human victims?

This hypothesis about the possible connection between the Great Famine and the Black Death was first made by William Chester Jordan.[23] I have developed this view in two separate studies, stressing the importance of the Great Cattle Pestilence as a missing link between the famine and the plague, but in both instances I could not take it beyond the framework of hypothesis because of a lack of empirical data.[24] Today, as of 2018, I am much more optimistic about the ability of interdisciplinary teams of historians, archaeologists, palaeopathologists, palaeomicrobiologists, and palaeo(epi)geneticists to put this hypothesis to the test. In particular, thanks to some recent ground-breaking advances in the stable isotope analysis of palaeo-diets and innovative methods in palaeogenetics, it is now possible to extract much more from skeletal remains than ever before. The ever-expanding goldmine of scientific possibilities related to skeletal assemblies from the Black Death cemeteries, such as East Smithfield, West Smithfield, and the Charterhouse burials in London, as well as the cemeteries of Hereford Cathedral, Thornton Abbey, and St John's Hospital, Cambridge, can shed much light on the impact of the famine on the health and diet of the post-famine generations.

[23] Jordan, *The Great Famine*, pp. 185–86.

[24] Slavin, 'The Great Bovine Pestilence', p. 1263; Slavin and DeWitte, 'Between Famine and Death', pp. 56–58.

Conclusions

The Great European Famine was a catastrophe of a remarkable magnitude. The crisis was initiated by a short-term weather anomaly, bringing about a very rare scenario of three back-to-back harvest failures in 1315, 1316, and 1317. Even though the adverse weather reduced the respective *gross* crop yields by 25, 35, and 14 per cent, respectively, and, thus, created a relative shortage, non-negligible proportions of food were still available for human consumption. It was not Nature, however, that spurred the transformation of shortage into famine; it was a combination of demographic and anthropogenic (institutional) factors. The crisis unfolded through a series of interconnected chain-links, leading gradually from the initial harvest failure to its tragic *finale* of famine, disease, and mortality. Food shortage created a tipping point, which put human decisions and institutional behaviour to the test, and they miserably failed. The disposal of *gross* harvests was managed in accordance with manorial customs, and a disproportional share of crops was invested as seed corn. The food crisis was aggravated further by the malfunctioning of local food markets, where segmentation, preferential trade, hoarding, and a very limited scale of foreign grain imports drove up prices to the point that *very few could afford purchasing crops*, even though *very many desperately needed to do so* to compensate for the calorific loss. The rising storage and transportation costs disrupted the food supply and exacerbated the crisis further. While storage costs (and, thus, spoilage rates) went up because of the adverse weather, transportation costs rose because of the combination of bad weather and violence (especially in the sea). The ongoing warfare also had a devastating and long-lasting impact on food supply, through floral and faunal destruction within the 'war zone', which cut local communities off from their natural resources. Furthermore, warfare reduced the access of commoners to food supplies through purveyance sales, extortion, and royal taxation, which could not have come at a worse time than 1315–1316.

The widespread disruption of the food supply, coupled with omnipresent poverty, implied that many had to starve and that some would die. Excess deaths were only one side of the demographic change; equally important was a decline in fertility rates and a rise in 'missing births'. Although our attempts to estimate the fall in population are far from being secure, all available evidence hints that England may have lost at least 15 per cent of her population (and most likely in the area of 15–20 per cent). Unfortunately, there is no way to come up with any comparable crude estimates for the Lordship of Ireland and south and east Wales, also ravaged by the famine, but it is possible that these regions may have suffered as badly, on the account of warfare. Southern and eastern Scotland, on the other hand, although undoubtedly suffering because of the combination of bad weather and warfare, seems to have got away with much lighter losses — partially thanks to her demographic, institutional, and dietary peculiarities.

The example of the Great Famine stresses the highly complex and eclectic nature of famine as a historical phenomenon. As we have seen, the famine was initiated, intensified, and let loose by a combination of exogenous, endogenous, and demographic factors. Taken together, Nature, demography, and institutions should be seen as three 'meta-structures', acting as a great three-headed hydra, with each head consisting of various destructive mechanisms and elements. It is only through understanding of the interaction between those three 'meta-structures' that one can endeavour to answer the most pressing question: What creates famine? Such an approach recognizes the need to analyse and weigh each 'meta-structure' in a meticulous manner according to specific contexts, and it stands in contrast with existing models of famine that are based on one 'meta-structure' or one factor. It would be incorrect to say that the existing theories and models of famine are faulty or wrong. Each of them has its own theoretical advantages and conceptual validity. But they also have limitations because they provide only a partial answer. Thus, the environmentalist approach explains *how famines are being initiated*, but not *why they happen*. The Malthusian interpretation stresses the connection between overpopulation, on the one hand, and land-hunger and poverty, on the other, but it ignores the connection between exogenous shocks and the *aggravation* of the same land-hunger and poverty, in the context of food crises. The view that market performance has an impact on famine is equally valid, but it tends to concentrate on market forces and fails to account for the link between reduced harvests and empty stalls, on the one hand, and the connection between poverty and food trade disruption, on the other. The biggest shortcoming of these theories is their exclusivist nature, which creates a false impression that they are incompatible with other models. Any attempt to explain famine by pointing to one 'meta-structure', let alone just one factor, is to reduce a historical argument to the oversimplified level of environmental, demographic, anthropogenic, or institutional determinism, which does not account for other, no less decisive, factors. It is time now to move away from those quasi-ideological frictions of contrasting scholarly views. Instead, a new approach to the study of famines is needed, one that breaks down the analysis into smaller and diversified variables, in accordance with economic and ecological complexity models.

Furthermore, the interaction between the three 'meta-structures' is neither universal nor uniform. Thus, while there were omnipresent signs of overpopulation, land pressure, and poverty in most parts of England, there is no evidence that the same situation prevailed in south and east Wales, the Lordship of Ireland, and Scotland. Nonetheless, both the Lordship of Ireland and south and east Wales suffered from famine, while Scotland seems to have escaped the catastrophe, in spite of the likely presence of a food shortage. To crack this paradox, one has to bear in mind that there is no such thing as a 'universal famine'. Even if a famine covers an entire country, its intensity will inevitably vary across regions — or spare some regions altogether. To understand why it is the case, it is essential to approach the famine regionally, taking into account natural, demographic, and institutional peculiarities of each region, as we did in Chapter 1. Because accounts and records from south and east Wales, the Lordship of Ireland, and Scotland are scarce, such an exercise was hardly possible in the present monograph. Still, whenever possible, I did my best to stress the experience of the crisis across different parts of the British Isles. Breaking down the analysis into smaller regional particulars is connected with the idea of 'complexity theory', which stresses the inability to reduce any explanation to a single or few variables and the importance of analysing regional, and even local, contexts. Recently, this model proved to be valuable for quantitative historical research; as Eoin Flaherty has shown, the geography of starvation during the Great Irish Famine of 1845–1852 can only be appreciated through internal variations of regional, and indeed local, contexts rather than singular macrodynamics.[1]

Neither can famine be fully appreciated through the prism of a single discipline. To determine the environmental, climatic, and palaeopathological contexts in which famines tend to occur, narrative and statistical sources need to be supplemented with palaeo-scientific data. In the case of the present study, the analysis of the weather anomaly of 1314–1316 has been supplemented by dendrochronology from England and Scotland. The deterioration in nutri-

[1] Flaherty, 'Assessing the Distribution'.

tional and calorific qualities of food and forage could not be understood properly without reliance upon related nutritional studies; the discussion of famine pathologies is corroborated by the growing corpus of palaeopathological data from several Black Death burial sites.

It would be appropriate to finish off by evaluating the overall importance of the Great Famine in the history of the fourteenth century. Indeed, as Ian Kershaw first noted in his 1973 paper, the famine (or, to be more precise, the agrarian crisis of 1315–1322, which also included the cattle pestilence of 1319–1320 and the disastrous harvest of 1321) was a watershed event, with far-reaching implications.[2] This view was strongly seconded by William Chester Jordan in his classic 1996 monograph on the Great Famine.[3] It appears as if all the new data brought together and analysed in the present study supports this view. The Great Famine may have been the second harshest mortality crisis, after the Black Death, unless the *pestis secunda* of 1361–1362 (whose mortality rates are yet to be established) was of greater proportions. Despite the heavy demographic losses, here established to have been in the area of 15–20 per cent, the population of England (and possibly other areas of the British Isles that were affected by the famine) resumed growing in the aftermath of the crisis, and it seems to have reached its pre-famine levels by the time of the arrival of the plague in 1348. However, as we have seen, the 'inter-disaster' generation appears to have been, at least in part, frail, because of the outbreak of the cattle pestilence in 1319–1320 and the subsequent crisis of dairy industry, but also because of the likely epigenetic complications.

When the famine struck, a large proportion of the English population was already living in misery, and there can be little doubt that the crisis aggravated the situation further. The inability of institutions — economic, social, and political alike — to cope with the famine meant that it was bound to become a formidable force in triggering what historians call the 'Fourteenth-Century Crisis'. After some two centuries of demographic and economic expansion, England (and possibly other parts of the British Isles) may have reached the limits of growth, having exhausted its resources and possibilities. The omnipresent inequality, poverty, and congestion meant that many of those who were incapable of securing cash to buy deficient food had to sacrifice their most important form of a capital — land. As a result, the proportion of rustics holding very small pieces of land or holding no land at all undoubtedly increased.

[2] Kershaw, 'The Great Famine and Agrarian Crisis in England'.

[3] Jordan, *The Great Famine*.

It was in these difficult conditions that they had to stand up for themselves and their households when facing the additional 'crises' of the pre-Black Death era, including the bovine pestilence, occasional outbreaks of sheep disease, and additional harvest failures.

In a wider perspective, there can be little doubt that the adjective 'great' is a befitting one for the famine of 1315–1317. As Campbell and Ó Gráda have established, three back-to-back harvest failures together had the probability of happening every two or three hundred years.[4] There were, of course, other harsh subsistence crises in the history of the British Isles. One may think about the crises of 1437–1439, 1481–1483, 1555–1556, 1595–1597, 1621–1623, 1727–1730, 1741–1742, and 1800–1801 in England, as well as the 1580s crisis, the Nine Years War Famine of 1594–1603, the Cromwellian Famine of 1649–1654, the Year of Slaughter Famine of 1740–1741, and the Potato Famine of 1845–1852 in Ireland.[5] But of these, only the Irish famines of 1740–1741 and 1845–1852 seem to be on par with the Great Famine of the early fourteenth century, in terms of their scale and impact. Indeed, the adverse combination of weather anomalies, a malfunctioning market, a bad transportation system, deficient storage facilities, ongoing warfare, extractive institutions, and omnipresent poverty makes the Great Famine reminiscent of the situation in the developing world, where famines, despite some optimistic predictions, continue to occur and torment local communities.

This brings us to the last point. Today, in the context of the twenty-first century, the term 'famine' is associated with countries of the 'developing world', while 'starvation' and the 'west' sound more like two mutually exclusive terms. To a large extent, this has to do with the accepted narrative of the 'rise of Europe as a success story of might and wealth', standing in contrast with the alleged 'stagnation and backwardness' of non-western civilizations. This narrative is perpetuated in the work of some notable economic historians, such as Eric Jones, David Landes, and Niall Ferguson, who speak about 'European exclusiveness'.[6] In contrast with this optimistic narrative, it is important to bear in mind that the late medieval British Isles (and indeed, all of pre-Industrial Europe) was, too, a developing world, struggling with various challenges related to food security. The Great European Famine of the early fourteenth

[4] Campbell and Ó Gráda, 'Harvest Shortfalls', p. 868.

[5] Campbell, 'Four Famines and a Pestilence'; Ó Gráda, 'Ireland'.

[6] Jones, *The European Miracle*; Landes, *The Wealth and Poverty of Nations*; Ferguson, *Civilization*.

century is an excellent reminder that in the event of a major harvest shortfall, a food crisis could unfold into a harsh famine through a combination of societal shortcomings, institutional failures, and costly wars. In many ways, the Great Famine resembles a 'typical' famine in the developing world. Despite chronological, cultural, and geographic differences, famines always have a number of common denominators. Perhaps the most obvious denominator is poverty. A large proportion of the early fourteenth-century inhabitants of the British Isles were poor and vulnerable — just as populations in today's Africa are. After all, famines are not just about people; they are about people who are poor, deprived, and often voiceless. It is our moral responsibility as historians to give these people a voice and hear their stories — recorded in manorial documents, gaol delivery rolls, purveyance accounts, credit registers, tree-rings, and skeletal remains — so that we can understand why famines still happen today, despite progress in eradicating them and the vain promises of globalization and democracy.

APPENDICES

Appendix 1. National Aggregated Gross Yields Calculation

Gross yields are the ratio between the number of grain or legumes sown in the previous year and the amount of the same crops harvested in the present year. Basically, there are two possible methods to calculate the yields. First, there is the 'auditor's method', whereby the yields have been calculated by medieval auditors and recorded on the margins of manorial accounts, under each crop's rubric. Second, there is the 'internal method', based on our own calculation of the ratio between the number of sown crops recorded in a previous year's account and the number of harvested crops recorded in the present year's account. To calculate the yields using this method, two consecutive account rolls are needed.[1] Although medieval auditors tended to be fairly accurate in their reckoning, there are instances when their calculations were not precise. Therefore, the 'internal method' is the preferred one among the two, and I have used it whenever possible.

Since, in the vast majority of cases, the account rolls do not record tithes paid from harvest, it is necessary to make an adjustment here accordingly. For instance, a 1300–1301 account records a 1300 harvest of 100 quarters of wheat. We also know from the preceding 1299–1300 account that 25 quarters of wheat were sown in the autumn of 1299. This implies a 'recorded yield' (not accounting for tithes, but recording seed corn needed for seeding) of 4:1. By deflating 4 by 0.9 to account for tithes, we arrive at 4.44, which is the gross yield. In other words, the total harvest, inclusive of the tithes, was 111.1, rather than 100 quarters, as recorded in the account.

As discussed in Chapter 2, averaging all yield observations can give misleading result, not truly reflecting the 'national' yields as a whole. First, because of the geography and the nature of the manorial regime, the majority of accounts

[1] Campbell, *Three Centuries of English Crop Yields*.

tend to be scattered unevenly towards the southern and eastern counties, while the western and northern counties have produced a comparatively small number of accounts. Second, because of local and regional circumstances, some counties performed better than others. For instance, crop yields were clearly higher in Kent, Norfolk, Suffolk, and Somerset (compared to the northern counties), located on higher altitudes, liable to higher precipitation levels, situated on less fertile soils, and suffering from ongoing warfare. To take those differences into account and eliminate the potential 'noise' as much as possible, I have grouped the accounts into seven regional groups, with each group assigned a different relative weight, based on its relative contribution to the national arable acreage. This method is very similar to that devised by Stephen Broadberry and his colleagues.[2] The regional groupings and their respective contribution are as follows:

South-eastern demesnes	(Essex, Hertfordshire, Middlesex, Surrey, Sussex, Kent) = 15.2%
South-central demesnes	(Berkshire, Hampshire, Wiltshire) = 8.9%
South-western demesnes	(Somerset, Dorset, Devon, Cornwall) = 10.7%
Midland demesnes	(Rutland, Leicestershire, Warwickshire, Gloucestershire, Northamptonshire, Bedfordshire, Buckinghamshire, Oxfordshire, Nottinghamshire) = 18.4%
Western demesnes	(Worcestershire, Herefordshire, Monmouthshire, Shropshire, Staffordshire, Cheshire) = 9.7%
Eastern demesnes	(Norfolk, Suffolk, Cambridgeshire, Huntingdonshire, Lincolnshire) = 22%
Northern demesnes	(Northumberland, Durham, Yorkshire (North, East, and West Ridings), Cumberland, Westmorland, Lancashire) = 15.8%

Within each block, I have calculated the weighted average rather than the average of all the available yield observations. Each regional weighted average was

[2] Broadberry and others, *British Economic Growth*, p. 88. Each county's weight derives from Campbell, *Three Centuries of English Crop Yields*.

then multiplied by the respective arable contribution share of each region, and the figures were added together to make up the national aggregated gross yields.

Finally, with each individual crop's yield estimated, we may convert these into a single, 'composite' yield, taking into account each crop's share in arable. This is done on the assumption that the respective share of each crop was 33 per cent for wheat, 7 per cent for rye and maslin, 2 per cent for winter barley, 2 per cent for beans, 3 per cent for peas, 1 per cent for vetches, 13 per cent for barley, 3 per cent for dredge, and 36 per cent for oats.[3] The 'national' yield of each crop is multiplied by its share within the total arable, thus rendering the national composite gross yield.

Appendix 2. Net Yields Calculation

To convert gross yields into their net equivalent, it is necessary to establish a proportion of each individual crop's harvest used as seed corn (S^i). To calculate the 'net' yields, the following formula was used:

$$Ny^i \text{ (net yield of individual crop)} = Gy^i - (Gy^i \times S^i)$$

where Gy^i, as we have seen in Chapter 2, is the gross yield of individual crop, while S^i is seed corn (namely, a share of the total individual crop receipt, used as seed). Gy^i calculated by dividing the quantities of seed corn by the total harvest. Let us suppose that in 1300, out of the total 111.1 quarters of wheat harvest, 25 quarters were invested in seeding. The S^i is, therefore, 0.23 ($25 \div 111.1 = 0.23$, or 23 per cent).

The S^i factor varied across crops, years, and manors, but, on average, in the pre-Black Death era, the composite figures stood at 24 per cent in non-famine years (20 per cent for wheat and maslin (wheat-rye mixture), 22 per cent for rye, 23 per cent for winter barley and spring barley, 20 per cent for dredge (barley-oat mixture), 27 per cent for legumes, and 29 per cent for oats).

Once the S^i is calculated, the net yield can be calculated by multiplying the gross yield by S^i. Let us suppose that wheat yields in 1300 were 4.44. Following that, the corresponding net yields stood at 3.42 ($4.44 \times 0.23 = 3.42$). During the famine, as discussed in Chapter 3, the seeding rates were considerably higher than in non-famine years, with the composite S^i standing at 0.38 in 1315 and 1316.

[3] Very similar figures are given by Broadberry and others, *British Economic Growth*, p. 89.

Appendix 3. Peasant Yields and the Subsistence Model During the Famine Years

Calorific values for wheat, rye, barley, oats, and peas (in kcal per bushel) *c.* 1300 are taken from Campbell's *Seigniorial Agriculture*.[4] The figures for maslin and dredge are assumed to have stood half-way between wheat and rye and between barley and oats, respectively. The values for beans and vetches are taken from 2002 version of McCance and Widdowson's *The Composition of Foods*.[5]

The figures are then converted into a 'composite bushel', which is, in effect, a weighted average of the calorific values of all crops, calculated on the basis of a respective share of each crop in total arable acreage, as in Appendix 1 (33 per cent for wheat, 7 per cent for rye and maslin, 2 per cent for winter barley, 2 per cent for beans, 3 per cent for peas, 1 per cent for vetches, 13 per cent for barley, 3 per cent for dredge, and 36 per cent for oats). This renders the figure of 67,105 kcal per bushel.

Around 1300, in 'normal' years, one acre would produce about 11.22 bushels of composite crops, gross of seed and tithe. After the deduction of tithe, the figure would decline to 10.2 bushels, and after a further deduction of seed (a further 24 per cent), it would fall to 7.75 bushels. This amount would have yielded about 515,000 kcal. Accounting for storage spoilage, we should subtract a further 10 per cent from this sum, arriving at some 463,500 kcal. The composite conversion of the remaining crops into food and drink would entail a further calorific loss of 38 per cent through composite extraction rates (20 per cent for milling and baking wheat, maslin, and rye; 70 per cent for malting and brewing small quantities of the same grains into ale; 18 per cent for cooking legume pottage; 70 per cent for malting and brewing barley, dredge, and oats into ale; 22 per cent for baking barley into flatbread; 33 per cent for baking dredge into bread; and 44 per cent for baking oats into bread). These figures closely correspond with Broadberry and his colleagues' 2015 study.[6] After the conversion of raw crops into bread, pottage, and ale, we arrive at about 287,368 kcal in processed form, ready to be consumed in the course of the year.

Suppose, an 'average' family of 4.25 members held a ten-acre arable plot, of which some two-thirds (6.67 acres) could be sown at any one point (with the remaining third, or 3.33 acres laying fallow). With the composite yields of

[4] Campbell, *Seigniorial Agriculture*, p. 215.

[5] McCance and Widdowson's 2002 version of *The Composition of Foods*.

[6] Broadberry and others, *British Economic Growth*, p. 282.

7.75 bushels per acre, 10 per cent spoilage rates, and the extraction rates of 38 per cent (as above), the acres sown would have rendered about 1,917,000 kcal (287,368 × 6.67 = 1,916,743). This figure would have endowed each family member with some 1238 kcal on a daily basis, which is below the approximate requirements of 1400 crop kcal per diem per capita (accounting for 70 per cent of the total 2000 kcal).

During the famine, per-acre yields, spoilage levels, and extraction rates were all different. As we have seen in Chapter 3, the composite per-acre yields, net of seed and tithe, were 5.44 and 4.38 bushels in 1315 and 1316, respectively. These amounts would render, respectively, about 361,640 and 291,207 kcal. I have assumed spoilage rates to have been in the area of 15 per cent in 1315–1316, but 10 per cent in 1316–1317 (see Chapter 6). This would depress the calorific estimates to about 307,394 and 262,086 kcal to be converted into food and drink over the course of 1315–1316 and 1316–1317, respectively. Because of the fact that more barley, dredge, and oats were baked rather than brewed during the crisis years, I have estimated a lower extraction rate of 28 per cent (via reducing the extraction rates for barley, dredge, and oats from 70 to 40 per cent, which still accounts for both baking and brewing). After the conversion of raw crops into bread, pottage, and ale, we arrive at about 221,324 and 188,702 kcal in processed form, ready to be consumed in the course 1315–1316 and 1316–1317. These figures would have allowed each family member to enjoy only 954 and 813 kcal on a daily basis.

A schematic representation of all three scenarios can be tabulated in the following manner (Table A.1):

Table A.1. Crop yields: their calorific equivalents and capacity to feed a family.

		Gross yield	**After tithing**	**After seeding**	**After spoilage**	**Processed**
***c.* 1300**	Bushel units	1.00	0.90	0.68	0.62	0.38
	Calories per bushel unit	67,105	60,395	45,900	41,310	25,612
	Bushels per acre	11.22	10.20	7.75	6.98	4.32
	Kcal per acre	752,918	677,626	514,996	463,496	287,368
	Bushels per 10 acres (6.7 sown acres)	75	68	52	47	29
	Kcal per 10 acres (6.7 sown acres)	5,021,964	4,519,767	3,435,023	3,091,521	1,916,743
	Kcal per family, per diem, 10 acres (6.7 sown acres)	13,787	12,408	9,430	8,487	5,262
	Kcal per person, per diem, 10 acres (6.7 sown acres)	3,244	2,920	2,219	1,997	1,238

		Gross yield	After tithing	After seeding	After spoilage	Processed
1315–1316	Bushel units	1.00	0.90	0.56	0.47	0.34
	Calories per bushel unit	67,105	60,395	37,445	31,828	22,916
	Bushels per acre	9.66	8.78	5.44	4.62	3.33
	Kcal per acre	648,100	583,290	361,640	307,394	221,324
	Bushels per 10 acres (6.7 sown acres)	64	59	36	31	22
	Kcal per 10 acres (6.7 sown acres)	4,322,828	3,890,545	2,412,138	2,050,317	1,476,228
	Kcal per family, per diem, 10 acres (6.7 sown acres)	11,868	10,681	6,622	5,629	4,053
	Kcal per person, per diem, 10 acres (6.7 sown acres)	2,792	2,513	1,558	1,324	954
1316–1317	Bushel units	1.00	0.90	0.56	0.50	0.36
	Calories per bushel unit	67,105	60,395	37,445	33,700	24,264
	Bushels per acre	7.78	7.07	4.38	3.94	2.84
	Kcal per acre	521,876	469,688	291,207	262,086	188,702
	Bushels per 10 acres (6.7 sown acres)	52	47	29	26	19
	Kcal per 10 acres (6.7 sown acres)	3,480,910	3,132,819	1,942,348	1,748,113	1,258,641
	Kcal per family, per diem, 10 acres (6.7 sown acres)	9,556	8,601	5,332	4,799	3,455
	Kcal per person, per diem, 10 acres (6.7 sown acres)	2,249	2,024	1,255	1,129	813

Appendix 4. Land and Sea Transportation Costs Calculation

In 1993, James Masschaele published an article on transportation costs in early fourteenth-century England, based on pre-Black Death sheriffs' accounts.[7] In this article, Masschaele laid out, in a detailed manner, his calculation methodology, which rendered the ratio in costs per ton-mile between land transport, river transport, and sea transport of about 8:4:1. According to Masschaele's calculations, early fourteenth-century land transport cost, was, on average, 1.5d per ton-mile. My methodological approach is similar to that of Masschaele, but several differences need to be pointed out. To begin with, Masschaele accounted

[7] Masschaele, 'Transport Costs in Medieval England', p. 273.

only for wheat carting, excluding other crop transportations. Conversely, I have included all crop transportations. Also, he assumed 1 bushel of wheat to have weighed, *c.* 1300, 48 lb rather than 53 lb. Following Bruce Campbell, I have accepted 53 lb to have been the weight of 1 bushel of wheat. The corresponding figures for other crops are as follows: 51 lb for rye, 52 lb for maslin (wheat-rye mixture), 46 lb for raw barley, 37 lb for malted barley, 36 lb for oats, 41 lb for dredge, 60 lb for beans and vetches, and 46 lb for peas. Also, I have included wheat flour in the sample. Wheat flour was accounted in 'tuns' (*dolea*), with each tun being, on average, 6.67 quarters.[8] Here, I have assumed flour to have been fine and well-sieved, as is indicated in Edward I's instructions to the sheriff of York on his Scottish campaign of 1298 and reflected in numerous sheriffs' accounts recording the conversion of ground grains (*farina*) into bran-less flour (*flora*).[9] The conversion of raw grains into *farina* and *flora* is outlined in Appendix 7.

Moreover, Masschaele measured the distances as the crow flies rather than on the basis of actual roads traversed. I have, as best as I could, estimated an approximate length of roads, using old Ordnance Survey maps. The calculated lengths agree, to a large degree, with the length of roads recorded in the sheriff's accounts, stated in leagues (*leuce*). One league corresponds here to 1.5 mile and not 1 mile, as Masschaele has assumed. Because of these methodological differences, based on different assumption, there are discrepancies between Masschaele's and my calculations. Fortunately, these discrepancies are insignificant: my calculations of land-carting costs are about 13 per cent lower than Masschaele's (1.3d per ton-mile).

Sea transport costs consist of the following elements: (1) portage from granaries or barges to the cargo compartment of ships; (2) wages paid to shipmasters for transporting goods from port to port; (3) wages paid to the delivery supervisor, usually the sheriff's clerk, and occasionally his assistant, for overseeing the shipping from port to port; (4) wages paid to lodemen (pilots) for navigating the ship through the sea; (5) cargo protection costs, or 'dunnage', usually related to covering grain sacks with canvas; and (6) unloading grain at the port of arrival. The distance between two ports has been calculated, with the help of the Sea-Distances Website.[10] Although the sheriffs' accounts break

[8] As indicated in the 1325 schedule of flour purveyance of the Bordeaux garrisons by Middlesex, Oxfordshire, and Hertfordshire (TNA, 101/571/12).

[9] Santiuste, *The Hammer of the Scots*, p. 90.

[10] *Sea-Distances.*

down the shipping expenses in a detailed manner, the wardrobe account books, recording grain transportation during the famine years, account for the total carting, storing, and shipping costs together. To disaggregate the shipping costs from the combined transportation and storage costs, a ratio between the former and latter needs to be established. A meticulous analysis of the sheriff's accounts makes this exercise possible, and it appears that the ratio between the combined transportation and storage costs, on the one hand, and the shipping costs, on the other, remained fairly constant across space and time, standing at 0.70:1.00. To estimate the costs of shipping by sea, mentioned in the wardrobe book accounts of the famine years, therefore, we may multiply the combined transportation plus the storage sum by the factor of 0.7. One potential pitfall in this methodology is the fact that it assumes that carting and storing costs rose on par with sea-shipping costs during the famine years. Unfortunately, there is only one famine-year account recording cart transportation (the sheriff of Devon's account for 1315–1316), and it does not indicate any rise in carting costs during that year. Obviously, this may not be reflective of the situation elsewhere. Furthermore, as discussed in Chapter 5, there is clear evidence that storage costs rose during the famine years because the inclement weather rendered barns into deficient state, which then required repairs. All the same, it is unlikely that the storage costs rose more than sea-shipping costs, as evidence from manorial accounts indicates. If, therefore, carting costs remained at the non-famine-year levels and storage costs rose slower than sea-shipping costs, this methodology underestimates, rather than overestimates, the sea-shipping costs during the famine years.

The calculations render the ratio between land and sea transport of 7:1, rather than the 8:1 suggested by Masschaele. This estimate closely agrees with Campbell and his colleagues' estimate in their 1993 study of London's grain supply.[11]

[11] Campbell and others, *A Medieval Capital*, pp. 193–98.

Appendix 5. Estimating the Financial Value of Crop-based Calories, 1315–1317

The present estimates are based on the following:

> The composite value of crops was 10s 2d per quarter in 1315–1316 and 10s 6d per quarter in 1316–1317 (based on the relative share of each crop in the total arable: wheat = 33%; maslin = 2%; rye = 5%; winter barley = 2%; beans = 2%; peas = 3%; vetches = 1%; spring barley = 13%; dredge = 3%; and oats = 36%)

The extraction rates during the famine would be different than in normal years because many of the crops meant to be turned into loaves. As in Appendix 3, I have estimated the composite extractions to have been 38 per cent *c.* 1300 and 28 per cent in 1315–1316 and 1316–1317.

According to this reckoning, one quarter of composite crops would have rendered about 389,000 kcal, worth 10s 2d in 1315–1316 and 10s 6d in 1316–1317. The corresponding figure for 1,000,000 kcal would have been 26s 2d in 1315–1316 and 26s 11d in 1316–1317.

Appendix 6. Estimated Death Rates, Birth Rates, and Population Decline Between September 1314 and September 1317

The point of departure for the following estimates regarding the population decline during the famine is that there were about 5.25 million people in England around the harvest of 1315. Obviously, this is by no means a secure estimate, but it represents a sort of middle ground between lower and higher estimates. If we assume that the annual population growth rates were about 0.5, then England's population may have grown between the harvests of 1314 and 1315 from about 5,223,881 to 5,250,000. This would be possible if the annual crude death rates (CDR) were 30/1000 (or 0.03), while the crude birth rates (CBR) were 35/1000 (0.035). Keeping these factors in mind, we may speculate that between the harvests of 1314 and 1315, about 156,716 people would have died (5,223,881 × 0.03), and about 182,836 people would have been born (5,223,881 × 0.035). With these rough and somewhat speculative estimates, we may now move to the famine demography.

As we have seen in Chapter 7, the indexed death rates (reflected in heriots recorded into manorial court rolls) have been estimated to be 3.80 and 5.02 their 1311–1315 level in 1315–1316 and 1316–1317, respectively, while the birth rates (reflected in merchets recorded into manorial court rolls) have been estimated to be 0.50 their 1311–1315 level in both 1315–1316 and 1316–1317. These are, however, by no means reflective estimates on a national level: heriots and merchets derive from court rolls on, respectively, forty-seven and nineteen manors, scattered primarily in East Anglia and south England. There is nothing to suggest mortality levels were the same in other regions. In particular, there is no evidence that more 'pastoral' and less densely populated counties, where arable played a lesser role and where overpopulation was not an issue, suffered from the crisis as badly. Taking these facts into account, in order to arrive at more secure estimates, it is necessary to convert 'manorial' excess rates into their 'comital' and then 'national' equivalents, based on comital weightings. This conversion is done on the basis that each county is assigned a different relative weight based on the share of arable land in its total cultivable acreage (as estimated by Broadberry and his colleagues) and its population size (readjusted from Broadberry and others).[12]

1. Converting manorial excess death rates into comital and national equivalents

The benchmark is Suffolk, where arable may have occupied about 60 per cent of its cultivable land; its excess mortality levels in 1315–1316 have been estimated to have been 3.80. At Hertfordshire, about 53 per cent of land was under the plough (0.88 per cent the Suffolk level); consequently, I have estimated excess mortality levels in this county to have been about 3.33. The same 'conversion' is then made for each county, for both 1315–1316 and 1316–1317.

This conversion, however, does not account for the war damage done within the 'war-zone' counties (Northumberland, Durham, North Yorkshire, Lancashire, Westmoreland, and Cumberland). These were primarily 'pastoral' counties, where arable farming did not play a paramount role. Nevertheless, there is univocal evidence that these counties were shrinking demographically, not only because of starvation caused by a combination of harvest failures and war-related environmental destruction, but also because in some cases, Scottish raiders would kill and take local parishioners captive. In some instances, families would emigrate to the south, fleeing the Scottish wrath. While it is impossible to estimate the mortality levels within these counties, I have assumed them to match at least those in East Anglia, estimated above. Thus, I have assigned, in

[12] Broadberry and others, *British Economic Growth*, pp. 25–26 and 66–67.

a speculative manner, each of these northern counties the respective figures of 3.80 and 5.02 for 1315–1316 and 1316–1317, respectively, to reflect the war crisis they were going through.

Once excess mortality rates have been calculated for each county, the 'national' rates may be estimated by calculating the weighted average of all counties. The average is weighted on the basis of each county's population size. This yields the respective figures of 2.94 and 4.26, for 1315–1316 and 1316–1317.

2. Converting manorial excess birth rates into comital and national equivalents

The conversion of 'manorial' excess birth rates into their 'comital' and then 'national' equivalents was done in the same manner. Just as with death rates, I have assumed a degree of correlation between agricultural specialism and birth rates, with more pastoral regions (except the 'war-zone' northern counties) to have been less prone to the crisis and, thus, have a lower reduction in birth rates. Once the figures have been estimated for each county, the weighted average for all counties could be calculated, rendering 'national' rates of 0.59 and 0.58 in 1315–1316 and 1316–1317, respectively — higher than the figure of 0.50 established on a 'manorial' level.

3. Estimating CDR and CBR for the famine years

If the CDR in 'normal' years in the early fourteenth century were in the area of 30/1000, or 0.03, the respective CDR in 1315–1316 and 1316–1317 would have been around somewhere around 88/1000 ($2.94 \times 0.03 = 0.088$) and 128/1000 ($4.26 \times 0.03 = 0.128$). Conversely, the CBR are estimated to have been about 35/1000, or 0.035, in non-famine years. Henceforth, the comparable figures for the famine years may have been in the area of 21/1000 ($0.59 \times 0.035 = 0.021$) in 1315–1316 and 20/1000 ($0.58 \times 0.035 = 0.020$) in 1316–1317.

4. Estimating the sheer deaths and births during the famine

With the estimated CDR and CBR, we may now endeavour to estimate the approximate numbers of deaths and births during the famine. If on the eve of the famine there were about 5,250,000 people living in England and if between the harvests of 1315 and 1316 the CDR stood at 88/1000, then we may speculate that about 462,000 people died in the course of that year (5,250,000

× 0.088 = 462,000). Conversely, if we assume the CBR to have been about 21/1000, then about 110,250 infants were born in the course of the same year (5,250,000 × 0.021 = 110,250). The estimated population change (Δ = CBR – CDR), then, would have been of –351,750 (110,250 births – 462,000 deaths). By the harvest of 1316, the population would have declined, according to this reckoning, to about 4,898,250 people.

Between the harvests of 1316 and 1317, approximately another further 626,976 people would have died (4,898,250 × 0.128) and only some 97,965 infants (4,898,250 × 0.02) would have been born. This implies the approximate Δ of –29,011 (97,965 births – 626,976 deaths) and an annual growth rate of –10.8 per cent. By the harvest of 1317, when the harvest would have been more or less over, the population would have declined to about 4,369,239 people.

5. Estimating population decline during the famine

Dividing the population estimate around the harvest of 1317 by that of the harvest of 1315 renders about 0.83 (4,369,239 ÷ 5,250,000), namely, a population loss of 17 per cent.

I am well aware that the estimated figure of 5.25 million around the harvest of 1315, serving as my benchmark, may well be either an under-estimate or overestimate and, as such, is by no means a convincing one. But it should be borne in mind that if we stick to the estimate of Broadberry and his colleagues (4.69 million), we eventually would end up with the same decline of 17 per cent.[13]

[13] Broadberry and others, *British Economic Growth*, p. 21.

Appendix 7. Multure Profits During the Famine Years

The following estimates assume a *by-weight* price ratio between raw wheat and wheat flour to have been in the area of 1:1.75, as it was the case in the late fifteenth and sixteenth centuries.[14] Unfortunately, I was not unable to identify any comparable records in the pre-Black Death period.

Suppose in 1316, a hypothetical tenant brought 1 quarter of wheat to be ground at his lord's mill, in accordance with the 'suit of mill' custom. One quarter of wheat would weigh 424 lb, and its financial equivalent would have been in the area of 15s. As a *multure* fee, the same peasant would have been charged a tiny fraction of his grain: say, 1/16, the median figure throughout the country south of the Humber. This would have entailed a loss of a half-bushel (= 0.625 quarters), namely, about 26.5 lb, worth about 11d. The remaining quantities to be milled by the same tenant would have amounted to 7.5 bushels (= 0.935 quarters), weighing about 397.5 lb and worth 14s 0.75d. The grinding of the raw grains would have increased the volume by approximately 16 per cent, thus rendering about 1.09 quarters of *farina*, consisting of flour and bran.

Having converted his grains into *farina*, the same tenant would have had two options: either to retain the *farina* as it was, ready to be baked into the coarsest type of whole-grain bread (*panis de toto blado*) without reducing the bakeable content, or to remove the bran and, thus, reduce the volume, weight, and calorific content by about 22, 15, and 10 per cent, respectively.[15] If the tenant chose the former option, he would then retain 397.5 lb of *farina*, valued at about 24s 7.5d. Perhaps this figure is somewhat inflated, given that the bran content would have dragged the value of *farina* down. It was nevertheless greater than the value of 337.9 lb of *flora*, valued about 20s 11d, if the tenant went with the latter option. This implies the respective profits of about 10s 6.5d (24s 7.5d – 14s 0.75d) or 6s 4.5d (20s 11d – 14s 0.75d = 6s 10.5d).

[14] Calculated from Rogers, *A History of Agriculture and Prices in England*, vol. III: *1401–1582*.

[15] For the reduction in volume, see Langdon, *Mills in the Medieval Economy*, p. 153; for the reduction in weight, see Campbell and others, *A Medieval Capital*, pp. 191–92; the calorific reduction estimates derive from the 2002 version of McCance and Widdowson's *The Composition of Foods*.

Appendix 8. Archival Shelf-Marks and Printed References of Statistical Sources Used in or Consulted for the Present Study

For Abbreviations, see pages xv–xviii

Manorial (aka, Demesne) Accounts

ENGLAND

Bedfordshire: Cranfield: TNA, SC 6/740/11; Millow: NorthantsRO, Spencer Roll 26; Shillington: TNA, SC 6/741/19

Berkshire: Ashcott: LH, 10766; Brightwell, Harwell, Wargrave, Culham, Waltham St Lawrence, Billingbear: HantsALS, 11M59/B1/69–73; Combe: KCA, COM/59; Windsor Park: TNA, SC 6/752/9

Buckinghamshire: Ivinghoe, Morton, West Wycombe: HantsALS, 11M59/B1/69–73; Langley-Marish: 1314–1318: TNA, SC 6/761/19; Beaumond: TNA, SC 6/759/1; Maids Moreton: BoL, DD Ch Ch27; Ibstone: MCA, 5077; Cheddington: MCA, 5546; Meesden: TNA, SC 6/867/4; Turweston: WAM, 7773–75

Cambridgeshire: Merton Hall in Cambridge (lands of Merton College): MCA, 4973–76; Harston: BL, Add. Ch. 13522; Gamlingay: MCA, 5370–73; Melbourne: BL, Add Ch 25866; Wisbech Barton: CUL, EDR D8/1/1–3; Newton: CUL, EDC 7/11B; Knapwell: TNA, SC 6/768/29

Cheshire: Frodsham: TNA, SC 6/801/12; Warmingham: BelCM, 1171

Derbyshire: Wirksworth: StaffRO (L), D30/4/5/54; Weston-on-Trent: StaffRO (S), D(W) 1734/J/1994

Dorset: Cranbourne: HaHM, Court Roll 1/10; Sydling St Nicholas: WCA, 18215

Durham: Billingham: DCA, Billingham accounts, 1315–1316, 1316–1317; Bewley: DCA, Bewley accounts, 1316–1317; Dalton: DCA, Dalton accounts, 1316; Elvethall: DCA, Hostillar Status 1318; Ferryhill: DCA, Ferryhill accounts, 1316–1317; Ketton: DCA, Ketton accounts, 1316–1317; Stillington: MCA, 5986

Essex: Birdbrook: WAM 25421–23; Feering: WAM, 25630, 25632, 25634; Kelvedon: WAM, 25800–4, HuntALS, M37/1/4; Middleton: CCA, DCc/BR/Middleton/27–30; Bocking: CCA, DCc/BR/Bocking/11; Thundersley: TNA, SC 6/847/26; Wivenhoe: ERO, T/B/122/1; Wickham: LMA MS 10467C

Gloucestershire: Bourton on the Hill: WAM, 8260–63; Ham (near Berkeley): BerCM, GAR 111; Berkeley: BerCM, GAR 291; Hinton: BerCM, GAR 163; Todenham: WAM, 25944–5, GRO, D1099/M30/5; Sutton-under-Brailes: GRO, D1099/M31/34; Minchinhampton: TNA, SC 6/856/18

Hampshire: Alresford, Ashmansworth, Beauworth, Bentley, Bishopstoke, Bishop's Waltham, Bitterne, Burghclere, Cheriton, Crawley, Droxford, East Meon, East Meon Church, Ecchinswell, Fareham, Hambledon, Havant, High Clere, Mardon, North Waltham, Overton, Sutton, Twyford, Weald, Woodhay: HantsALS, 11M59/B1/69–73; Crondall, Long Sutton, Hannington, Wooton St Lawrence, Whitchurch, Hurstbourne Tarrant, Wonston, Houghton, Martyr Worthy, Chilbolton, Michelmersh, Easton, Silkstead, Littleton, Barton Priors: HantsALS, DC/J1/12–15; Cosham and Werrore: QCA, 246, 251; Heckley: QCA, 200–03; Padwell: QCA, 276–77; Rowner (in Gosport): CRO, PB 4/143

Herefordshire: Tyberton: HCA, 359a

Hertfordshire: Amwell: BL, Add. Roll 66734; Kinsbourne: WAM, 8803–04; King's Langley: TNA, SC 6/866/20; Stevenage: TNA, SC 6/871/5–6, LMA, MS 25404; Wheathampstead: HertALS, D/Elw/M130–1; Wymondley: HertALS, 57524

Huntingdonshire: Houghton: TNA, SC 6/878/20; King's Ripton: TNA, SC 6/882/28

Isle of Wight: Calbourne: HantsALS, 11M59/B1/69–73; Swanston: TNA, SC 6/987/13

Kent: Adisham: CCA, DCc/BR/Adisham/5; Agney (Old Romney): CCA, DCc/BR/Agney/29–30; Appledore: CCA, DCc/BR/Appledore/16–19; Barksore: CCA, DCc/BR/Barksore/21–23; Barton Carrucate (Northgate, Canterbury): CCA, DCc/BR/Bartone Carrucate/5, 12, 45; Chartham: CCA, DCc/BR/Chartham/9, 11–13; Cliffe: CCA, DCc/BR/Cliffe/28; Copton (Preston): CCA, DCc/BR/Copton/9–11; East Farleigh: CCA, DCc/BR/East Farleigh/10–11; East Peckham: CCA, DCc/BR/East Peckham/28–31; Eastry: CCA, DCc/BR/Eastry/40, 42–46; Ebony: CCA, DCc/BR/Ebony/27–30; Fairfield: CCA, DCc/BR/Fairfield/8–13; Great Chart: CCA, DCc/BR/Great Chart/28–30; Ham: CCA, DCc/BR/Ham/18–20; Hollingbourne: CCA, DCc/BR/Hollingbourne/26–28; Ickham: CCA, DCc/BR/Ickham/20–22; Leysdown: CCA, DCc/BR/Leysdown/16–17; Loose: CCA, DCc/BR/Loose/30–32; Lydden: CCA, DCc/BR/Lydden/34–37; Meopham: CCA, DCc/BR/Meopham/59–63; Mersham: CCA, DCc/BR/Mersham/7–8; Monkton: CCA, DCc/BR/Monkton/44, 46, 50; Orpington: CCA, DCc/BR/Orpington/14–18; Ruckinge: CCA, DCc/BR/Ruckinge/4–6; Seasalter: CCA, DCc/BR/Seasalter/1; West Farleigh: CCA, DCc/BR/West Farleigh/31–34; Westwell: CCA, DCc/BR/Westwell/11–16; Maidstone: LPA, ED 659; Otford: LPA, ED 832; Aldham and 'le Woodland': ESRO, SAS/G/45/2; Westerham: WAM, 26411–2, 26415; Wouldham: WCA, C834

Lancashire: Lytham St Annes: DCA, Lytham 1317–1318 account

Leicestershire: Blackfordby: SalopRO, 972/1/1/595; Kibworth: MCA, 6219–21

Lincolnshire: Benniworth: TNA, SC 6/1144/2; Rippingale: LA, Anc 3/1; Temple Bruer: TNA, DL 29/242/3882

Middlesex: Eyebury: TNA, SC 6/919/13–16; Knightsbridge: WAM, 16403–8; Teddington: TNA, SC/6/918/14, WAM, 27176

Norfolk: Brandiston: MagCA, 166/10; Burnham Thorpe: NRO, WAL 63 269x1; Calthorpe: NRO, NCR Case 24, Shelf C; East Lexham: NRO, Kimberley, 1/8/1; Easton: NRO, DCN 60/8/17; Eaton (Norwich Cathedral Priory sacrist's manor): BoL, Norfolk Rolls 24–25; Gateley: NRO, DCN 60/13/17; Gnatingdon: NRO, DCN 60/14/16; Hemsby: NRO, DCN 60/15/10; Hindolveston: NRO, DCN 60/18/20; Hindringham: NRO, DCN 60/20/18; Kempstone: NRO, WIS 2/163x1; Martham: TNA, DCN 60/23/15; Monks Granges: TNA, DCN 60/26/16; Newton: CUL, Ch(H)/4/4/30; North Elmham: NRO, DCN 60/10/16; Scratby: NRO, DCN 60/30/4–6; Sedgeford: NRO, DCN 60/33/19; Taverham: NRO, DCN 60/35/18; Thornham: NRO, DCN 60/37/12; West Tofts: NRO, Philips 500; Worstead: NRO, DCN 60/39/6; Wymondham: NRO, NRS 8811 21 E4 and NRS 18524 33 D4; Great Raynham: RHM, Great Raynham Accounts, 1315–1316; Stanhoe: RHM, Stanhoe Accounts, 1314–1315

Northamptonshire: Boroughbury: NorthantsRO, FM 248; Werrington: NorthantsRO, FM 244

Northumberland: Pontilond: MCA, 5986

Nottinghamshire: Gringley: TNA, SC 6/953/18; Clipston: TNA, 953/8; Wheatley: TNA, SC 6/954/21–23

Oxfordshire: Adderbury, Witney: HantsALS, 11M59/B1/69–73; Bicester: TNA, SC 6/955/20; Chastleton: BoL, Ch Ch 6; Cuxham: MCA, 5838–41; Caversfield: TNA, SC 6/957/16; Launton: WAM 15325–7; Little Tew: BoL, DD Ch Ch, 104–05

Rutland: Oakham: WAM, 20239–42

Shropshire: Adderley: TNA, SC 6/965/2

Somerset: Holway, Hull, Nailsbourne, Poundisford, Rimpton, Staplegrove, Taunton: HantsALS, 11M59/B1/69–73; Baltonsborough, Doulting, Glastonbury, Greinton, High Ham, Hunstrete, Marksbury, Milton Podimore, Pilton, Shapwick, South Brent, East Brent, Street, Walton, Westonzoyland, Middlezoy, Othery, Wrington: LH, 10766

Suffolk: Downham (Newhall): SuffRO (BSE), 651.35/1; Eye Grange, Hadleigh, Thorndon: TNA, SC 6/996/14; Fornham St Martin: SuffRO (BSE), BE 3.15/9/2/16(b); Hinderclay: UCL, Bacon Rolls 445–47; Sapiston: SuffRO (I), HD 1538/338/15; Westley: SuffRO (BSE), E3.15/15/2/1

Surrey: Byfleet: TNA, SC 6/1011/4–7; Charlwood: CCA, DCc/BR/Charlwood/33–34; Cheam: CCA, DCc/BR/Cheam/30–34; Esher, Farnham: HantsALS, 11M59/B1/69–73; Farleigh: MCA, 4825–30; Horsley (royal manor): TNA, SC 6/1013/4; Horsley (manor of Canterbury Cathedral Priory): CCA, DCc/BR/Horsley/7; Malden: MCA, 4656–57; Thorncroft (in Leatherhead): MCA, 5745–48; Sheen: TNA, SC 6/1014/2–6; Walworth: CCA, DCc/BR/Walworth/6–9

Wiltshire: Bishopstone, Downton, Fonthill, Knoyle, Upton Knoyle: HantsALS, 11M59/B1/69–73; Longbridge Deverill: LH, 9645, 10688, 9647; Monkton Deverill: LH, 9763–66; Westwood, Stockton, Wroughton, Overton, Alton Priors, Enford: HantsALS, DC/J1/12–15; Cowesfield (in Whiteparish): WSHC, 192/31

Worcestershire: Chaddesley Corbett: WAM, 21066–8; Grimley: WCA, C582; Castlemorton with Longdon manor: WAM, 21047 and TNA, SC 6/1109/5; Oldington: TNA, SC 6/1070/21–23; Pershore: WAM, 22101–4, 22107

Yorkshire: Bolton, Stede, Halton, Riddings, How, Ingthorpe, Malham, Angram, Kildwick, Rither: *The Bolton Priory Compotus, 1286–1325: Together with a Priory Account Roll for 1377–1378*, ed. by I. Kershaw and D. M. Smith (Woodbridge: Boydell and Brewer, 2000); Meltham: TNA, SC 6/1085/11; Tickhill Castle: TNA, SC 6/1088/1; Patrington, Burton, Skidby, Elloughton, Wilton, Wetwang, Sherburn, 'Swythmund', Archbishop of York's grange near York: TNA, SC 6/1144/2

WALES

Glamorganshire: Llanblethian, Llantwit-juxta-Neath, Neath, Roath (near Cardiff): TNA, SC 6/1202/7–9

Monmouthshire: Caerleon: TNA, SC 6/921–24; Llangwm: TNA, SC 6/922/31; Llangeston: TNA, SC 6/923/6; Llantrissent: TNA, SC 6/923/26–27; Treigruk: TNA, SC 6/925/15–16; Trelleck: TNA, SC 6/925/32; Troy: TNA, SC 6/926/18–19; Usk: TNA, SC 6/927/18, 22

Manorial Court Rolls

ENGLAND

Bedfordshire: Rowlands Manor (in Cople): BARS, Russell Box 212 (R8/14/1/1–10); Sewell: BARS, Russell Box 100; Wroxhill: NorthantsRO, E(S) 1190 mm 2,4,7,8

Berkshire: Bray: SGCA, SGC XV.58.D.1, 5, 7; Brightwalton: TNA, SC 2/153/67; Combe: KCA, BEC/7, 13–15; Hungerford: KCA, BEC/07, 13–15; Inkpen: BRO, D/EC/M53; Wantage: KCA, BEC/14–15

Buckinghamshire: Bledlow: KCA, BEC/07, 13–15; Cheddington: MCA, 5641–44; Chesham Bois: CBS, D/BASM/18/6; Chicheley: NorthantsRO, SS2398; Great Horwood: NCA, 3913; Ibstone: MCA, 5216; Northall: BL, Add Ch 67975–77

Cambridgeshire: Balsham: LMA, ACC/1876/MR/02/001; Chesterton: BoL, Gough Camb 1; Cottenham, Dry Drayton and Oakington: CUL, QC 11; Foxton: CTCA, Box 27/4–5; Littleport: CambrRO, R93/96; Methwold: CrCA, Box 39; Mildreth: LMA, H01/ST/E/79/1; Newton-cum-Hauxton: CUL, EDC 7/11A; Ovington Bozomes: CUL, CUA, D.XXIII.73; Sutton-in-the-Isle: CUL, EDC 7/4A

Dorset: Gillingham: DHC, D/FAN, 10 Edward II; Milbourne Stileham: KCA, BEC/13–15; Povington: KCA, BEC/07, 13, 15; Tarrant: TNA, SC 2/170/51

Essex: Dunton Waylett: KCA, BEC/13, 15; Great Birch: ERO, D/Dr/M2; Great Bromley: ERO, D/DRg 1/30; Great Waltham: ERO, D/DTu 257; Little Dunmow: TNA, SC 2/171/63; Matching: ESO, D/DeW M1; Messing: ERO, D/DH/X1; Rickling: TNA, SC 2/173/51–52; Roding Marcie: UCA, E3/CR1/1

Gloucestershire: Hawkesbury: TNA, SC 2/175/42; Mitchinhampton: TNA, SC 2/175/83; Shipton Moyne: GRO, D1571/M18

Hampshire: Anne Bec (in Monxton): KCA, BEC/7, 13–15; Ellingham: EaCA, ECR 17/61; Hound and Netley: NYRO, ZJX 3/10/6 Roll 12; Quarley: KCA, BEC/7, 13–15; Southwick: HantsALS, M53/C1–27

Herefordshire: Canon Pyon: HCA, R1013; Clehonger, Eggleton, Holme Lacy and Marden: HCA, R1113–14; R1091–3; TNA, C115/95/6834/5; Pencombe: HereARC, A63/I/1/2–4

Hertfordshire: Ashwell: TNA, SC 2/176/28–30; Branfield: HertALS, 40702; Cassio: BL, Add. MS 40626; Chipping Barnet and East Barnet: BL, Add. MS 40167; Croxley (in Rickmansworth): BL, Add. MS 5834 and Add. MS 6057; Hexton: HertALS, DE/B/2355/M/111; Newland Squillers: BoL, MS Gough Herts 1; Norton: HertALS, 65498–65518; Park manor (in St Albans, near St Stephen's parish): BL, Add. MS 40625; Ridge: HertALS, DE/B/2067/B/M1; Sandridge: HertALS, 40700; Westmill: TNA, SC 2/178/74

Kent: Hoo St Werburgh: TNA, SC 2/181/28

Leicestershire: Barkby: MCA, 6500–05

Lincolnshire: Alsacoe [5 parishes: Atterby, Grayingham, Harpswell, Helmswell and Snitterby]: LA, KR 53, 59, 62, 71, 76, 81, 84, 90, 94; Manley [7 parishes: Ashby, Brumby, Burringham, Redbourne, Scunthorpe, Stainton, Winterton]: LA, KR 47, 57, 65, 69, 80, 87, 92, 96, 101; Corringham [10 parishes: Aisby, Blyton, Great Corringham, Heapham, Little Corringham, Morton, Northorpe, Somerby, Springthorpe, Walkerith]: LA, KR 43, 46, 56, 60, 64, 68, 79, 82, 85, 91, 95, 99; Crowle: LA, CM 1/1–8; Edwinstowe: LA, DC Linc A 4/4/3 and 15; Fulstowe: LA, Anc 3/5/2, 3, 5 and 13; Halton and Bourton-on-the-Stour: WAM, 14545–46; Ingoldmells: LA, MM 8/3/5, 15–16

Middlesex: Edgware: TNA, SC 2/188/54; Harrow: LMA, ACC 76/2410; Isleworth: LMA, Acc 1379/2; Ruislip: KCA, BEC/7, 13, 15; Tottenham: HALH: MT/1/1–2

Norfolk: Aldeby: NRO, DCN 60/1/6–7; Blackworth: NRO, FEL 317; Brandiston: MagCA, 89/2; Bressingham: NRO, WAR 3–11; Burnham Thorpe: NRO, WAL 1635; Coltishall: KCA, COL/363; East Raynham: RHM, East Raynham court rolls (attic; boxes 27, 31, 33, 35); East Wretham: KCA, BEC/7, 13–15; Gressenhall: NRO, ING 31–46; Hargham: CUL, Buxton Papers, Box 77;

Heacham: NRO, Lest DA/6–8; Hindolveston: NRO, DCN 60/19/3; Hingershale: NRO, WIS 1; Horsham St Faith: NRO, NRS 19505; Hunstanton: NRO, ING/227; Kempstone: NRO, WIS 1; Kipton: CUL, Weasenham Box 3; Langham: RHM, Langham Court Rolls (boxes 32 and 48); Lessingham: KCA, BEC/7, 13–15; Marham: NRO, Hare 2180, 194x4; Sedgeford: NRO, DCN 60/32/8–9; Tattersett, RHM, Tattersett court rolls (box 52); Tivetshall: NRO, WAL 444, 274x2, WAL 1231, 1690, 1692–94; Walsoken: WFM, WK/2; West Rudham: RHM, West Rudham court rolls (attic boxes); Wimbotsham: NRO, Hare 4237; Wymondham: NRO, NRS 18475 and 18481

Northamptonshire: Brigstock: NorthantsRO, Montague, X364A/22–31

Oxfordshire: Cottisford: KCA, BEC/7, 13, 15; Cuxham: MCA, 5908; Launton: WAM, 15424–27; Newington: CCA/U15/26/50–67; Swyncombe: KCA, BEC/7, 13, 15

Shropshire: Claverley: SalopRO, 5586/1/7

Somerset: Abbotts Leigh: SARS, DD\SAS/C795/PD/79; Fairfield: SARS, DD\AH/66/5/1–2; Hillfarrance: SARS, DD\AH/66/7/1

Staffordshire: Alrewas: StaffTO (S), D(W)0/3/8; Essington: StaffRO (S), D1790/C/70, 73–74; Haywood: StaffRO (S), D(W) 1734/2/1/426 and D(W) 1734/2/1/598; Hilton Park: StaffRO (S), D1790/C/2; Pattingham: StaffRO (S), D(W)1807/1; Tamworth: KUL, Tamworth Court Rolls, EDII/8–22

Suffolk: Drinkstone: SuffRO (BSE); E7/10/1.1–2; Flixton: SuffRO (L), HA12/C3/2; Fornham All Saints: SuffRO (BSE), E3/15.9/1.2 and 1.4; Fornham St Martin and St Genevieve: SuffRO (BSE), E3/15.9/1.1–1.2; Great Barton: SuffRO (BSE), E18/151/1; Great Blakenham: EaCA, ECR 29/122–134a; Hinderclay: UCL, Bacon Roll 119–20; Horham: SuffRO (I), HA 68/484/86; Lakenheath: CUL, EDC 7/16A; Melton and Kingston: CUL, EDC 7/17A; Palgrave: SuffRO (I), HA 411/2/1/12/1/3–4; Preston: KCA, COL/15; Redgrave: UCL, Bacon Roll 11–14; Rickinghall: BL, Add Ch 63422–32; South Elmham: SuffRO (L), HA12/C2/3–8; Walsham-le-Willows: *Court Rolls of Walsham le Willows, 1303–50*, ed. by Ray Lock (Woodbridge: Boydell and Brewer, 1998), pp. 34–61; Worlingworth: SuffRO (I), HA 116/3/19/1/2

Surrey: Addington: LPA, TH 19–23; Farleigh: MCA, 4909–13; Lambeth: LPA, ED 1902; Leatherhead: MCA, 5744–50; Merstham: CCA/U15/22/10, 16–18, 20; Milton: CCCA (O), Ga/e, f, i, j, k; Pleshney: TNA, DL 30/62/875; Shalford: TNA, SC 2/173/66; Streatham: KCA, COL/13, 15; Tooting Bec: KCA, COL/13, 15

Sussex: Udimore: ESRO, RAY 3/1/1–3/1/6; Wartling, Cowbeech and Socknersh: BL, Add Roll 32617–20, 31502–4

Warwickshire: Alveston: WCA, E6–9, R1162; Atherstone: KCA, BEC/7, 13–15, ATH/26; Middleton: UNL, MiM 131/4–11; Monkspath (in Tanworth): BoL, MS Top. Warwick c. 1

Wiltshire: Brixton Deverill: KCA, BEC/7, 13–15; Chisenbury: KCA, BEC/7, 13–15; Lavington Rector: MagCA, 78/; Longbridge Deverill and Monkton Deverill: LH, 9660–65, 8063 and 9770; Ogbourne: KCA, BEC/13–15

Worcestershire: Blackwell, Broadwas, Cleeve Prior, Cropthorne and Netherton, Hallow and Grimley, Harvington, Himbleton, Moor, Newnham, Overbury, Sedgeberrow, Teddington, Tibberton and Wolverley: WCA, E6–9, R1162; Bockleton: StaffRO (S), D(W)1788/P39.B10; Eldersfield: WAAC, 007:134/BA1531/69a/1; Halesowen [4 townships, on the Worcestershire-Shropshire border: Hales parish, Hales borough, Romsley and Warley]: Birmingham Archives, Heritage and Photography Service (BAHPS), 346233–346239; 350357–350360; Powick: WAAC, BA 845:705.93, parcel 4; Stoke Prior: WCA, E198–200, 202–03; Tardebigge: WAAC, BA 1188:705.128, box 12

Yorkshire: Gargrave: YMA, AA 9.14; Wakefield: *Court Rolls of the Manor of Wakefield*, vol. III: *1313 to 1316, and 1286*, ed. by John Lister, The Yorkshire Archaeological Society, 57 (Leeds: Printed for the Yorkshire Archaeological Society, 1917); *Court Rolls of the Manor of Wakefield*, vol. IV: *1315 to 1317, and 1286*, ed. by John Lister, The Yorkshire Archaeological Society, 78 (Leeds: Printed for the Yorkshire Archaeological Society, 1930); LUL, MD 225/1/37–48; BL, Add Ch 54408

WALES

Denbighshire: Dyffryn Clwyd: TNA, SC 2/215/71–74, 216/1–3

Monmouthshire: Porthcasseg: NLW, Badminton 1642–1643; Troy: TNA, SC 2/192/11

Tithe Receipts

Durham: Billingham: DCA, Billingham account rolls, 1305, 1315, 1320, 1323; Ferryhill: DCA, Ferryhill account rolls 1320, 1324; Newton Bewley: DCA, Bewley account roll 1320; Spennymoor: DCA, Ferryhill account roll 1316; Wolviston: DCA, Bewley account rolls, 1303, 1320, 1325

Essex: Feering rectory: WAM, 25602, 25604, 25607, 25609, 25611, 25613, 25617, 25619, 25621, 25623, 25625, 25627, 25631, 25633, 25635, 25637, 25642, 25644, 25647, 25469; TNA, SC 6/841/5

Gloucestershire: Todenham rectory: GRO, D1099/M30/4

Hampshire: East Meon and Hambledon: HantsALS, 11M59/B1/57–78; Wooton St Lawrence, HantsALS, DC/J1/6–17

Hertfordshire: Wheathampstead: WAM, 8919

Kent: Eastry: CCA, DCc/BR/Eastry/28–31, 34, 36, 39, 41; Monkton and Birchington: CCA, DCc/BR/Monkton/22, 30, 33, 37, 38, 40, 47, 50, 54

Middlesex: Hendon rectory: WAM, 32624–6

Norfolk: Scratby: NRO, DCN 60/30/4–8; Sedgeford: NRO, LEST/IC/2, LEST/IB/79–80

Northumberland: Norham parish and Islandshire [5 parishes]: DCA, Proctor of Norham accounts, 1300–1301, 1314–1315, 1315–1316, 1317–1321

Wiltshire: Overton, Alton Priors, Enford: HantsALS, DC/J1/6–17

Yorkshire: Skipton and Kildwick: *The Bolton Priory Compotus, 1286–1325: Together with a Priory Account Roll for 1377–1378*, ed. by I. Kershaw and D. M. Smith (Woodbridge: Boydell and Brewer, 2000)

Rentals and Extents

Feering extent (Essex), 1289: CUL, MS Kk 5.29, fols 95^{v}–111^{v}

Monkton extent (Kent): CCA, Register C, fols 128^{r}–137^{r}

Sedgeford extent (Norfolk): BL, Add MS 57973, fols 2^{r}–18^{v}

Port Customs Accounts

Boston: TNA, E 122/6/3

Bristol: TNA, E 122/15/1

Exeter: Maryanne Kowaleski, *Local Customs Accounts of the Port of Exeter, 1266–1321*, Devon and Cornwall Record Society, 36 (Exeter, 1993), pp. 130–52

Ipswich: TNA, E 122/50/3

King's Lynn: E 122/93/17

Scarborough: TNA, E 122/56/7, 10; E 122/34/1–2

Southampton: TNA, E 122/36/8

Winchelsea: TNA, 122/32/6; SC 6/1031/26

Lay Subsidy Rolls

Bedfordshire: TNA, E 179/71/10–13 (1310, 1315, 1316, and 1333)

Cambridge, city of: TNA, E 179/81/5 (1315)

Cambridgeshire: TNA, E 179/81/7 (1315)

Hertfordshire: TNA, E 179/20/8–11 (1308, 1314, 1317, and 1323)

Oxfordshire: TNA, E 179/8 and 10 (1316 and 1307)

Oyer and Terminer Rolls

Gloucestershire: TNA, JUST 1/296 (1324)
Herefordshire: TNA, JUST 1/316 (1325)
Lancashire: TNA, JUST 1/425 (1324)
Norfolk: TNA, JUST 1/598 (1325)
Suffolk: TNA, JUST 1/853 (1324–1325)
Yorkshire: TNA, JUST 1/1117 (1324)

Sheriffs' Accounts (Including Purveyance Accounts)

Cambridgeshire and Huntingdonshire: TNA, E 101/552/11–15 and 50
Cumberland: TNA, E 101/554/9, 11–12
Devon: TNA, E 101/555/3–4 and 7
Essex and Hertfordshire: TNA, E 101/556/6–7, 9–10, 14–15, 17
Gloucestershire: TNA, E 101/559/5, 8–9
Hampshire: TNA, E 101/561/5, 9–10
Kent: TNA, E 101/566/2–3, 9–10, 13–14
Lincolnshire: TNA, E 101/568/3, 7–10, 17–19
London and Middlesex: TNA, E 101/571/10, 12 and 18
Norfolk and Suffolk: TNA, E 101/574/22–25, 27–29 and 32
Northamptonshire: TNA, E 101/577/8–9
Nottinghamshire and Derbyshire: TNA, E 101/580/21
Oxfordshire and Berkshire: TNA, E 101/582/9
Somerset and Dorset: TNA, E 101/585/5–6, 9, 12–13
Surrey and Sussex: TNA, E 101/588/7, 10 and 12
Wiltshire: TNA, E 101/593/5–7
Worcestershire: TNA, E 101/592/8
Yorkshire: TNA, E 101/597/17–19
Central purveyance account of all sheriffs, for 1324–1327: TNA, E 358/3

Royal Wardrobe Accounts

1302–1303 account: BL, Add MS 17360

1310–1311 account: BL, Nero, C viii

1310–1311 account related to Berwick-upon-Tweed: BoL, Tanner 197

1315–1316 account: TNA, E 101/376/7

1316–1317 account: SAL, MS 120

1317–1318 account: SAL, MS 121

1324–1327 account: SAL, MS 122

Gaol Delivery Rolls and Justiciar of Ireland's Rolls

Bedfordshire: TNA, JUST 3/1/1–3

Buckinghamshire: TNA, JUST 3/5/2–3

Essex: TNA, JUST 3/19/1–4

Herefordshire: TNA, JUST 3/222

Hertfordshire: TNA, JUST 3/23/2 and JUST 3/22/2

Kent: TNA, JUST 3/26/2 and JUST 3/27/2–3

Middlesex and London: TNA, JUST 3/40/2, JUST 3/41/1, and JUST 3/42/1

Norfolk: TNA, JUST 3/48 and JUST 3/49/1

Northamptonshire: TNA, JUST 3/51/2, JUST 3/51/1, and JUST 3/51/3

Suffolk: TNA, JUST 3/63/4

Surrey and Sussex: TNA, JUST 3/112

Yorkshire: TNA, JUST 3/75 and JUST 3/76

Various counties: TNA, JUST 3/110

Justiciar of Ireland (Cork, Dublin, Kildare, Kilkenny, Limerick, Louth, Meath, Tipperary, and Waterford): NAI, KB 2/1–12

Coroners' Rolls

Bedfordshire: TNA, JUST 2/6 and 2/272

Kent: TNA, JUST 2/47

London and Middlesex: TNA, JUST 2/94a–94b

Northamptonshire: TNA, JUST 1/640, JUST 2/106–108b

Wallingford Castle (Berkshire): BRO, W/JCm/1–4

Ecclesiastical Valuations

The 1291–1292 *Taxatio ecclesiastica*: *Taxatio Ecclesiastica Angliae et Walliae Auctoritate*, ed. by Thomas Astle, Samuel Ayscough, and John Caley (London: Record Commission, 1802)

***Nova taxatio* of Yorkshire, Durham, Northumberland, Cumberland, Westmoreland, and Lancashire (1318–1328)**: TNA, E 179/67/9

***Nova taxatio* of Irish provinces of Meath, Cloyne, Ross, Limerick, Emly, Waterford, Cashel, and Cork (1318)**: TNA, E 101/233/21

***Nova taxatio* of Ossory province**: RCBCIA, The Red Book of Ossory

Episcopal Registers

Bath and Wells: *The Register of John de Drokensford, Bishop of Bath and Wells (AD 1309–1329)*, ed. by Rev. Bishop Hobhouse (Bath: Printed by the Somerset Record Society, 1887)

Canterbury: LPA, Register Reynolds

Carlisle: *The Register of John de Halton, Bishop of Carlisle, AD 1292–1324*, vol. 2, ed. by W. N. Thompson (London: Issued for the Canterbury and York Society, 1913)

Coventry and Lichfield: *The Register of Walter Langton, Bishop of Coventry and Lichfield, 1296–1321*, vol. 1, ed. by J. B. Hughes (Woodbridge: Boydell and Brewer, Printed for the Canterbury and York Society, 2001)

Hereford: *Registrum Ricardi de Swinfield Episcopi Herefordensis, AD MCCLXXXIII–MCCCXVII*, ed. by William W. Capes (London: Issued for the Canterbury and York Society, 1909); *The Register of Adam de Oreleton, Bishop of Hereford (AD 1317–1327)*, ed. by Rev. A. T. Bannister (Hereford: Wilson and Phillips, 1907)

Lincoln: LA, REG/2

Norwich: NRO, DN/Reg/1

Salisbury: *The Registers of Roger Martival, Bishop of Salisbury, 1315–1330*, vol. I: *The Register of Presentations and Institutions to Benefices*, ed. by Kathleen Edwards (Canterbury: Canterbury and York Society, 1959); *The Registers of Roger Martival, Bishop of Salisbury, 1315–1330*, vol. II: *The Register of Divers Letters*, ed. by C. R. Elrington (Canterbury: Canterbury and York Society, 1963)

Winchester: *The Registers of John de Sandale and Rigaud de Asserio, Bishops of Winchester (AD 1316–1323)*, ed. by Francis Joseph Baigent (London: Simpkin and co., 1897)

Worcester: *The Register of Thomas de Cobham, Bishop of Worcester, 1317–1327*, ed. by Ernest Harold Pearce (London: Mitchell Hughes and Clarke, 1930); *The Register of the Diocese of Worcester during the Vacancy of the See*, ed. by J. W. Willis Bund (Oxford: James Parker and co., 1893); Worcestershire Record Office, b. 716–093-B.A.2648/1

Feet of Fines

Bedfordshire: TNA, CP 25/1/3/40–47

Berkshire: TNA, CP 25/1/9/39–42, CP 25/1/10/43–45

Buckinghamshire: TNA, CP 25/1/18/61–72

Cambridgeshire: TNA, CP 25/1/26/52–53, CP 25/1/27/54–63

Cornwall: TNA, CP 25/1/32/16–18

Cumberland: TNA, CP 25/1/35/8

Derbyshire: TNA, CP 25/1/37/24–28; Calendared: Charles Kerry, 'A Calendar of the Fines for the County of Derby, from Their Commencement in the Reign of Richard I', *Derbyshire Archaeological and Natural History Society*, 15 (1893), 1–19; *Derbyshire Archaeological and Natural History Society*, 17 (1895), 95–113; *Derbyshire Archaeological and Natural History Society*, 18 (1896), 1–17

Dorset: TNA, CP25/1/49/28–32

Devon: CP 25/1/42/37–40, CP 25/1/43/41; Calendared: *Devon Feet of Fines, 1 Edward I–43 Edward III: 1272–1369*, vol. 2, ed. by Oswald J. Reichel, F. B. Prideaux, and H. Tapley-Soper (Exeter: Devon and Cornwall Record Society, 1939)

Essex: TNA, CP 25/1/60/123–26; CP 25/1/61/127–41, CP 25/1/62/142–48; Calendared: *Feet of Fines for Essex*, vol. 2, ed. by Ernest F. Kirk (Colchester: Published by the Society at the Museum in the Castle, 1913–1928)

Gloucestershire: TNA, CP 25/1/76/44–51; Calendared: *Abstracts of Feet of Fines relating to Gloucestershire 1300–1359*, ed. by Christopher Robin Elrington (Bristol: Bristol and Gloucestershire Archaeological Society, Gloucestershire Record Series, 2006)

Hampshire: TNA, CP 25/1/205/16–19

Herefordshire: TNA, CP 25/1/82/29–34

Hertfordshire: TNA, CP 25/1/87/53–64

Huntingdonshire: Calendared: *A Calendar of the Feet of Fines Relating to the County of Huntingdon Levied in the King's Court from the Fifth Year of Richard I to the End of the Reign of Elizabeth 1194–1603*, ed. by G. J. Turner (Cambridge: Cambridge Antiquarian Society, 1913)

Kent: TNA, CP/25/1/100/87–96; CP/25/1/101/97–108; Calendared: James Greenstreet, 'Abstract of Kent Fines, Anno 1 to Anno 6 Edward II', *Archaeologia Cantiana*, 11 (1877), 305–58; James Greenstreet, 'Kent Fines, Anno 7 Edward II', *Archaeologia Cantiana*, 12 (1878), 289–308; James Greenstreet, 'Kent Fines, Temp. Edward II', *Archaeologia Cantiana*, 13 (1880), 289–320; James Greenstreet, 'Kent Fines, 10–15 Edward II', *Archaeologia Cantiana*, 14 (1882), 241–80

Lancashire: TNA, CP 25/1/119/18–21; Calendared: *Final Concords of the County of Lancaster from the Original Chirographs, or Feet of Fines Preserved in the Public Record Office London, Part 2: Edward II and Edward III. AD 1307 to AD 1377*, ed. by William Farre, Lancashire and Cheshire Record Society, 46 (London: Printed for the Record Society 1903)

Leicestershire: TNA, CP 25/1/124/46–52

Lincolnshire: TNA, CP 25/1/135–37

Norfolk: TNA, CP 25/1/162/27–139; Calendared: *A Short Calendar of the Feet of Fines for Norfolk. Part 2, Comprising the Fines of the Reigns of Edward II, Edward III, Richard II, Henry IV, Henry V, Henry VI, Edward IV, and Richard III*, ed. by Walter Rye (Norwich: A. H. Goose, 1886)

Northamptonshire: TNA, CP 25/1/175/63–70

Northumberland: TNA, CP/25/1/181/10; Calendared: *Feet of Fines, Northumberland AD 1273–AD 1346*, ed. by E. S. Scroggs and Charles Johnson (Newcastle upon Tyne: Newcastle upon Tyne Records Committee, 1934)

Nottinghamshire: TNA, CP 25/1/184

Oxfordshire: TNA, CP 25/1/189/14–16

Rutland: TNA, CP 25/1/192/5

Shropshire: TNA, CP 25/1/194/8–10

Somerset: TNA, CP 25/1/198/15–18; Calendared: *Pedes Finium Commonly Called Feet of Fines for the County of Somerset, 1 Edward II to 20 Edward III., AD 1307 to AD 1346*, ed. by Emanuel Green, Somerset Record Society, 12 (London: Harrison and Sons, 1898)

Staffordshire: TNA, CP 25/1/209/10–12; Calendared: *Final Concords, Edward I and Edward II, From the Transcripts in the William Salt Library and from the Publica Record Office*, ed. by the William Salt Archaeological Society (London: Harrison and Sons, 1911), pp. 27–111

Suffolk: TNA, CP 25/217/52–62

Surrey: TNA, CP 25/1/228/31–41; Calendared: *Pedes Finium, or Fines relating to the County of Surrey, Levied in the King's Court, from the Seventh Year of Richard I to the End of the Reign of Henry VII*, ed. by Frank B. Lewis (Guildford: Surrey Archaeological Society, 1894)

Sussex: TNA, CP 25/1/236/42–48, CP 25/1/237/49–50; Calendared: *An Abstract of Feet of Fines relating to the County of Sussex, from 1 Edward II to 24 Henry VII*, ed. by L. F. Salzmann (London: Printed for the Sussex Record Society, 1916)

Warwickshire: TNA, CP 25/1/245/40–44, CP 25/1/246/45; Calendared: *Warwickshire Feet of Fines*, vol. 2: *13 Ed. I (1284)–18 Ed. III (1345)*, ed. by Ethel Stokes and Lucy Drucker (London: Printed for the Dugdale Society, 1939)

Wiltshire: TNA, CP 25/1/253/32–37; Calendared: *Abstracts of Feet of Fines relating to Wiltshire for the Reigns of Edward I and Edward II*, ed. by R. B. Pugh (Devizes: Printed for the Wiltshire Archaeological and Natural History Society, 1939)

Westmorland: TNA, CP 25/1/249/6

Worcestershire: TNA, CP 25/1/259/14–17

Yorkshire: TNA, CP 25/1/270/85–92, CP 25/1/271/93–95

Monastic and Hospital Accounts

Canterbury Cathedral Priory: **(a) Almoner's accounts**: CCA, DCc/Almoner 17, 18, 21, 20, 5, 23, 4, 6, 26; **(b) Granger's accounts**: CCA, DCc/Granger 8–10; **(c) Bartoner's accounts**: CCA, DCc/Bartoner 5–8

Norwich Cathedral Priory: **(a) Master of the Cellar's accounts**: NRO, DCN 1/1/22–28; **(b) Almoner's accounts**: NRO, DCN 1/6/9, 12–13

St John's Hospital (Winchester) Accounts: HantsALS, W/H1/2–6

The Great Hospital (Nowich) Accounts: NRO, NCR 24A

Clyst Gabriel Hospital Accounts: ExCA, 5206–12

Credit Registers

***Condempnaciones Register* of Dean and Chapter of York Cathedral**: YML, M2/la, fols 1^{r}–27^{v}

Works Cited

Printed Primary Sources (Excluding Those Listed in Appendix 8)

Annála Connacht: The Annals of Connacht, ed. by A. Martin Freeman (Dublin: Dublin Institute for Advance Studies, 1944); an online edition is available courtesy of the CELT Project: <http://www.ucc.ie/celt/published/T100011/index.html> [accessed April 2018]

Annala Uladh: Annals of Ulster otherwise Annala Senait, Annals of Senat: A Chronicle of Irish Affairs from AD 431 to AD 1540, ed. and trans. by W. M. Hennessy and B. Mac Carthy, 4 vols (Dublin: Printed for Her Majesty's Stationery Office 1887–1901, reprinted, with fresh introduction and bibliography, Dublin, 1998); an online edition is available courtesy of the CELT Project: <http://www.ucc.ie/celt/published/T100001B/index.html> [accessed April 2018]

Annales Londonienses, in *Chronicles of the Reigns of Edward I and Edward II*, vol. i, ed. by William Stubbs, Rolls Series, 76:1 (London: Longmans 1882)

Annales Monasterii de Bermundeseia, in *Annales Monastici*, ed. by H. L. Luard, Rolls Series, 36:3 (London: Longmans, 1866)

Annales Monasterii de Osenia, in *Annales Monastici*, ed. by H. L. Luard, Rolls Series, 36:3 (London: Longmans, 1866)

Annales Paulini, in *Chronicles of the Reigns of Edward I and Edward II*, vol. i, ed. by William Stubbs, Rolls Series 76:1 (London: Longman, 1882)

Annales Prioratus de Wigornia, in *Annales Monastici*, ed. by H. L. Luard, Rolls Series, 36:3 (London: Longmans, 1866)

The Annals of Clonmacnoise, ed. by Denis Murphy (Dublin: Royal Society of Antiquaries of Ireland, 1896)

The Annals of Loch Cé: A Chronicle of Irish Affairs from AD 1014 to AD 1590, ed. and trans. by William M. Hennessy (London: Longman, 1871); an online edition is available courtesy of the CELT Project: <http://www.ucc.ie/celt/online/T100010A/> [accessed April 2018]

The Black Death, ed. and trans. by Rosemary Horrox (Manchester: Manchester University Press, 1994)

The Bruce by John Barbour, ed. by W. M. Mackenzie (London: Adam and Charles Black, 1909)

Brut y Tywysogyon, or the Chronicle of the Princes. Peniarth MS. 20 Version, ed. and trans. by Thomas Jones (Cardiff: University of Wales Press, 1952)

Calendar of Close Rolls Preserved in the Public Record Office: Edward II, AD *1313–1318* (London: Her Majesty's Stationery Office, 1893)

Calendar of Documents Relating to Scotland, vol. III: *1307–57*, ed. by Joseph Bain (Edinburgh: General Register House, 1887)

Calendar of Inquisitions Miscellaneous (Chancery) Preserved in the Public Record Office, vol. II (London: His Majesty's Stationery Office, 1916)

Calendar of Letter-Books of the City of London: E, 1314–1337, ed. by Reginald R Sharpe (London: His Majesty's Stationery Office, 1903)

Calendar of Patent Rolls Preserved in the Public Record Office: Edward II, AD *1313–1317* (London: (London: Her Majesty's Stationery Office, 1898)

Calendar of Patent Rolls Preserved in the Public Record Office: Edward II, AD *1317–1321* (London: His Majesty's Stationery Office, 1903)

Calendar of Patent Rolls Preserved in the Public Record Office: Edward II, AD *1321–1324* (London: His Majesty's Stationery Office, 1904)

Castleford Chronicle, or 'The Boke of Brut', ed. Carline D. Eckhardt (Oxford: Oxford University Press, 1996)

The Cellarers' Rolls of Battle Abbey, 1275–1513, ed. by Eleanor Searle, Sussex Record Society, 65 (Sydney: Sydney University Press, 1967)

Chronica Monasterii de Melsa, ed. by E. Bond, Rolls Series, 43:2 (London: Longmans, 1867)

'The Chronicle of John Somer, OFM', ed. by Linne Ruth Mooney and Jeremy Catto, in *Chronology, Conquest and Conflict in Medieval England: Camden Miscellany 34* (Cambridge: Cambridge University Press, 1997), pp. 197–285

The Chronicle of Lanercost, trans. by Sir Herbert Maxwell (Glasgow: James Maclehose and Sons, 1913)

The Chronicle of Robert Greystanes, in *Historiae Dunelmensis Scriptores Tres*, ed. by James Raine, The Publications of the Surtees Society, 9 (Edinburgh: Laing and Forbes, 1839)

The Chronicle of St Mary's Abbey, York, From Bodley MS. 39, ed. by H. H. E. Craster and M. E. Thornton, The Publications of the Surtees Society, 148 (London: Andrews and Co., 1934)

Chronicon Abbatiae de Parco Ludae: The Chronicle of Louth Park Abbey, ed. by E. Venables (Horncastle: Morton 1891)

Chronicon Henrici Knighton vel Cnitthon Monachi Leycestrensis, vol. I, ed. by Joseph Rawson Lumby (London: Printed for Her Majesty's Stationary Office, 1889)

Chronique de Sepringham, in *Le Livere de Reis de Britannie et Le Livere de Reis de Engleterre*, ed. by John Glover (London: Longmans, 1865)

Concilia Magnae Britanniae et Hibernia, a Synodo Verolamiensi, AD CCCCXLVI *ad Londinensem* AD MDCCXVII, ed. by David Wilkins (London: Gosling and others, 1737)

Corpus orationum. Tomus XII, *Calendarium liturgicum*, ed. by Eugène Moeller, Jean-Marie Clément, and Bertrand Coppieters 't Wallant, Corpus Christianorum Series Latina, 160 K (Turnhout: Brepols, 2001)

The Correspondence, Inventories, Account Rolls and Law Proceedings of the Priory of Coldingham, ed. by James Raine (Durham: Surtees Society, 1841)

The English Manor, c. 1200–c. 1500, ed. by Mark Bailey (Manchester: Manchester University Press, 2002)

Ellis, William, *The Country Housewife's Family Companion* (Totnes: Prospect Books, 2000)

Extracts from the Account Rolls of the Abbey of Durham from the Original MSS, ed. by Joseph Thomas Fowler (Durham: Published for the Society for Andrews and Co., 1898–1901)

Flores Historiarum, ed. by H. R. Luard, Rolls Series, 84:3 (London: Longmans, 1889)

Galbraith, V. H., 'Extracts from the Historia Aurea and a French "Brut" (1317–47)', *English Historical Review*, 43 (1928), 203–17

Gesta Edwardi de Carnarvan, in *Chronicles of the Reigns of Edward I and Edward II*, vol. II, ed. by William Stubbs, Rolls Series, 76:2 (London: Longmans 1883)

Histoire de l'abbey de Sainte Croix de Quimperlé, ed. by Dom Placide le Duc and R. F. le Men (Quimperlé: Th. Claret, 1880)

Historic and Municipal Documents of Ireland, 1172–1320: From the Archives of the City of Dublin, ed. by J. T. Gilbert (London: Longmans, 1870)

Household Accounts from Medieval England, ed. by Christopher Woolgar (Oxford: Oxford University Press, 1992)

Jacob's Well: An English Treatise on the Cleansing of Man's Conscience, ed. by Arthur Brandeis, Early English Text Society, 115 (London: Paul Kegan, 1900)

Jacques Legrand, *Archiloge sophie et livre des bonnes moeurs*, ed. by Evencio Beltran (Paris, Champion: 1986)

Joannis Długosz senioris canonici Cracoviensis Opera omnia, vol. VII, ed. by A. Przezdziecki (Cracow: Ex typographia Kirchmajeriana, 1876)

Johannis de Trokelowe Annales, ed. by H.T. Riley, Rolls Series, 28:3 (London: Longman, 1866)

De legibus et consuetudinibus regni Angliae, ed. by George E. Woodbine (New Haven: Yale University Press, 1932)

The Leofric Missal, as Used in the Cathedral of Exeter, ed. by F. E. Warren (Oxford: Clarendon Press, 1883)

Liber sacramentorum Romanae Ecclesiae, ed. by Henry Austin Wilson (Oxford: Clarendon Press, 1894)

Le livere de Reis de Brittanie e le livere de Reis de Engleterre, ed. by John Glover, Rolls Series, 42 (London: Longmans, 1865)

Local Customs Accounts of the Port of Exeter, 1266–1321, ed. by Maryanne Kowaleski (Exeter: Devon and Cornwall Record Society, 1993)

Medieval English Political Writings, ed. by James M. Dean (Kalamazoo: Medieval Institute Publications, 1996); an online edition is available, *TEAMS Middle English Text Series*, Medieval Institute of the University of Rochester, <https://d.lib.rochester.edu/teams/publication/dean-medieval-english-political-writings> [accessed April 2018]

The Minor Poems of the Vernon Ms, ed. by Carl Horstmann (London: Early Texts Society, 1892)

The Missal of Robert of Jumièges, ed. by H. A. Wilson (London: Harrison and Sons, 1896)
Missale ad usum insignis Ecclesiæ Eboracensis, vol. II, ed. by W. Henderson (Durham; Andrews & co., 1874)
Nekrasov, Nikolay A., *Polnoe Sobranie Sochineniy v 15 Tomakh* (Moscow: Nauka, 1981)
'The NLW Peniarth 32 Latin Chronicle', ed. by Diana Luft, *Studia Celtica*, 44 (2010), 47–70
Open-Field Farming in Medieval England, ed. by Warren O. Ault (London: Allen and Unwin, 1972)
A Repertorium of Middle English Prose Sermons, ed. by Veronica O' Mara and Suzanne Paul (Turnhout: Brepols, 2007)
Robert of Brunne's 'Handlyng Synne, AD 1303', ed. by Frederick J. Furnivall (London: Published for the Early Text Society, 1910)
Scotichronicon by Walter Bower in Latin and English, vol. VI, ed. by D. Watt (Aberdeen: Aberdeen University Press, 1991)
Scotichronicon by Walter Bower in Latin and English, vol. VII, ed. by D. Watt (Aberdeen: Aberdeen University Press, 1996)
The Statutes of the Realm, vol. I, ed. by John Raithby (London: Record Commission, 1810)
Thomae Walsingham Historia Anglicana, ed. by H. T. Riley, Rolls Series, 28:1 (London: Longman, 1863–1864)
Thorney Annals, 963–1412, ed. by Cyril Hart (Lewiston, NY: Mellen Press, 1997)
Valor Ecclesiasticus Temp. Henr. VIII. Auctoritate Regia Institutus, vol. V (London: Printed by Command of His Majesty King George III, 1825)
Vita Edwardi Secundi, ed. by Wendy R. Childs (Oxford: Oxford University Press, 2005)
Vita et Mors Edwardi Regis Angliae, in *Chronicles of the Reigns of Edward I and Edward II*, vol. II, ed. by William Stubbs, Rolls Series, 76:2 (London: Longmans 1883)
Walter of Henley and Other Treatises on Estate Management and Accounting, ed. by Dorothea Oschinsky (Oxford: Clarendon Press, 1971)
William Shakespeare, *Macbeth: Fully Annotated* (Chicago: R. R. Donnelley and Sons, 2005)
Wynnere and Wastoure and the Parlement of the Thre Ages, ed. by Warren Ginsberg (Kalamazoo: Medieval Institute Publications, 1992)

Secondary Sources

Aberth, John, *From the Brink of the Apocalypse: Confronting Famine, War, Plague, and Death in the Later Middle Ages* (New York: Routledge, 2001)
Adamets, Serguei, 'Famine in Nineteenth and Twentieth Century Russia: Mortality by Age, Cause and Gender', in *Famine Demography: Perspectives from Past and Present*, ed. by Tim Dyson and Cormac Ó Gráda (Oxford: Oxford University Press, 2002), pp. 158–80
Akkerman, Abraham, 'Winter Cities and Mood Disorder: Observations from European City-Form at the End of the Little Ice Age', *Trames*, 18 (2014), 19–37

Alfani, Guido and Cormac Ó Gráda eds, *Famine in European History* (Cambridge: Cambridge University Press, 2017)
Allen, Martin, 'The First Sterling Area', *Economic History Review*, 70 (2017), 79–100
——, 'Silver Production and the Money Supply in England and Wales, 1086–*c.* 1500', *Economic History Review*, 64 (2011), 114–31
Antoine, Daniel, and Simon Hillson, 'The Great Famine, Black Death and Health in 14th Century London', *Archaeology International*, 9 (2005), 26–28
Antoine, D., and others, 'Using Growth Structures in Teeth from Victims of the Black Death to Investigate the Effects of the Great Famine of AD 1315–1317', *American Journal of Physical Anthropology*, 126 (2005), 65
Appleby, Andrew B., *Famine in Tudor and Stuart England* (Liverpool: Liverpool University Press, 1978)
Arnoux, Mathieu, 'Famine, blés marocains, marins génois et pirates français à Sandwich (1316–1318)', in *Orbis disciplinae: Hommages en l'honneur de Patrick Gautier Dalché*, ed. by N. Bouloux, A. Dan, and G. Tolias (Turnhout: Brepols, 2017), pp. 507–23
Arthur, Paula, 'The Black Death and Mortality: A Reassessment', *Fourteenth Century England VI*, ed. by Chris Given-Wilson (Woodbridge: Boydell and Brewer, 2010), pp. 49–72
Aston, Trevor, and C. H. E. Philpin, eds, *The Brenner Debate: Agrarian Class Structure and Economic Development in Pre-Industrial Europe* (Cambridge: Cambridge University Press, 1985)
Ault, W. O., 'By-Laws of Gleaning and the Problems of Harvest', *Economic History Review*, 14 (1961), 210–17
Bailey, Mark, 'Peasant Welfare in England, 1290–1348', *Economic History Review*, 51 (1998), 223–51
——, 'The Rabbit and the Medieval East Anglian Economy', *Agricultural History Review*, 36 (1988), 1–20
Barber, J., and A. Dzeniskevich, eds, *Life and Death in Besieged Leningrad, 1941–1944* (New York: Palgrave Macmillan, 2005)
Bardsley, Sandy, *Women's Roles in the Middle Ages* (Westport: Greenwood Press, 2007)
Barrow, Geoffrey W. S., *Kingship and Unity: Scotland, 1000–1306* (Edinburgh: Edinburgh University Press, 1981)
——, *Robert Bruce and the Community of the Realm of Scotland*, 4th edition (Edinburgh: Edinburgh University Press, 2005)
Bassino, Jean-Pascal, 'Market Integration and Famines in Early Modern Japan, 1717–1857' (Paul Valéry University (Montpellier III) working paper, <http://federation.ens.fr/ydepot/semin/texte0708/BAS2007MAR.pdf> [accessed April 2018])
Beckles, Diane M., and Maysaya Thitisaksakul, 'How Environmental Stress affects Starch Composition and Functionality in Cereal Endosperm', *Starch — Stärke*, 66 (2004), 58–71
Bennett, Judith, *Ale, Beer and Brewsters in England: Women's Work in a Changing World, 1300–1600* (Oxford: Oxford University Press, 1996)
——, *A Medieval Life: Cecilia Penifader of Brigstock, c. 1295–1344* (New York: McGraw-Hill, 1998)

——, 'The Village Ale-Wife: Women and Brewing in Fourteenth-Century England', in *Women and Work in Preindustrial Europe*, ed. by Barbara A. Hanawalt (Bloomington: University of Indiana Press, 1986), pp. 20–36

——, *Women in the Medieval English Countryside: Gender and Household in Brigstock before the Plague* (Oxford: Oxford University Press, 1987)

Bennett, Judith A., *Natives and Exotics: World War II and Environment in the Southern Pacific* (Honolulu: University of Hawai'i Press, 2009)

Bois, Guy, *The Crisis of Feudalism: Economy and Society in Eastern Normandy, c. 1300–1550* (Cambridge: Cambridge University Press, 1984)

Boriak, Hennadii, 'Sources and Resources on the Famine in Ukraine's Archival System', *Harvard Ukrainian Studies*, 27 (2004–2005), 117–47

Boucher, Andy, Luke Craddock-Bennett, and Tegan Daly, *Death in the Close: A Medieval Mystery* (Hereford: Archaeological Investigations, 2015)

Boyle, Phelim P., and Cormac Ó Gráda, 'Fertility Trends, Excess Mortality, and the Great Irish Famine', *Demography*, 23 (1986), 543–62

Braid, Robert, 'Economic Behavior, Markets and Crises: The English Economy in the Wake of Plague and Famine in the 14th Century', in *Le interazioni fra economia e ambiente biologico nell'Europa preindustriale, secc. XIII–XVIII/ Economic and Biological Interactions in Pre-Industrial Europe from the 13th to 18th Centuries*, ed. Simonetta Cavaciocchi, Fondazione Istituto Internazionale di Storia Economica 'F. Datini', Prato, Seri II: Atti delle 'Settimane de Studi' e altri Convegni no. 41 (Florence: Firenze University Press, 2010), pp. 335–72

Brady, Lisa, *War upon the Land: Military Strategy and the Transformation of Southern Landscapes during the American Civil War* (Athens: University of Georgia Press, 2012)

Brady, Niall D. K., 'The Sacred Barn: Barn-Building in Southern England, 1100–1550: A Study of Grain Storage Technology and Its Cultural Context' (unpubl. doctoral dissertation, Cornell Univ., 1996)

Brand, Paul, 'Inge, Sir William', in *Oxford Dictionary of National Biography* (Oxford: Oxford University Press, 2004); online edition, Jan. 2008, <http://www.oxforddnb.com/view/article/37587> [accessed April 2018]

Brantz, Dorothee, 'Environments of Death: Trench Warfare on the Western Front, 1914–18', in *War and the Environment: Military Destruction in the Modern Age*, ed. by Charles E. Closmann (College Station: Texas A & M University Press, 2009), pp. 68–91

Brenner, Robert, 'Agrarian Class Structure and Economic Development in Pre-Industrial Europe', *Past and Present*, 70 (1976), 30–74

Bridbury, A. R., *England and the Salt Trade in the Later Middle Ages* (Oxford: Clarendon Press, 1955)

Briggs, Chris, *Credit and Village Society in Fourteenth-century England* (Oxford: Oxford University Press, 2009)

——, 'Creditors and Debtors and their Relationships at Oakington, Cottenham and Dry Drayton (Cambridgeshire), 1291–1350', in *Credit and Debt in Medieval England, c. 1180–c. 1350*, ed. by P. R. Schofield and N. J. Mayhew (Oxbow: Oxford, 2002), pp. 127–48

——, 'Empowered or Marginalized? Rural Women and Credit in Later Thirteenth- and Fourteenth-Century England', *Continuity and Change*, 19 (2004), 13–43
——, 'Money and Rural Credit in the Later Middle Ages Revisited', in *Money, Prices and Wages: Essays in Honour of Professor Nicholas Mayhew*, ed. by Martin Allen and D'Maris Coffman (Basingstoke: Palgrave Macmillan, 2015), pp. 129–42
——, 'Taxation, Warfare, and the Early Fourteenth Century "Crisis" in the North: Cumberland Lay Subsidies, 1332–1348', *Economic History Review*, 58 (2005), 639–72
Britnell, Richard, *Britain and Ireland, 1050–1530* (Oxford: Oxford University Press, 2004)
——, 'Forstal, Forsetalling and the Statute of Forestallers', *English Historical Review*, 102 (1987), 89–102
——, *Growth and Decline in Colchester, 1300–1525* (Cambridge: Cambridge University Press, 1986)
——, 'La commercializzazione dei ceriali in Inghilterra (1250–1350)', *Quaderni Storici*, 96 (1997), 631–61
——, 'Morals, Laws and Ale in Medieval England', in *Le Droit et sa perception dans la littérature et les mentalités médiévales*, ed. by U. Müller, F. Hundsnurcher, and C. Sommer (Göppingen: Kümmerle Verlag, 1993), pp. 21–29
——, 'Price-Setting in English Borough Markets, 1349–1500', *Canadian Journal of History*, 31 (1996), 2–15
——, 'The Proliferation of Markets in England, 1200–1349', *Economic History Review*, 34 (1981), 209–21
Britton, Edward, *The Community of the Vill: A Study in the History of the Family and Village Life in Fourteenth-Century England* (Toronto: Macmillan of Canada, 1977)
Broadberry, Stephen, and others, *British Economic Growth, 1270–1870* (Cambridge: Cambridge University Press, 2015)
Brown, Michael, *Bannockburn: The Scottish War and the British Isles, 1307–1323* (Edinburgh: Edinburgh University Press, 2008)
Burger, G. C. E., H. R. Sandstead, and J. C. Drummond, eds, *Malnutrition and Starvation in Western Netherlands, September 1944 to July 1945: Part I and II* (The Hague: General State Printing Office, 1948)
Bush, Ray, 'Food Riots: Poverty, Power and Protest', *Journal of Agrarian Change*, 10 (2010), 119–29
Caffyn, Douglas J. M., 'River Transport, 1189–1600' (unpubl. doctoral thesis, Univ. of Sussex, 2010)
Cage, A. G., and W. E. N. Austin, 'Marine Climate Variability during the Last Millennium: The Loch Sunart Record, Scotland, UK', *Quaternary Science Reviews*, 29 (2010), 1633–47
Cai, Y., and W. Feng, 'Famine, Social Disruption, and Involuntary Fetal Loss: Evidence from Chinese Survey Data', *Demography*, 42 (2005), 301–22
Campbell, Bruce M. S., 'The Agrarian Problem in the Early Fourteenth Century', *Past and Present*, 188 (2005), 3–70
——, 'Agricultural Progress in Medieval England: Some Evidence from Eastern Norfolk', *Economic History Review*, 36 (1983), 26–46

——, 'Agriculture in Kent in the High Middle Ages', in *Later Medieval Kent, 1220–1540*, ed. by Sheila Sweetinburgh (Woodbridge: Boydell and Brewer, 2010), pp. 25–53
——, 'Arable Productivity in Medieval England: Some Evidence from Norfolk', *Journal of Economic History*, 43 (1983), 379–404
——, 'Benchmarking Medieval Economic Development: England, Wales, Scotland, and Ireland circa 1290', *Economic History Review*, 61 (2008), 896–948
——, *English Seigniorial Agriculture, 1250–1450* (Cambridge: Cambridge University Press, 2000)
——, 'Four Famines and a Pestilence: Harvest, Price, and Wage Variations in England, 13th to 19th Centuries', in *Agrarhistoria på manga sätt; 28 studier om manniskan och jorden. Festskrift till Janken Myrdal på hans 60-årsdag*, ed. by B. Liljewall and others (Stockholm: KSLAB, 2009), pp. 43–48
——, *The Great Transition: Climate, Disease and Society in the Late-Medieval World* (Cambridge: Cambridge University Press, 2016)
——, 'Matching Supply to Demand: Crop Production and Disposal by English Demesnes in the Century of the Black Death', *Journal of Economic History*, 57 (1997), 827–58
——, 'Nature as Historical Protagonist: Environment and Society in Pre-Industrial England (The 2008 Tawney Memorial Lecture)', *Economic History Review*, 63 (2010), 281–314
——, 'Panzootics, Pandemics and Climatic Anomalies in the Fourteenth Century', in *Beiträge zum Göttinger Umwelthistorischen Kolloquium 2010–2011*, ed. by Berndt Herrmann (Göttingen: Universitätsverlag Göttingen, 2011), pp. 177–215
——, 'Population Pressure, Inheritance, and the Land Market in a Fourteenth Century Peasant Community', in *Land, Kinship and Lifecycle*, ed. by R. M. Smith (Cambridge: Cambridge University Press, 1984), pp. 87–134
Campbell, Bruce M. S., and Lorraine Barry, 'The Population Geography of Great Britain, *c.* 1290: A Provisional Reconstruction', in *Population, Welfare and Economic Change in Britain, 1290–1834*, ed. by Chris Briggs, P. M. Kitson, and S. J. Thompson (Woodbridge: Boydell and Brewer, 2014), pp. 43–78
Campbell, Bruce M. S., and Ken Bartley, *England on the Eve of the Black Death: An Atlas of Lay Lordship, Land and Wealth, 1300–49* (Manchester: Manchester University Press, 2006)
Campbell, Bruce M. S., and Cormac Ó Gráda, 'Harvest Shortfalls, Grain Prices, and Famines in Preindustrial England', *Journal of Economic History*, 71 (2011), 859–86
Campbell, Bruce M. S., and others, *A Medieval Capital and its Grain Supply: Agrarian Production and Its Distribution in the London Region c. 1300* (London: Centre for Metropolitan History, 1993)
Campbell, Bruce M. S., and Mark Overton, 'A New Perspective on Medieval and Early Modern Agriculture: Six Centuries of Norfolk Farming *c.* 1250-*c.* 1850', *Past and Present*, 141 (1993), 38–105
Campbell-Jackson, Louise, and Antje Horsch, 'The Psychological Impact of Stillbirth on Women: A Systematic Review', *Illness, Crisis and Loss*, 22 (2014), 237–56

Carlin, Martha, 'Cheating the Boss: Robert Carpenter's Embezzlement Instructions (1261×1268) and Employee Fraud in Medieval England', in *Commercial Activity, Markets and Entrepreneurs in the Middle Ages: Essays in Honour of Richard Britnell*, ed. by B. Dodds and C. Liddy (Woodbridge: Boydell and Brewer, 2011), pp. 183–98

Cessford, C., and others, 'The St John's Hospital Cemetery and Environs, Cambridge: Contextualizing the Medieval Urban Dead', *Archaeological Journal*, 172 (2015), 52–120

Cheng, Weichung, *War, Trade and Piracy in the China Seas (1622–1683)* (Leiden: Brill, 2013)

Chereji, Rodica, Iuliana Cretescu, and Rodica Cărpiță, 'The Influence of Alpha Amylase on the Quality of Bread', *Lucrări științifice Zootehnie și Biotehnologii*, 41 (2008), 30–34

Claridge, Jordan, 'England's Sacred Beast?', *History Today*, 63 (June 2013), 1–6

Claridge, Jordan, and John Langdon, 'The Composition of Famuli Labour on English Demesnes, *c.* 1300', *Agricultural History Review*, 63 (2015), 195–210

——, 'Storage in Medieval England: The Evidence from Purveyance Accounts, 1295–1349', *Economic History Review*, 64 (2011), 1242–65

Clark, Elaine, 'Debt Litigation in a Late Medieval English Vill', in *Pathways to Medieval Peasants*, ed. by J. A. Raftis (Toronto: PIMS, 1981), pp. 247–79

——, 'Medieval Debt Litigation: Essex and Norfolk, 1270–1490' (unpubl. doctoral dissertation, Univ. of Michigan, 1977)

Clark, William A., *Crime and Punishment in Soviet Officialdom: Combating Corruption in the Political Elite, 1965–1990* (New York: M. E. Sharpe, 1993)

Clay, Jason W., and Bonnie K. Holcomb, *Politics and the Ethiopian Famine: 1984–1985* (Cambridge, MA: Cultural Survival Inc., 1986)

Closmann, Charles E., ed., *War and the Environment: Military Destruction in the Modern Age* (College Station: Texas A & M University Press, 2009)

Coale, Ansley J., and Paul Demeny, *Regional Model Life Tables and Stable Populations*, 2nd edn (New York: Academic Press, 1983)

Cohn, Samuel K., *Popular Protest in Late Medieval English Towns* (Cambridge: Cambridge University Press, 2013)

Comet, Georges, 'Technology and Agricultural Expansion in the Middle Ages: The Example of France North of the Loire', in *Medieval Farming and Technology: The Impact of Agricultural Change in Northwest Europe*, ed. by Grenville Astill and John Langdon (Leiden: Brill, 1997), pp. 11–40

Connell, Brian, and others, *A Bioarchaeological Study of Medieval Burials on the Site of St Mary Spital: Excavations at Spitalfields Market, London E1, 1991–2007* (London: Museum of London Archaeology, 2012)

Cooper, R. J., and others, 'A Tree-Ring Reconstruction of East Anglian (UK) Hydroclimate Variability over the Last Millennium', *Climate Dynamics*, Online First, 20 March 2012, DOI: 10.1007/s00382-012-1328-x

Cooter, R., 'Of War and Epidemics: Unnatural Couplings, Problematic Conceptions', *Journal of the Society for the Social History of Medicine*, 16 (2003), 283–302

Coss, Peter, *The Origins of the English Gentry* (Cambridge: Cambridge University Press, 2003)

Cox, Catherine, *Negotiating Insanity in the Southeast of Ireland, 1820–1900* (Manchester: Manchester University Press, 2012)

Crum-Cianflone, Nancy F., 'Bacterial, Fungal, Parasitic, and Viral Myositis', *Clinical Microbiological Reviews*, 21 (2008), 473–94

Crump, J. A., S. P. Luby, and E. D. Mintz, 'The Global Burden of Typhoid Fever', *Bulletin of World Health Organization*, 82 (2004), 346–53

Cullen, Karen J., *Famine in Scotland: The 'Ill' Years of the 1690s* (Edinburgh: Edinburgh University Press, 2010)

Cussans, G. W., and others, 'A Survey of the Incidence of the *Bromus* Species as Weeds of Winter Cereals in England, Wales and parts of Scotland', *Weed Research*, 34 (2006), 361–68

Davies, Mike, and Jonathan Kissock, 'The Feet of Fines, the Land Market and the English Agricultural Crisis of 1315 to 1322', *Journal of Historical Geography*, 30 (2004), 215–30

Davies, Rees R., *Lordship and Society in the March of Wales, 1282–1400* (Oxford: Clarendon Press, 1978)

Davies, Wendy, *Wales in the Early Middle Ages* (Leicester: Leicester University Press, 1982)

Davis, James, 'Baking for the Common Good: A Reassessment of the Assize of Bread in Medieval England', *Economic History Review*, 57 (2004), 465–502

——, *Medieval Market Morality: Life, Law and Ethics in the English Marketplace, 1200–1500* (Cambridge: Cambridge University Press, 2012)

Davis, Mike, *Late Victorian Holocausts: El Niño Famines and the Making of the Third Word* (New York: Verso, 2001)

Dawson, A. G., and others, 'Greenland (GISP2) Ice Core and Historical Indicators of Complex North Atlantic Climate Changes during the Fourteenth Century', *Holocene*, 17 (2007), 427–34

Demarcq, Gérard, *Dunkerque 1 000 ans d'histoire* (Cambrai: Le Téméraire, 1993)

Desai, Meghnad, 'The Agrarian Crisis in Medieval England', *Bulletin of Economic Research*, 43 (1991), 223–58

Devereux, Stephen, 'The Malawi Famine of 2002', *IDS Bulletin*, 33 (2002), 70–78

——, 'Sen's Entitlement Approach: Critiques and Counter-critiques', in *The New Famines: Why Famines Persist in an Era of Globalization*, ed. by Stephen Devereux (London: Routledge, 2007), pp. 66–90

——, *Theories of Famine* (Hemel Hempstead: Harvester Wheatsheaf, 1993)

——, 'Why does Famine Persist in Africa?', *Food Security*, 1 (2009), 1–25

De Waal, Alexander, *Famine Crimes: Politics and the Disaster Relief Industry in Africa* (Bloomington: Indiana University Press, 1997)

——, *Famine that Kills: Darfur, Sudan* (Oxford: Oxford University Press, 2005)

DeWitte, Sharon N., and James W. Wood, 'Selectivity of Black Death Mortality with Respect to Preexisting Health', *PNAS*, 105 (2008), 1436–41

Dick, Henry C., and others, 'Detection and Characterisation of Black Death Burials by Multi-Proxy Geophysical Methods', *Journal of Archaeological Science*, 59 (2015), 132–41

Didwania, N., and M. Joshi, 'Mycotoxins: A Critical Review on Occurrence and Significance', *International Journal of Pharmacy and Pharmaceutical Sciences*, 5 (2013), 1014–19

Digby, William, *The Famine Campaign in Southern India: Madras and Bombay Presidencies and Province of Mysore, 1876–1878* (London: Longmans, 1878)

Dobson, Mary J., *Contours of Death and Disease in Early Modern England* (Cambridge: Cambridge University Press, 1997)

Dodds, Ben, *Peasants and Production in the Medieval North-East: The Evidence from Tithes, 1270–1536* (Woodbridge: Boydell and Brewer, 2007)

——, 'Tithe and Agrarian Output between the Tyne and Tees' (unpubl. doctoral thesis, Univ. of Durham, 2002)

Down, Kevin, 'Colonial Society and Economy', in *A New History of Ireland*, vol. II: *Medieval Ireland 1169–1534*, ed. by Art Cosgrove (Oxford: Oxford University Press, 1987), pp. 439–91

Dragoa, Filippo, and others, 'Acute Low Doses of Melatonin Stimulate Rat Sex Behavior: The Role of Serotonin Neurotransmission', *European Journal of Pharmacology*, 385 (1999), 1–6

Drèze, Jean, and Amartya Sen, *Hunger and Public Action* (Oxford: Oxford University Press, 1991)

Driscoll, S., *Alba: The Gaelic Kingdom of Scotland AD 800–1124* (Edinburgh: Birlinn, 2002)

DuBoulay, Francis Robin H., *The Lordship of Canterbury: An Essay on Medieval Society* (London: Barnes and Noble, 1966)

Duffy, Seán, 'The Bruce Invasion of Ireland: A Revised Itinerary and Chronology', in *Robert the Bruce's Irish Wars: The Invasions of Ireland, 1306–1329*, ed. by Seán Duffy (Stroud: Tempus, 2002), pp. 9–43

Dugdale, William, *Monasticon Anglicanum* (London: James Bohn, 1846)

Dunning, Chester S. L., *Russia's First Civil War: The Time of Troubles and the Founding of the Romanov Dynasty* (University Park: Penn State University Press, 2010)

Dyer, Christopher, 'Changes in Diet in the Late Middle Ages: The Case of Harvest Workers', *Agricultural Historical Review*, 36 (1988), 77–99

——, 'Did Peasants Need Markets and Towns? The Experience of Late Medieval England', in *London and Beyond: Essays in Honour of Derek Keene*, ed. by Matthew Davies and James Galloway (London: London Institute of Historical Research 2012), pp. 25–47

——, 'Did the Peasants Really Starve in Medieval England?', in *Food and Eating in Medieval Europe*, ed. by Martha Carlin and Joel Rosenthal (London: Hambledon Press, 1998), pp. 53–72

——, "England's Economy in the Fifteenth Century', *The Fifteenth Century XIII: Exploring the Evidence: Commemoration, Administration and Economy*, ed. by Linda Clark (Woodbridge: Boydell and Brewer, 2014), 201–26

——, 'Gardens and Garden Produce in Late Medieval England', in *Food in Medieval England: Diet and Nutrition*, ed. by C. M. Woolgar, D. Serjeantson, and T. Waldron (Oxford: Oxford University Press, 2006), pp. 27–40

——, 'The Hidden Trade of the Middle Ages: Evidence from the West Midlands of England', *Journal of History Geography*, 18 (1992), 283–303

——, *Lords and Peasants in a Changing Society: The Estates of the Bishopric of Worcester, 680–1540* (Cambridge: Cambridge University Press, 1980)

——, *Making a Living in the Middle Ages: The People of Britain, 850–1520* (New Haven: Yale University Press, 2002)

——, 'The Material World of English Peasants, 1200–1540: Archaeological Perspectives on Rural Economy and Welfare', *Agricultural History Review*, 62 (2014), 1–22

——, 'Poverty and its Relief in Late Medieval England', *Past and Present*, 216 (2012), 41–78

——, 'Seasonal Patterns in Food Consumption in the Later Middle Ages', in *Food in Medieval England: Diet and Nutrition*, ed. by Chris Woolgar, Dale Serjeantson, and Tony Waldron (Oxford: Oxford University Press, 2006), pp. 201–14

——, *Standards of Living in the Later Middle Ages: Social Change in England c. 1200–1520* (Cambridge: Cambridge University Press, 1989)

Dyer, Christopher, Peter Coss, and Chris Wickham, eds, *Rodney Hilton's Middle Ages: An Exploration of Historical Themes* (Oxford: Oxford University Press, 2007)

Dyson, Tim, 'Famine in Berar, 1896–97 and 1899–1900: Echoes and Chain Reactions', *Famine Demography: Perspectives from Past and Present*, ed. by Tim Dyson and Cormac Ó Gráda (Oxford: Oxford University Press, 2002), pp. 93–112

——, 'On the Demography of South Asian Famines Part I', *Journal of Population Studies*, 45 (1991), 5–25

——, 'On the Demography of South Asian Famines Part II', *Journal of Population Studies*, 45 (1991), 279–97

Dyson, Tim and Cormac Ó Gráda, 'Introduction', in *Famine Demography: Perspectives from Past and Present*, ed. by Tim Dyson and Cormac Ó Gráda (Oxford: Oxford University Press, 2002), pp. 1–18

Eddison, Jill, *Medieval Pirates: Pirates, Raiders and Privateers, 1204–1453* (Stroud: The History Press, 2013)

Edwards, J. F., and B. P. Hindle, 'The Transportation System of Medieval England and Wales', *Journal of Historical Geography*, 17 (1991), 123–34

Ehrlich, Paul R., *The Population Bomb* (New York: Ballantine Books, 1968)

Eitinger, Leo, *Concentration Camp Survivors in Norway and Israel* (The Hague: Martinus Nijhoff, 1972)

Engler, Steven, and others, 'The Irish Famine of 1740–1741: Famine Vulnerability and "Climate Migration"', *Climate of the Past*, 9 (2013), 1161–79

Enrnæs, Mette, Karl Gunnar Persson, and Søren Rich, 'Feeding the British: Convergence and Market Efficiency in the Nineteenth-Century Grain Trade', *Economic History Review*, 61 (2008), 140–71

Epstein, Steven A., *Wage Labor and Guilds in Medieval Europe* (Chapel Hill: University of North Carolina Press, 1991)

Fagan, Brian, *The Little Ice Age: How Climate Made History, 1300–1850* (New York: Basic Books, 2000)

Farmer, David, 'Prices and Wages', in *The Agrarian History of England and Wales*, vol. II: *1042–1350*, ed. by H. E. Hallam (Cambridge: Cambridge University Press, 1988), pp. 715–817

——, 'Prices and Wages', in *The Agrarian History of England and Wales*, vol. III: *1348–1500*, ed. by Edward Miller (Cambridge: Cambridge University, 1991), pp. 431–525

Ferguson, Niall, *Civilization: The West and the Rest* (London: Allen Lane, 2011)

Flaherty, Eoin, 'Assessing the Distribution of Social-Ecological Resilience and Risk: Ireland as a Case Study of the Uneven Impact of Famine', *Ecological Complexity*, 19 (2014), 35–45

Fogel, Robert, *The Escape from Hunger and Premature Death, 1700–2100* (Cambridge: Cambridge University Press, 2004)

Forbes, T. R., *Chronicle from Aldgate* (New Haven: Yale University Press, 1971)

Galley, Chris, 'The Stillbirth Rate in Early Modern England', *Local Population Studies*, 81 (2008), 75–83

Galloway, James A., 'One Market or Many? London and the Grain Trade of England', in *Trade, Urban Hinterlands and Market Integration c. 1300–1600*, ed. by James A. Galloway (London: Centre for Metropolitan History 2000), pp. 23–42

Garenne, Michel, 'An Atypical Urban Famine: Antananarivo, Madagascar 1985–1986', in *The New Famines: Why Famines Persist in an Era of Globalization*, ed. by Stephen Devereux (Abingdon: Routledge, 2007), pp. 178–96

Garnsey, Peter, *Famine and Food Supply in the Graeco-Roman World: Responses to Risk and Crisis* (Cambridge: Cambridge University Press, 1988)

Geary, Frank, and Tom Stark, 'Trends in Real Wages during the Industrial Revolution: A View from across the Irish Sea', *Economic History Review*, 57 (2004), 362–95

Geltner, Guy, *The Medieval Prison: A Social History* (Princeton: Princeton University Press, 2008)

Goodspeed, Tyler, *Famine and Finance: Credit and the Great Famine of Ireland* (London: Palgrave Macmillan, 2017)

Gorski, Richard, *The Fourteenth-Century Sheriff: English Local Administration in the Late Middle Ages* (Woodbridge: Boydell and Brewer, 2003)

Green, Reginald Herbold, 'The Course of the Four Horsemen: The Costs of War and its Aftermath in Sub-Saharan Africa. An Overview', in *War and Hunger*, ed. by J. Macrae and A. Zwi (London: Zed Books, 1994), pp. 37–49

Gregg, Joan Young, 'The Narrative Exempla of *Jacob's Well*: A Source Study of with an Index for *Jacob's Well* to Index Exemplorum' (unpubl. doctoral dissertation, New York Univ., 1973)

Griffiths, David, 'Medieval Coastal Sand Inundation in Britain and Ireland', *Medieval Archaeology*, 59 (2015), 103–21

Guinnane, Timothy W., *The Vanishing Irish: Households, Migration, and the Rural Economy in Ireland, 1850–1914* (Princeton: Princeton University Press, 1997)

Haines, Roy Martin, *King Edward II: His Life, His Reign, and Its Aftermath, 1284–1330* (Montreal: McGill-Queen's Press, 2003)

Hallam, H., 'Population Movements in England, 1086–1350', in *The Agrarian History of England and Wales*, vol. II: *1042–1350*, ed. by H. Hallam (Cambridge: Cambridge University Press, 1988), pp. 508–93

Hanawalt, Barbara, *Crime and Conflict in English Communities, 1300–1348* (Cambridge, Mass.: Harvard University Press, 1979)

——, *The Ties That Bound: Peasant Families in Medieval England* (Oxford: Oxford University Press, 1986)

Harvey, Barbara, *Living and Dying in England, 1100–1540: The Monastic Experience* (Oxford: Oxford University Press, 1993)

Harvey, Paul D. A., *Medieval Oxfordshire Village: Cuxham, 1240 to 1400* (Oxford: Oxford University Press 1965)

Hatcher, John, *Rural Economy and Society in the Duchy of Cornwall, 1300–1500* (Cambridge: Cambridge University Press, 1970)

Healey, Jonathan, 'Famine and the Female Mortality Advantage: Sex, Gender and Mortality in Northwest England, *c.* 1590–1630', *Continuity and Change*, 30 (2015), 153–92

Heebøll-Holm, Thomas K., *Ports, Piracy and Maritime War: Piracy in the English Channel and the Atlantic, c. 1280-c. 1330* (Leiden: Brill, 2013)

Heijmans, T. Bastiaan, and others, 'Persistent Epigenetic Differences Associated with Prenatal Exposure to Famine', *PNAS*, 105 (2008), 17046–17049

Hendrix, Cullen (with Henk-Jan Brinkman), 'Food Insecurity and Conflict Dynamics: Causal Linkages and Complex Feedbacks', *Stability: International Journal of Security and Development* (2013) (online publication: <http://www.stabilityjournal.org/articles/10.5334/sta.bm/> [accessed April 2018])

Hernández-Julián, Rey, Hani Mansour, and Christina Peters, 'The Effects of Intrauterine Malnutrition on Birth and Fertility Outcomes: Evidence From the 1974 Bangladesh Famine', *Demography*, 51 (2014), 1775–96

Hilton, Rodney, *Class Conflict and the Crisis of Feudalism: Essays in Medieval Social History* (London: Hambledon Press, 1985)

Hionidou, Violetta, 'Black Market, Hyperinflation and Hunger, Greece 1941–1944', *Food and Foodways*, 12 (2004), 81–106

——, *Famine and Death in Occupied Greece, 1941–1944* (Cambridge: Cambridge University Press, 2006)

——, '"Send us Food or Coffins': The 1941–43 Famine on the Aegean Island of Syros', in *Famine Demography: Perspectives from Past and Present*, ed. by Tim Dyson and Cormac Ó Gráda (Oxford: Oxford University Press, 2002), pp. 181–203

Howard-Hassmann, Rhoda E., 'State-Induced Famine and Penal Starvation in North Korea', *Genocide Studies and Prevention: An International Journal*, 7 (2012), 147–65

Howe, Paul, and Stephen Devereux, 'Famine Intensity and Magnitude Scales: A Proposal for an Instrumental Definition of Famine', *Disasters*, 38 (2004), 353–72

Hudson, William, 'The Prior of Norwich's Manor of Hindolveston', *Norfolk Archaeology*, 20 (1917–1919), 179–214

Hurst, J. D., 'Fuel Supply and the Medieval Salt Industry in Droitwich', *Transactions of the Worcestershire Archaeological Society*, 19 (2004), 111–32

Hybel, Nils, 'The Grain Trade in Northern Europe before 1350', *Economic History Review*, 55 (2002), 219–47

Jackson, Henry, *The Arithmetical Catechism, Compiled from Various Authors* (Portsmouth: Miller and Gray, 1824)

James, Francis Godwin, 'Charity Endowments as Sources of Local Credit in Seventeenth- and Eighteenth-Century England', *Journal of Economic History*, 8 (1948), 153–70

Jenks, Stuart, 'The Lay Subsidies and the State of the English Economy (1275–1334)', *Vierteljahrschrift für Sozial- und Wirtschaftsgeschichte*, 85 (1998), 1–39

Jones, Andrew, 'Harvest Customs and Labourers' Perquisites in Southern England, 1150–1350: The Hay Harvest', *Agricultural History Review*, 25 (1977), 98–107

Jones, E. T., 'River Navigation in Medieval England', *Journal of Historical Geography*, 26 (2000), 60–75

Jones, Eric, *The European Miracle: Environments, Economies and Geopolitics in the History of Europe and Asia* (Cambridge: Cambridge University Press, 1981)

Jones, W. R., 'Purveyance for War and the Community of the Realm in Late Medieval England', *Albion: A Quarterly Journal Concerned with British Studies*, 7 (1975), 300–16

Jordan, William Chester, 'The Gleaners', in *Boundaries in the Medieval and Wider World: Essays in Honour of Paul Freedman*, ed. by Thomas W. Barton and others (Turnhout: Brepols, 2017), pp. 201–20

——, 'The Great Famine: 1315–1322 Revisited', in *Ecologies and Economies in Medieval and Early Modern Europe: Studies in Environmental History for Richard C. Hoffmann*, ed. by Scott G. Bruce (Leiden: Brill, 2010), pp. 45–62

——, *The Great Famine: Northern Europe in the Early Fourteenth Century* (Princeton, N. J.: Princeton University Press, 1996)

Jørgensen, Dolly, 'Local Government Responses to Urban River Pollution in Late Medieval England', *Water History*, 2 (2010), 35–52

Jurkowski, Maureen, C. Smith, and D. Crook, *Lay Taxes in England and Wales, 1188–1688* (Kew: Public Record Office, 1998)

Kaati, Gunnar, and others, 'Transgenerational Response to Nutrition, Early Life Circumstances and Longevity', *European Journal of Human Genetics*, 15 (2007), 784–90

Kacki, Sacha, 'Influence de l'état sanitaire des populations anciennes sur la mortalité en temps de peste: Contribution à la paléoépidémiologie' (unpubl. doctoral thesis, Univ. of Bordeaux, 2016)

Kaminsky, Rina G., Renato Valenzuela Castillo, and Coralia Abrego Flores, 'Growth Retardation and Severe Anemia in Children with Trichuris Dysenteric Syndrome', *Asian Pacific Journal of Tropical Biomedicine*, 5 (2015), 591–97

Kershaw, Ian, *Bolton Priory: The Economy of a Northern Monastery, 1268–1325* (Oxford: Oxford University Press, 1973)

——, 'The Great Famine and Agrarian Crisis in England 1315–1322', *Past and Present*, 59 (1973), 3–50

——, 'A Note on the Scots in the West Riding, 1318–19', *Northern History*, 17 (1981), 231–39

Keys, A., and others, *The Biology of Human Starvation* (Minneapolis: University of Minnesota Press, 1950)

Kim, Woon Keun, Hyunok Lee, and Daniel A. Sumner, 'Assessing the Food Situation in North Korea', *Economic Development and Cultural Change*, 46 (1998), 519–35

King, Gary, and Charlotte Henderson, 'Living Cheek by Jowl: The Pathoecology of Medieval York', *Quaternary International*, 30 (2013), 1–12

Kirschenbaum, Lisa A., *The Legacy of the Siege of Leningrad, 1941–1995: Myth, Memories, and Monuments* (Cambridge: Cambridge University Press, 2006)

Kislitsyn, N., *Leningrad ne sdayotsya* [*Leningrad Won't Surrender*] (Moscow: Progress, 1991)

Kitsikopoulos, Harry, 'Standards of Living and Capital Formation in Pre-plague England: A Peasant Budget Model', *Economic History Review*, 53 (2000), 237–61

Klein, Axel, 'The Moral Economy of Somali Piracy — Organised Criminal Business or Subsistence Activity', *Global Policy*, 4 (2013), 94–100

Kotloff, Karen, and others, 'Burden and Aetiology of Diarrhoeal Disease in Infants and Young Children in Developing Countries (the Global Enteric Multicenter Study, GEMS): A Prospective, Case-control Study', *Lancet*, 382(2013), 209–22

Kowaleski, Maryanne, 'The Grain Trade in Fourteenth-Century Exeter', in *The Salt of Common Life: Individuality and Choice in the Medieval Town, Countryside and Church: Essays Presented to J. Ambrose Raftis*, ed. by Edwin B. DeWindt (Kalamazoo: Medieval Institute Publications, 1993), pp. 1–52

Krug, Ilana, 'Purveyance and Peasants at the Beginning of the Hundred Years War: Maddicott Revisited', in *The Hundred Years War (Part II): Different Vistas*, ed. by Don Kagay and Andrew Villalon (Leiden: Brill, 2008), pp. 345–65

——, 'Wartime Corruption and Complaints of the English Peasantry', in *Noble Ideals and Bloody Realities: Warfare in the Middle Ages*, ed. by Niall Christie and Maya Yazigi (Leiden: Brill, 2006), pp. 177–93

Labuschagne, M. T., O. Elago, and E. Koen, 'The Influence of Extreme Temperatures during Grain Filling on Protein Fractions, and Its Relation to Some Quality Characteristics in Bread, Biscuit and Durum Wheat', *Cereal Chemistry*, 86 (2009), 61–66

——, 'The Influence of Temperature Extremes on Quality and Starch Characteristics in Bread, Biscuit and Durum Wheat', *Journal of Cereal Science*, 49 (2009), 184–89

Lamb, Elizabeth H., 'The Impact of Previous Perinatal Loss on Subsequent Pregnancy and Parenting', *Journal of Perinatal Education*, 11 (2002), 33–40

Lambert, Craig L., *Shipping the Medieval Military: English Maritime Logistics in the Fourteenth Century* (Woodbridge: Boydell and Brewer, 2011)

Landes, David, *The Wealth and Poverty of Nations: Why Some are So Rich and Some So Poor* (New York: Norton, 1998)

Langdon, John, 'Inland Water Transport in Medieval England', *Journal of Historical Geography*, 19 (1993), pp. 1–11

——, Lordship and Peasant Consumerism in the Milling Industry of Early Fourteenth-Century England', *Past and Present*, 145 (1994), 3–46

——, *Mills in the Medieval Economy: England, 1300–1540* (Oxford: Oxford University Press, 2004)

Langdon, John, and Jordan Claridge, 'Transport in Medieval England', *History Compass*, 9 (2011), 864–75

Langdon, John, Jill Walker, and John R. Falconer, 'Boom and Bust: Building Investment on the Bishop of Winchester's Estate in the Early Fourteenth Century', in *The Winchester Pipe Rolls and Medieval English Society*, ed. by Richard H. Britnell (Woodbridge: Boydell and Brewer, 2003), pp. 139–55

Laragy, Georgina, 'Narratives of Poverty in Irish Suicides between the Great Famine and the First World War, 1845–1914', in *Poverty and Sickness in Modern Europe: Narratives of the Sick Poor, 1780–1938*, ed. by Andreas Gestrich, Elizabeth Hurren, and Steven King (London: Continuum Press, 2012)

Larking, Lambert B., ed., *The Knights Hospitaller in England: Being the Report of Prior Philip de Thame to the Grand Master Elyan de Villanova for AD 1338*, Camden Society, 65 (London: Printed for the Camden Society, 1857)

Latimer, Paul, 'Money and the English Economy in the Twelfth and Thirteenth Centuries', *History Compass*, 9 (2011), 246–56

Lawlor, Hugh Jackson, 'Calendar of the Liber Ruber of the Diocese of Ossory', *Proceedings of the Royal Irish Academy, Section C: Archaeology, Celtic Studies, History, Linguistics, Literature*, 27 (1908/1909), pp. 159–208

Leaning, Jennifer, and Debarati Guha-Sapir, 'Natural Disasters, Armed Conflict, and Public Health', *New England Journal of Medicine*, 369 (2013), pp. 1836–42

Linder, Amnon, *Raising Arms: Liturgy in the Struggle to Liberate Jerusalem in the Late Middle Ages* (Brepols: Turnhout, 2003)

Livingstone, Marilyn R., 'The *Nonae*: The Records of the Taxation of the Ninth in England 1340–41' (unpubl. doctoral thesis, Queen's University of Belfast, 2003)

Lloyd, T. H., *England and the German Hanse, 1157–1611: A Study in Their Trade and Commercial Diplomacy* (Cambridge: Cambridge University Press, 1991)

Lucas, Henry S., 'The Great European Famine of 1315, 1316 and 1317', *Speculum*, 5 (1930), 343–77

——, 'John Crabbe: Flemish Pirate, Merchant and Adventurer', *Speculum*, 20 (1945), 334–50

Lust, Kersti, 'Providing for the Hungry? Famine Relief in the Russian Baltic Province of Estland, 1867–69', *Social History*, 40 (2015), 15–37

Lydon, J., 'The Impact of the Bruce Invasion, 1315–27', in *A New History of Ireland*, vol II: *Medieval Ireland, 1169–1534*, ed. by A. Cosgrove (Oxford: Clarendon Press, 1987), pp. 275–302

Maddicott, John R., *The English Peasantry and the Demands of the Crown, 1294–1341*, Past and Present Supplement no. 1 (Oxford: Oxford University Press, 1975)

Malanima, Paolo, 'Wages, Productivity and Working Time in Italy 1300–1913', *Journal of European Economic History* 36 (2007), 127–74

Malthus, Thomas R., *An Essay on the Principle of Population* (New York: Norton and Co., 1976)

Marshall, Garry, *A View of Medieval Chesham* (Chesham: The Chess Valley Archaeological and Historical Society, 2013)

Martins, V. J. B., and others, 'Long-Lasting Effects of Undernutrition', *International Journal of Environmental Research and Public Health*, 8 (2011), pp. 1817–46

Marvin, Julia, 'Cannibalism as an Aspect of Famine in Two English Chronicles', in *Food and Eating in Medieval Society*, ed. by M. Carlin and J. T. Rosenthal (London: Hambledon, 1998), pp. 73–86

Marx, Karl, *Capital: Critique of Political Economy*, vol. I, trans. by Ben Fowkes (London: Penguin Classics, 1990)

Mary, Yannick, and others, 'High Frequency Environmental Changes and Deposition Processes in a 2kyr-Long Sedimentological Record from the Cap Breton Canyon (Bay of Biscay)', *Holocene*, 25 (2015), 348–65

Masschaele, James, *Peasants, Merchants, and Markets: Inland Trade in Medieval England, 1150–1350* (New York: St Martin's Press, 1997)

——, 'Transport Costs in Medieval England', *Economic History Review*, 46 (1993), 266–79

Maxwell, Daniel, and Merry Fitzpatrick, 'The 2011 Somalia Famine: Context, Causes, and Complications', *Global Food Security*, 1 (2012) 5–12

Mayer, Peter, *Suicide and Society in India* (London: Routledge, 2010)

Mayhew, Nicholas J., 'Modelling Medieval Monetisation', in *A Commercializing Economy: England 1086 to c. 1300*, ed. by Richard H. Britnell, and Bruce M. S. Campbell (Manchester: Manchester University Press, 1995), pp. 55–77

——, 'Silver Production and the Money Supply in England and Wales, 1086–*c.* 1500', *Economic History Review*, 64 (2011), 114–31

McCloskey, Donald N., 'English Open Fields as Behavior towards Risk', *Research in Economic History*, 1 (1976), 124–71

——, 'The Prudent Peasant: New Findings on Open Fields', *Journal of Economic History*, 51 (1991), 343–55

McCormick, Finbar, 'The Decline of the Cow: Agricultural and Settlement Change in Early Medieval Ireland', *Peritia*, 20 (2008), 210–25

McIntosh, Marjorie Keniston, *Poor Relief in England, 1350–1600* (Cambridge: Cambridge University Press, 2012)

McMahon, Richard, *Homicide in Pre-Famine and Famine Ireland* (Liverpool, Liverpool University Press, 2013)

——, '"A Violent Society"? Homicide Rates in Ireland, 1831–1850', *Irish Economic and Social History*, 36 (2009), 1–20

McLean, Teresa, *Medieval English Gardens* (London: Barrie and Jenkins, 1989)

McNamee, Colm, *The Wars of the Bruces: Scotland, England and Ireland, 1306–1328* (East Linton: Tuckwell Press, 1997)

Mendelsohn, Robert, and Ariel Dinar, 'Climate, Water, and Agriculture', *Land Economics* 79 (2003), 328–41

Mendelsohn, Robert, Ariel Dinar, and Arne Dalfelt, *Climate Change Impacts on African Agriculture* (New Haven: Yale University Press, 2003)

Moore, John S., '"Quot homines?" The Population of Domesday England', *Anglo-Norman Studies*, 19 (1997), 307–34

More, Alexander F., and others, 'Next Generation Ice Core Technology Reveals True Minimum Natural Levels of Lead (Pb) in the Atmosphere: Insights from the Black Death', *GeoHealth*, 1 (2017) <DOI: 10.1002/2017GH000064> ('an early view' version published 31 May 2017)

Morgan, Hiram, *Tyrone's Rebellion: The Outbreak of the Nine Years War in Tudor Ireland* (Woodbridge: Boydell and Brewer, 1993)
Muldrew, Craig, *Food, Energy and the Creation of Industriousness: Work and Material Culture* (Cambridge: Cambridge University Press, 2011)
Mullan, John, *Mortality, Gender and the Plague of 1361–2 on the Estate of the Bishop of Winchester*, Cardiff Historical Papers (Cardiff: Cardiff University Press, 2007–2008)
Munro, John, 'The Medieval Origins of the Financial Revolution: Usury, *Rentes*, and Negotiability', The *International History Review*, 25 (2003), 505–62
——, 'The "New Institutional Economics" and the Changing Fortunes of Fairs in Medieval and Early Modern Europe: The Textile Trades, Warfare, and Transaction Costs', *Vierteljahrschrift für Sozial- und Wirtschaftsgeschichte*, 88 (2001), 1–47
Muscheler, Raimund, and others, 'Solar Activity during the Last 1000 yr Inferred from Radionuclide Record', *Quaternary Science Reviews*, 26 (2007), 82–97
Myllyntaus, Timo, 'Summer Frost: A Natural Hazard with Fatal Consequences in Preindustrial Finland', in *Natural Disasters, Cultural Responses: Case Studies Toward a Global Environmental History*, ed. by Christof Mauch and Christian Pfister (Lanham: Lexington Books, 2009), pp. 77–102
Naylor, R. T., *Wages of Crime: Black Markets, Illegal Finance, and the Underworld Economy* (Ithaca: Cornell University Press, 2004)
Neville, Cynthia, *Land, Law and People in Medieval Scotland* (Edinburgh: Edinburgh University Press, 2010)
Newfield, Timothy, 'A Cattle Panzootic in Early Fourteenth-Century Europe', *Agricultural History Review*, 57(2009), 155–90
——, 'The Contours of Disease and Hunger in Carolingian and Early Ottonian Europe (*c.* 750-*c.* 950 CE)' (unpubl. doctoral dissertation, McGill Univ., 2010)
——, 'A Great Destruction of Cattle: The Impact and Extent of Epizootic Disease in Early Fourteenth-Century Northwestern Europe' (unpubl. MA dissertation, Univ. of Toronto, 2005)
Newman, Edward I., 'Medieval Sheep-Corn Farming: How Much Grain Yield Could Each Sheep Support?', *Agricultural History Review*, 50 (2002), 164–80
Nicholls, Kenneth W., *Gaelic and Gaelicised Ireland* (Dublin: Gill and Macmillan, 1972)
Nightingale, Pamela, 'The English Parochial Clergy as Investors and Creditors in the First Half of the Fourteenth Century', in *Credit and Debt in Medieval England, c. 1180–c. 1350*, ed. by P. R. Schofield and N. J. Mayhew (Oxbow: Oxford, 2002), pp. 89–105
Norse, David, 'Multiple Threats to Regional Food Production: Environment, Economy, Population?', *Food Policy*, 19 (1994), 133–48
North, Douglass, *Structure and Change in Economic History* (New York: Norton 1981)
Ogilvie, Sheilagh, *Institutions and European Trade: Merchant Guilds, 1000–1800* (Cambridge: Cambridge University Press, 2011)
Ó Gráda, Cormac, *Black '47 and Beyond: The Great Irish Famine in History, Economy, and Memory* (Princeton: Princeton University Press, 2000)
——, *Eating People is Wrong, and Other Essays on Famine, Its Past and Its Future* (Princeton: Princeton University Press, 2015)

——, *Famine: A Short History* (Princeton: Princeton University Press, 2009)
——, *The Great Irish Famine* (Cambridge: Cambridge University Press, 1995)
——, 'Ireland', in *Famine in European History*, ed. by Guido Alfani and Cormac Ó Gráda (Cambridge: Cambridge University Press, 2017), pp. 166–84
——, *Ireland Before and After the Famine: Explorations in Economic History, 1800–1925*, 2nd edition (Manchester: Manchester University Press, 1993)
——, *Ireland: A New Economic History 1780–1939* (Oxford: The Clarendon Press, 1995)
——, 'Markets and Famines: Evidence from Nineteenth Century Finland', *Economic Development and Cultural Change*, 49 (2001), 575–90
——, 'Markets and Famines in Pre-Industrial Europe', *Journal of Interdisciplinary History*, 36 (2005), 143–66
——, 'Neither Feast, Nor Famine: England before the Industrial Revolution', in *Institutions, Innovation, and Industrialization: Essays in Economic History and Development*, ed. by Avner Greif, Lynne Kiesling, and John V. C. Nye (Princeton: Princeton University Press, 2015), pp. 7–32
Ó Gráda, Cormac, and Jean-Michel Chevet, 'Famine and Market in Ancien *Régime* France', *Journal of Economic History*, 62 (2002), 706–33
Ó Gráda, Cormac, and Andrés Eiríksson, eds, *Ireland's Great Famine: Interdisciplinary Perspectives* (Dublin: University College Dublin Press, 2006)
Ó Gráda, Cormac, and Joel Mokyr, 'What Do People Die of During Famines: The Great Irish Famine in Comparative Perspective', *European Review of Economic History*, 6 (2002), 339–63
Olsson, Lennart, 'On the Causes of Famine — Drought, Desertification and Market Failure in the Sudan', *Ambio*, 22 (1983), 395–404
Oppenheimer, Clive, *Eruptions that Shook the World* (Cambridge: Cambridge University Press, 2011)
Oram, Richard, 'Bruce, Balliol and the Lordship of Galloway: South-West, Scotland and the Wars of Independence', *Transactions of the Dumfriesshire and Galloway Natural History and Antiquarian Society*, 67 (1992), 29–47
——, 'The Worst Disaster Suffered by the People of Scotland in Recorded History': Climate Change, Dearth and Pathogens in the Long Fourteenth Century', *Proceedings of the Society of Antiquaries of Scotland*, 144 (2014), 223–44
Oram, Richard, and W. Paul Adderley, 'Lordship and Environmental Change in Central Highland Scotland *c.* 1300–*c.* 1400', *Journal of the North Atlantic*, 1 (2008), 74–84
Orme, Nicholas, 'A Medieval Almhouse for the Clergy: Clyst Gabriel Hospital near Exeter', *Journal of Ecclesiastical History*, 39 (1988), 1–15
——, 'Mortality in Fourteenth-Century Exeter', *Medical History*, 32 (1988), 195–203
Özmucur, Süleyman, and Şevket Pamuk, 'Real Wages and Standards of Living in the Ottoman Empire, 1489–1914', *Journal of Economic History*, 62 (2002), 293–321
Page, Mark, 'The Peasant Land Market on the Estate of the Bishopric of Winchester before the Black Death', in *The Winchester Pipe Rolls and Medieval English Society*, ed. by Richard Britnell (Woodbridge: Boydell and Brewer, 2003), pp. 61–80

Paping, R., E. Vanhaute, and C. Ó Gráda, eds, *When the Potato Failed: Causes and Effects of the Last European Subsistence Crisis, 1845–1850*, Corn Publication Series, 9 (Turnhout: Brepols, 2007).

Parker, Geoffrey, *Global Crisis: War, Climate Change and Catastrophe in the Seventeenth Century* (New Haven: Yale University Press, 2013)

Pearson, Chris, *Mobilizing Nature: The Environmental History of War and Militarization in Modern France* (Manchester: Manchester University Press, 2012)

Penman, Michael, *Robert the Bruce: King of the Scots* (New Haven: Yale University Press, 2014)

Pfizenmaier, Sam, ed., *Charterhouse Square: Black Death Cemetery and Carthusian Monastery, Meat Market and Suburb* (London: Museum of London Archaeology, 2016)

Phillips, Seymour, *Edward II* (New Haven: Yale University Press, 2010)

Pitkänen, Kari J., 'Famine Mortality in Finland: Is there a Sex Bias?', in *Famine Demography: Perspectives from Past and Present*, ed. by Tim Dyson and Cormac Ó Gráda (Oxford: Oxford University Press, 2002), pp. 65–92

Podberscek, Anthony L., 'Good to Pet and Eat: The Keeping and Consuming of Dogs and Cats in South Korea', *Journal of Social Issues*, 65 (2009), 615–32

Poirteir, Cathal, *Famine Echoes — Folk Memories of the Great Irish Famine: An Oral History of Ireland's Greatest Tragedy* (Dublin: Gill and Macmillan, 1995)

Poos, Lawrence, *A Rural Society after the Black Death: Essex, 1350–1525* (Cambridge: Cambridge University Press, 1991)

Poos, Lawrence R., and Richard M. Smith '"Legal Windows onto Historical Populations?" Recent Research in Demography and the Manor Court in Medieval England', *Law and History Review*, 2 (1984), 128–52

——, 'Shades Still on the Window: A Reply to Zvi Razi', *Law and History Review*, 3 (1985), 409–29

Poos, Lawrence R., Zvi Razi, and Richard M. Smith, 'The Population History of Medieval English Villages: A Debate on the Use of Manor Court Records', in *Medieval Society and the Manor Court*, ed. by Zvi Razi and Richard Smith (Oxford: Clarendon Press, 1996), pp. 298–368

Popkin, Samuel L., *The Rational Peasant: The Political Economy of Rural Society in Vietnam* (Los Angeles: University of California Press, 1979)

Postan, Michael M., *Essays on Medieval Agriculture and General Problems of the Medieval Economy* (Cambridge: Cambridge University Press, 1973)

——, *The Famulus: The Estate Labourer in the* XII*th* *and* XIII*th* *Centuries* (Cambridge: Cambridge University Press, 1954)

——, 'Village Livestock in the Thirteenth Century', *Economic History Review*, 15 (1962), 219–49

Postan, Michael M., and Jan Titow, 'Heriots and Prices on Winchester Manors', *Economic History Review*, 11 (1959), 392–411

Praschak-Rieder, Nicole, and others, 'Seasonal Variation in Human Brain Serotonin Transporter Binding', *Archives of General Psychiatry*, 65 (2008), 1072–78

Prestwich, Michael, *Armies and Warfare in the Middle Ages: The English Experience* (New Haven: Yale University Press, 1996)

——, 'Gilbert de Middleton and the Attack on the Cardinals, 1317', in *Warriors and Churchmen in the Middle Ages: Essays Presented to Karl Leyser*, ed. by Timothy Reuter (London: Hambledon Press, 1992), pp. 179–94

——, *War, Politics, and Finance Under Edward I* (Totowa, NJ: Rowman and Littlefield, 1972)

Pribyl, Kathleen, 'Weather in Late Medieval Norfolk: Agricultural Practices and their Climatological Significance' (unpubl. doctoral thesis, Univ. of Bern, 2011)

Proctor, C. J., A. Baker, and W. L. Barnes, 'A Three Thousand Year Record of North Atlantic Climate', *Climate Dynamics*, 19 (2002), 449–54

Proctor, C. J., and others, 'A Thousand Year Speleothem Proxy Record of North Atlantic Climate from Scotland', *Climate Dynamics*, 16 (2000), 815–20; the dataset is available online at <https://www.ncdc.noaa.gov/paleo/study/5418> [accessed April 2018]

Raftis, Ambrose J., *Peasant Economic Development within the English Manorial System* (Montreal: McGill University Press, 1996)

Rahmato, Dessalegn, *Famine and Survival Strategies: A Case Study from Northeast Ethiopia* (Upsala: Nordiska Afrikainstitutet, 1991)

Ravallion, Martin, *Markets and Famines* (Oxford: Oxford University Press, 1987)

Rawcliffe, Carole, *Medicine for the Soul: The Life, Death and Resurrection of an English Medieval Hospital* (Southampton: Sutton, 1999)

——, *Urban Bodies: Communal Health in Late Medieval English Towns and Cities* (Woodbridge: Boydell and Brewer, 2013)

Razi, Zvi, 'Family, Land and the Village Community in Later Medieval England', *Past and Present*, 93 (1981), 3–36.

——, *Life, Marriage and Death in a Medieval Parish* (Cambridge: Cambridge University Press, 1980)

——, 'The Myth of the Immutable English Family', *Past and Present*, 140 (1993), 3–44

Reid, Anna, *Leningrad: Tragedy of a City under Siege, 1941–44* (London: Bloomsbury Publishing)

Reitsema, L. J., 'Beyond Diet Reconstruction: Stable Isotope Applications to Human Physiology, Health, and Nutrition', *American Journal of Human Biology*, 25 (2013), 445–56

Richardson, Gary, 'The Prudent Village: Risk Pooling Institutions in Medieval English Agriculture', *Journal of Economic History*, 65 (2005), 386–413

Richardson H. G., and G. O. Sayles, 'Irish Revenue, 1278–1384', *Proceedings of the Royal Irish Academy, Section C: Archaeology, Celtic Studies, History, Linguistics, Literature*, 62 (1961–1963), pp. 87–100

Rogers, James E. Thorold, *A History of Agriculture and Prices in England*, vol. III: *1401–1582* (Oxford: Clarendon Press, 1882)

Rubin, Olivier, 'A Collapse of Entitlements and Democratic Responsiveness', *Journal of Asian and African Studies*, 44 (2009), 279–98

——, 'The Entitlement Approach — A Case for Framework Development rather than Demolition', *Journal of Development Studies*, 45 (2009), 622–41

Rubin, Miri, *Charity and Community in Medieval Cambridge* (Cambridge: Cambridge University Press, 2002)

Rutledge, Elizabeth, 'Immigration and Population Growth in Early Fourteenth-Century Norwich: Evidence from the Tithing Roll', *Urban History Yearbook*, 15 (1988), 15–30
Rydval, Miloš, and others, 'Reconstructing 800 Years of Summer Temperatures in Scotland from Tree Rings', *Climate Dynamics*, 48 (2017), 1–24
Ryn, Z., 'Suicides in the Nazi Concentration Camps', *Suicide and Life-Threatening Behavior*, 16 (1986), 419–33
Samatar, Abdi Ismail, Mark Lindberg, and Basil Mahayni, 'The Dialectics of Piracy in Somalia: The Rich versus the Poor', *Third World Quarterly*, 31 (2010), 1377–94
Santín-Montanyá, Inés, Encarnación Zambrana-Quesada, and José Luis Tenorio-Pasamón, 'Weed Management in Cereals in Semi-Arid Environments: A Review', in *Herbicides — Current Research and Case Studies in Use*, ed. by Andrew J. Price and Jessica A. Kelton (Rijeka: Intech, 2013), pp. 133–52; <http://www.intechopen.com/books/herbicides-current-research-and-case-studies-in-use> [accessed April 2018]
Santiuste, David, *The Hammer of the Scots: Edward I and the Scottish Wars of Independence Pen and Sword Books* (Barnsley: Pen and Sword Books, 2015)
Schofield, Philipp R., 'Access to Credit in the Early Fourteenth-Century', in *Credit and Debt in Medieval England, c. 1180–c. 1350*, ed. by Phillipp R. Schofield and Nicholas J. Mayhew (Oxbow: Oxford, 2002), pp. 106–26
——, 'Approaching Poverty in the Medieval Countryside', in *Poverty and Prosperity in the Middle Ages and Renaissance*, ed. by A. Scott and C. Kosso (Turnhout: Brepols, 2012), pp. 95–111
——, 'Dealing in Crisis: External Credit and the Early Fourteenth-century English Village', in *Medieval Merchants and Money: Essays in Honour of James L. Bolton*, ed. by Martin Allen and Matthew Davies (London: Institute of Historical Research, 2016), pp. 253–70
——, 'Dearth, Debt and the Local Land Market in a Late Thirteenth-Century Village Community', *Agricultural History Review*, 45 (1997), 1–17
——, 'The Market, Economic Growth and Famine in the Medieval English Countryside in the Early Fourteenth Century', *Quaestiones Medii Aevi Novae*, 20 (2016), 269–84
——, 'The Social Economy of the Medieval Village in the Early Fourteenth Century', *Economic History Review*, 61 (2008), 38–63
——, 'Wales and the Great Famine of the Early Fourteenth Century', *Welsh History Review*, 29 (2018), 143–67
Schofield P. R. and N. J. Mayhew, eds, *Credit and Debt in Medieval England, c. 1180–c. 1350* (Oxbow: Oxford, 2002)
Scott S. and C. J. Duncan, 'The Mortality Crisis of 1623 in North-West England', *Local Population Studies*, 58 (1997), 14–25
Scott S., S. R. Duncan, and C. J. Duncan, 'Infant Mortality and Famine: A Study in Historical Epidemiology in Northern England', *Journal of Epidemiology and Community Health*, 49 (1995), 245–52
Scudder, Thayer, *Global Threats, Global Futures: Living with Declining Living Standards* (Cheltenham: Edward Elgar Publishing, 2010)
Scully, Terence, *The Art of Cookery in the Middle Ages* (Woodbridge: Boydell and Brewer, 1995)

Seabourne, Gwen, *Royal Regulation of Loans and Sales in Medieval England: Monkish Superstition and Civil Tyranny* (Woodbridge: Boydell and Brewer, 2003)

Sen, Amartya Kumar, *Poverty and Famines: An Essay on Entitlement and Deprivation* (Oxford: Clarendon Press, 1981)

Serjeantson, Dale, *Farming and Fishing in the Outer Hebrides, AD 600 to 1700* (Southampton: The Highfield Press, 2013)

Sharp, Buchanan, *Famine and Scarcity in Late Medieval and Early Modern England: The Regulation of Grain Marketing, 1256–1631* (Cambridge: Cambridge University Press, 2016)

——, 'Royal Paternalism and the Moral Economy in the Reign of Edward II: The Response to the Great Famine', *Economic History Review*, 66 (2013), 628–47

Slavin, Philip, *Bread and Ale for the Brethren: The Provisioning of Norwich Cathedral Priory, 1260–1536* (Hatfield: University of Hertfordshire Press, 2012)

——, 'Climate and Famines: A Historical Reassessment', *WIRES Climate Change*, 7 (2016), 433–47

——, 'Crusaders in Crisis: Towards the Re-assessment of the Origins and Nature of the "People's Crusade" of 1095–6', *Imago Temporis*, 4 (2010), 175–99

——, 'Ecology, Famine and Religious Violence: The Case of the Popular Crusading Movement, 1095–1320', in *Crisis a l'Edat Mitjana: Models, explicacions i representacions*, ed. by Pere Benito i Monclús (Lleida: Milenio, 2013), pp. 173–90

——, '*Feeding the Brethren: Grain Provisioning of Norwich Cathedral Priory, c. 1280–1370*' (unpubl. doctoral dissertation, Univ. of Toronto, 2008)

——, 'Flogging a Dead Cow: Coping with Animal Panzootics on the Eve of the Black Death', in *Coping with Crisis: Re-Evaluating the Role of Crises in Economic and Social History*, ed. by Alex Brown, Andy Burn, and Rob Doherty (Woodbridge: Boydell and Brewer, 2015), pp. 111–35

——, 'Food Security, Safety, and Crises', in *A Cultural History of Food in the Renaissance, 1300–1600*, ed. by Ken Albala (London: Berg, 2012), pp. 63–82

——, 'The Great Bovine Pestilence and Its Economic and Environmental Consequences in England and Wales, 1318–50', *Economic History Review*, 65 (2012), 1239–66

——, 'Late-Medieval Goose Farming: Evidence from Eastern England, *c.* 1250–1400', *Agricultural History Review*, 58:1 (2010), 1–29

——, 'Landscapes of Panzootics: Sheep Scab and Rural Environment in England in 1279–1280' *Landscapes*, 17 (2016), 156–70

——, 'Market Failure during the Great Famine in England and Wales (1315–1317)', *Past and Present*, 222 (2014), 9–49

——, 'Peasant Livestock Husbandry in Late-Thirteenth-Century Suffolk: Economy, Environment and Society', in *Peasants and Lords in Medieval English Economy: Essays in Honour of Bruce M. S. Campbell*, ed. by Maryanne Kowaleski, John Langdon, and Phillipp R. Schofield (Turnhout: Brepols, 2015), pp. 3–26

——, 'Warfare and Ecological Destruction in Early Fourteenth-Century British Isles', *Environmental History*, 19 (2014), 528–50

Slavin, Philip, and Sharon DeWitte, 'Between Famine and Death: Physiological Stress and Dairy Deficiency in England on the Eve of the Black Death (1315–50): New Evidence from Paleoepidemiology and Manorial Accounts', *Journal of Interdisciplinary History*, 44 (2013), 37–60

Smith, Adam, *An Inquiry into the Nature and Causes of the Wealth of Nations: A Selected Edition*, ed. by Kathryn Sutherland (Oxford: Oxford Paperbacks, 2008)

Smith, Brendan, *Colonisation and Conquest in Medieval Ireland: The English in Louth, 1170–1330* (Cambridge: Cambridge University Press)

Smith, Richard, 'English Peasant Life-cycles and Socio-economic Networks: A Quantitative Geographical Case Study' (unpubl. doctoral thesis, Univ. of Cambridge, 1974)

——, 'Further Models of Medieval Marriage: Landlords, Serfs and Priests in Rural England, *c.* 1290–1370', in *Georges Duby L'ecriture de l'Histoire*, ed. by Claudie Duhamel-Amado and Guy Lobrichon (Brussels: De Boeck Université, 1996), pp. 160–73

——, 'Plagues and People: The Long Demographic Cycle, 1250–1670', in *The Peopling of Britain: The Shaping of a Human Landscape*, ed. by Paul Slack and Ryk Ward (Oxford: Oxford University Press, 2002), pp. 177–209

——, 'Some Issues concerning Families and Their Property in Rural England 1250–1800', in *Land, Kinship and Lifecycle*, ed. by Richard M. Smith (Cambridge: Cambridge University Press, 1984), pp. 1–86

Snyder, Timothy, *Bloodlands: Europe Between Hitler and Stalin* (London: The Bodley Head, 2010)

Söderberg, Johan, 'Long-term Trends in Real Wages of Labourers', in *Historical Monetary and Financial Statistics for Sweden: Exchange Rates, Prices, and Wages, 1277–2008*, ed. by Rodney Edvinsson, Tor Jacobson, and Daniel Waldenström (Stockholm: Ekerlids Förlag, 2010), pp. 453–78

Solar, Peter M., 'The Crisis of the Late 1840s: What Can Be Learned From Prices?', in *When the Potato Failed: Causes and Effects of the Last European Subsistence Crisis, 1845–1850*, ed. by R. Paping, E. Vanhaute, and C. Ó Gráda, Corn Publication Series, 9 (Turnhout: Brepols, 2007), pp. 79–94

Sorokin, Pitirim, *Golod kak Faktor* (Petrograd: Kolos, 1921; repr. Moscow: Academia, 2003)

Stathakopoulos, Dionysios Ch., *Famine and Pestilence in the Late Roman and Early Byzantine Empire: A Systematic Survey of Subsistence Crises and Epidemics* (Aldershot: Ashgate, 2004)

Stern, Derek V., *A Hertfordshire Demesne of Westminster Abbey: Profits, Productivity and Weather* (Hatfield: University of Hertfordshire Press, 2000)

Stevens, Matthew Frank, 'The Great Famine in Dyffryn Clwyd, 1315–1322', *Denbighshire Historical Society Transactions*, 63 (2015), 13–35

——, *Urban Assimilation in Post-conquest Wales: Ethnicity, Gender and Economy in Ruthin, 1282–1348* (Cardiff: University of Wales Press, 2010)

Stone, David J, 'The Consumption and Supply of Birds in Late Medieval England', in *Food in Medieval England: Diet and Nutrition*, ed. by Chris Woolgar, Dale Serjeantson, and Tony Waldron (Oxford: Oxford University Press, 2006), pp. 148–61

——, 'The Impact of Drought in Early Fourteenth-Century England', *Economic History Review*, 67 (2014), 435–62

Susser, Ezra, and others, 'Schizophrenia after Prenatal Famine', *Archives of General Psychiatry*, 53 (1996), 25–31

Susser, Mervyn, and Zena Stein, 'Timing in Prenatal Nutrition: a Reprise of the Dutch Famine Study', *Nutrition Reviews*, 52 (1994), 84–94

Swanson, Robert N., 'Indulgences for Prayers for the Dead in the Diocese of Lincoln in the Early Fourteenth Century', *Journal of Ecclesiastical History*, 52 (2001), 197–219

——, 'A Universal Levy: Tithes and Economy Agency', in *Agriculture and Rural Society After the Black Death: Common Themes and Regional Variations*, ed. by Ben Dodds and Richard Britnell (Hatfield: University of Hertfordshire Press, 2008), pp. 89–112

Taylor, Alice, 'Homo Ligius and Unfreedom in Medieval Scotland', in *New Perspectives on Medieval Scotland, 1093–1286*, ed. by Matthew Hammond (Woodbridge: Boydell and Brewer, 2013), pp. 85–116

Thaxton, Ralph, *Catastrophe and Contention in Rural China: Mao's Great Leap Forward Famine and the Origins of Righteous Resistance in Da Fo Village* (Cambridge: Cambridge University Press, 2008)

Thomason, W. E., and others, 'Understanding Pre-Harvest Sprouting of Wheat', *Virginia Cooperative Extension*, Publication 424–060 (2009), 1–4

Thompson, E. P., 'The Moral Economy of the English Crowd in the Eighteenth Century', *Past and Present*, 50 (1971), 76–136

Titow, Jan Z., 'Evidence of Weather in the Account Rolls of the Bishopric of Winchester 1209–1350', *Economic History Review*, 12 (1960), 360–407

——, 'Some Evidence of the Thirteenth Century Population Increase', *Economic History Review*, 13 (1961), 218–24

Trethowan, R. M., 'Preharvest Sprouting Tolerance', in *Application of Physiology in Wheat Breeding*, ed. by M. P. Reynolds, J. I. Ortiz-Monasterio, and A. McNab (Mexico City: CIMMYT, 2001), pp. 145–47

Tubach, Frederic C., *Index Exemplorum: A Handbook of Medieval Religious Tales* (Helsinki: Suomalainen Tiedeakatemia, 1969)

Tucker, Richard P., 'The Impact of Warfare on the Natural World: A Historical Survey', in *Natural Enemy, Natural Ally: Toward an Environmental History of War*, ed. by Richard P. Tucker and Edmund Russell (Corvallis: Oregon State University Press, 2004), pp. 15–41

Tucker, Richard P., and Edmund Russell, eds, *Natural Enemy, Natural Ally: Toward an Environmental History of War* (Corvallis: Oregon State University Press, 2004)

Tufanova, Olga A., 'Izobrazhenie goloda 1601–1603 gg. v Russkikh i inostrannykh istochnikakh o Smutnom Vremeni', [Depiction of 1601–1603 Famine years in Russian and Foreign Sources about the Time of Troubles], *Vestnik Slavyanskikh Kul'tur*, 24 (2012), 77–83

Tyson, Diana B., 'Against the King's Taxes — The Second Manuscript', *Nottingham Medieval Studies*, 54 (2010), 73–92

Udry, Christopher, 'Credit Markets in Northern Nigeria: Credit as Insurance in a Rural Economy', *The World Bank Economic Review*, 4 (1990), 251–69

Van Haute, Eric, and Thijs Lambrecht, 'Famine, Exchange Networks and the Village Community: A Comparative Analysis of the Subsistence Crises of the 1740s and the 1840s in Flanders', *Continuity and Change*, 26 (2011), 155–86

Van Werveke, Hans, 'La famine de l'an 1316 en Flandre et dans les régions voisines', *Revue du Nord*, 41 (1959), 5–14

Várdy, Steven Bela, and Agnes Huszár Várdy, 'Cannibalism in Stalinist Russia and Mao's China', *East European Quarterly*, 41 (2007), 223–38

Veenendaal, Marjolein V. E., and others, 'Transgenerational Effects of Prenatal Exposure to the 1944–45 Dutch Famine', in *The Fetal Origins of Adult Disease, the Evidence and Mechanisms*, ed. by Marjolein Veenendaal (Rotterdam: Legatron Electronic Publishing, 2012), pp. 96–104

Vere Cruess, William, *Home and Farm Food Preservation* (Bedford, MA: Applewood Books, 1918)

Waites, Bryan, *Monasteries and Landscape of the North York Moors and Wolds* (Stroud: Tempus, 2007)

Walsh, Oonagh, 'Gender and Insanity in Nineteenth Century Ireland', in *Sex and Seclusion, Class and Custody: Perspectives on Gender and Class in the History of British and Irish psychiatry*, ed. by Jonathan Andrews and Anne Digby (Amsterdam: Rodopi, 2004), pp. 69–93

——, *Insanity, Power and Politics in Nineteenth-Century Ireland: The Connaught District Lunatic Asylum* (Manchester: Manchester University Press, 2013)

Waters, Robert E. C., *Genealogical Memoirs of the Extinct Family of Chester of Chicheley, Their Ancestors and Descendants* (London: Robson and Sons 1878)

Watson, Fiona, *Under the Hammer: Edward I and Scotland, 1286–1307* (East Linton: Tuckwell Press, 1998)

White, Henry, 'Account of the Making of Cheshire Cheese', *British Farmer's Magazine*, 9 (1845), 281–93

Wilson, Rob, and others, 'A Millennial Long March–July Precipitation Reconstruction for Southern-central England', *Climate Dynamics*, 40: 3–4 (2012), 997–1017, DOI: 10.1007/s00382-012-1318-z

Winchester, Angus, *The Harvest of the Hills: Rural Life in Northern England and the Scottish Borders, 1400–1700* (Edinburgh: Edinburgh University Press, 2000)

Woods, R., 'The Measurement of Historical Trends in Fetal Mortality in England and Wales', *Population Studies*, 59 (2005), 147–62

Wrigley, E. A., 'Some Reflections on Corn Yields and Prices in Pre-Industrial Economies', in *Famine, Disease and the Social Order in Early Modern Society*, ed. by John Walter and Roger Schofield (Cambridge: Cambridge University Press, 1989), pp. 235–78

Wrigley, E. A., and others, *English Population History from Family Reconstruction, 1580–1837* (Cambridge: Cambridge University Press, 1997)

Wrigley, E. A., and R. S. Schofield, *The Population History of England 1541–1871: A Reconstruction* (Cambridge, MA: Harvard University Press, 1981)

Wright, J. R., *The Church and the English Crown, 1305–1334: A Study based on the Register of Archbishop Walter Reynolds* (Toronto: PIMS, 1980)

Yates, Margaret 'The Market in Freehold Land, 1300–1509: The Evidence of Feet of Fines', *Economic History Review*, 66 (2013), 579–600

Yaussy, S. L., S. N. DeWitte, and R. R. Redfern, 'Frailty and Famine: Patterns of Mortality and Physiological Stress among Victims of Famine in Medieval London', *American Journal of Physical Anthropology*, 160 (2016), 272–83

Yeliseyeva, I. I., and Ye. I. Gribova, eds, *Sankt-Peterburg: 1703–2003: Yubileynyi statisticheskiy sbornik* [Saint Petersburg: 1703–2003: The Jubilee Statistical Yearbook] (Saint Peterburg: Sudostroyenie, 2003)

Young, A. 'The Earls and Earldom of Buchan in the Thirteenth Century', in *Medieval Scotland: Crown, Lordship and Community: Essays Presented to G. W. S. Barrow*, ed. by A. Grant and K. J. Stringer (Edinburgh: Edinburgh University Press, 1993), pp. 174–202

Zamenhof, S., E. van Marthens, and L. Grauel, 'DNA (cell number) in Neonatal Brain: Second Generation (F2) Alteration by Maternal (F0) Dietary Protein Restriction', *Science*, 172 (1971), 850–51

Zierler, David, *The Invention of Ecocide: Agent Orange, Vietnam, and the Scientists Who Changed the Way We Think about the Environment* (Athens: University of Georgia Press, 2011)

Online Databases and Websites

Africa Watch, 'Sudan, Nationwide Famine: Culpable Negligence in the Management of Food Security, War, and the Use of Food as a Political Weapon', 7 November 1990, <http://www.hrw.org/reports/pdfs/s/sudan/sudan90n.pdf> [accessed April 2018]

Campbell, Bruce M. S., *Three Centuries of English Crop Yields, 1211–1491*, 2007, <http://www.cropyields.ac.uk> [accessed April 2018]

CBC News, 'Somali Food Aid Stolen, Sold in Markets', 15 Augst 2011, <http://www.cbc.ca/news/world/somalia-food-aid-stolen-sold-in-markets-1.1033052> [accessed April 2018]

CELT: Corpus of Electronic Texts <http://www.ucc.ie/celt/> [accessed April 2018]

CIRCLE: A Calendar of Irish Chancery Letters, c. 1244–1509 <https://chancery.tcd.ie/roll/9-Edward-II/close> [accessed April 2018]

Cooper, R. J., and others, 'East Anglia 1,100 Year Spring–Summer Precipitation Reconstruction', in *NOAA/NCDC Paleoclimatology Data Contribution Series #2012–043* <https://www.ncdc.noaa.gov/paleo/study/12896> [accessed April 2018]

Famine Early Warning Systems Network <https://www.fews.net/> [accessed April 2018].

Hoffman, Philip T., 'Paris 1380–1870' [commodity prices data set], in *Global Price and Income History Group Datasets* <http://gpih.ucdavis.edu/Datafilelist.htm> [accessed April 2018]

Holdomor 1932–33: Famine Genocide in Ukraine, 'Holdomor Survivor Accounts and Memoirs', <http://holodomorct.org/holodomor-information-links/accounts-memoirs/> [accessed April 2018]

Kausmally, T., 'East Smithfield Black Death Cemetery Summary', in *Collections, The Museum of London*, 28 March 2007, <https://www.museumoflondon.org.uk/collections/

other-collection-databases-and-libraries/centre-human-bioarchaeology/osteological-database/medieval-cemeteries/east-smithfield-black-death-medieval> [accessed April 2018]

Letters, S., *Gazetteer of Markets and Fairs in England and Wales* <http://www.history.ac.uk/cmh/gaz/gazweb2.html> [accessed April 2018]

McCance and Widdowson, *The Composition of Foods*, 2002 version, < https://www.gov.uk/government/publications/composition-of-foods-integrated-dataset-cofid> [accessed April 2018]

MET Office, *UK Climate — Historic Station Data* <http://www.metoffice.gov.uk/public/weather/climate-historic/#?tab=climateHistoric> [accessed April 2018]

Munro, John H. A., 'Revisions of David Farmer's series of English "National Agrarian Wages"', <https://www.economics.utoronto.ca/wwwfiles/archives/munro5/ResearchData.html> [accessed April 2018]

——, 'Revisions of the Phelps Brown and Hopkins "Basket of Consumables" Commodity Price Series, 1264–1700', 2006, <http://www.economics.utoronto.ca/munro5/ResearchData.html> [accessed June 2015]

NASA, 'Severe Drought Causes Famine in East Africa', <https://earthobservatory.nasa.gov/NaturalHazards/view.php?id=51411> [accessed April 2018]

Poynder, Nicholas, *Database of England Monthly Grain Prices, 1270–1955* <http://www.iisg.nl/hpw/data.php#united> [accessed on April 2018]

——, 'Database of French (Parisian) Monthly Grain Prices, 1549–1698', in *Global Price and Income History Group Datasets* <http://gpih.ucdavis.edu/Datafilelist.htm> [accessed April 2018]

Radnedge, Aidan, 'Somalia Drought Victims Attempting Suicide to Escape Worst Ever Famine', *Metro News*, 21 July 2011, <http://metro.co.uk/2011/07/21/somalia-drought-victims-attempting-suicide-to-escape-worst-ever-famine-94056/#ixzz4XfL2rFEd> [accessed April 2018]

Records of the Parliaments of Scotland to 1707 (RPS) <http://www.rps.ac.uk/> [accessed April 2018]

Sea-Distances <http://www.sea-distances.org/> [accessed April 2018]

SKULD, 'Transportation of Wheat', 12 July 2017, <https://www.skuld.com/topics/cargo/solid-bulk/agricultural-cargoes/transportation-of-wheat/ > [accessed April 2018]

Soilscapes database, Cranfield Soil and Agrifood Institute, University of Cranfield, <http://www.landis.org.uk/soilscapes/> [accessed April 2018]

SWI, 'Food Speculation Blamed for East Africa Famine', 22 November 2011, < http://www.swissinfo.ch/eng/politics/foreign_affairs/Food_speculation_blamed_for_East_Africa_famine.html?cid = 31607472> [accessed April 2018]

Taxatio Database <https://www.hrionline.ac.uk/taxatio/> [accessed April 2018]

TEAMS Middle English Text Series, Medieval Institute of the University of Rochester, <https://d.lib.rochester.edu/teams> [accessed April 2018]

University of Minnesota, 'Nutritional Needs of Horses', <http://www.extension.umn.edu/distribution/livestocksystems/components/0480_03.html> [accessed April 2018]

Wilson, R., and others, 'Southern-Central England 1000 Year March–July Precipitation Reconstruction', in the *NOAA/NCDC Paleoclimatology Data Contribution Series #2012–054* <https://www.ncdc.noaa.gov/paleo/study/12907> [accessed April 2018]

World Food Programme, 'Drought Affects Millions in the Sahel', <https://www.wfp.org/stories/drought-returns-sahel> [accessed April 2018]

Index

Environmental Histories of the North Atlantic World

All volumes in this series are evaluated by an Editorial Board, strictly on academic grounds, based on reports prepared by referees who have been commissioned by virtue of their specialism in the appropriate field. The Board ensures that the screening is done independently and without conflicts of interest. The definitive texts supplied by authors are also subject to review by the Board before being approved for publication. Further, the volumes are copyedited to conform to the publisher's stylebook and to the best international academic standards in the field.

Title in Series

Thomas Finan, *Landscape and History on the Medieval Irish Frontier: The King's Cantreds in the Thirteenth Century* (2016)

Environment, Colonization, and the Baltic Crusader States: Terra Sacra I, ed. by Aleksander Pluskowski (2019)

Ecologies of Crusading, Colonization, and Religious Conversion in the Medieval Baltic: Terra Sacra II, ed. by Aleksander Pluskowski (2019)